MATHEMATICAL METHODS FOR OCEANOGRAPHERS

MATHEMATICAL METHODS FOR OCEANOGRAPHERS

An Introduction

EDWARD LAWS
University of Hawaii, Honolulu

JOHN WILEY & SONS, INC.
New York • Chichester • Weinheim • Toronto • Singapore • Brisbane

Library of Congress Cataloging-in-Publication Data

Laws, Edward A., 1945–
Mathematical methods for oceanographers : an introduction / Edward Laws.
p. cm.
ISBN 0-471-16221-3 (cloth : alk. paper)
1. Oceanography—Mathematics. I. Title.
GC10.4.M36L38 1997
551.46′0015′1—dc20 96-31998

CONTENTS

PREFACE

People who are successful scientists are able to recognize patterns in data. This statement is true of both the natural and the social sciences, and it is certainly true of oceanography. The ability to recognize patterns is, in fact, a characteristic of successful persons in many walks of life, including business, linguistics, medicine, government service, and many other professions. Thousands of years ago Polynesians sailed over the Pacific Ocean. They were able to chart their course over vast areas of the ocean based primarily on their ability to recognize patterns in the stars and, in particular, by their recognition of a relationship between the position of certain stars and latitude. They also recognized that there was a relationship between the movement of wind-driven waves and direction, and were able to use this fact to navigate during the day and during times when the stars were hidden by clouds. They realized, for example, that in the Southern Hemisphere the swell created by the trade winds came out of the southeast, and by orienting their canoes relative to that swell they could maintain their course even on starless nights.

In physics, the birth of quantum mechanics can be traced to the fact that certain scientists were able to recognize patterns in data which could not be explained by the laws of Newtonian physics. One of the most important developments in the early history of quantum mechanics, for example, was the discovery of a systematic pattern in the wavelengths of light in the emission spectrum of atomic hydrogen. The wavelengths in the visible region of the spectrum can all be described by the equation

$$\frac{1}{\lambda} = R\left(\frac{1}{2^2} - \frac{1}{n^2}\right)$$

where λ is the wavelength of light, R is a constant, and n is an integer greater than 2. The recognition of this pattern by J. J. Balmer in 1885 led to the concept of quantization of energy levels, the integers 2 and n representing different energy levels within the atom. Balmer's accomplishment was due to his recognition of a pattern in the emission spectrum of atomic hydrogen as observed on a photographic plate, a spectrum many people might have seen as nothing more than a hodgepodge of emission lines.

Two important skills associated with pattern recognition are (1) the ability to separate the signal from the noise when a pattern exists and (2) the ability to recognize the absence of convincing evidence that a pattern exists when confronted with nothing but noise. The ability of Polynesian voyagers to recognize patterns in wind-generated waves is a good example of the former. Local storms, for example, could easily impose a considerable component of noise on the systematic swell created by the trade winds. The navigational skills of the Polynesians required that they be able to somehow filter out that noise and navigate according to the pattern of swells generated by the trade winds.

The ability to avoid being misled by spurious patterns requires some knowledge of or at least sense of statistics. This is a skill with which every scientist should be equipped. As an assistant professor of oceanography about 20 years ago, I still recall being asked to read a graduate student's dissertation concerned with the influence of temperature on the $\delta^{18}O$ of the calcium carbonate produced by marine coccolithophores. The student had grown several species under controlled conditions at a series of temperatures and had plotted the $\delta^{18}O$ results against temperature. He had fit linear regression lines to the data for each species and had then wanted to know whether there was any difference in the slopes of the regression lines. Since he had no knowledge of statistics, he had simply looked at the data and decided, based on visual inspection of the results, that there was no difference in the slopes of the regression lines. This sort of subjective data analysis should never be allowed to pass muster. What particularly struck me in this case was that none of the student's other committee members apparently knew anything about statistics, either.

That experience, and others like it, motivated me to begin teaching a course concerned with data analysis to graduate students in the University of Hawaii Oceanography Department. Most of the chapters in this book closely follow the presentation of subject matter in that course. The book assumes that the reader is familiar with calculus at least through first-order ordinary differential equations and has had at least an introductory course in statistics. The first chapter is a review of fundamental concepts of calculus relevant to the remaining chapters. It is intended to be a review and not an introduction to the subject. Statistical concepts are introduced where appropriate, but with the exception of analysis of variance, the presentation is again assumed to constitute a review and not an introduction.

The remaining chapters of the book are a systematic introduction to some of the basic mathematical methods whose use frequently arises in the analysis of oceanographic data and in the search for patterns. These include model I linear regression (Chapter 2), correlation analysis (Chapter 3), model II linear regression (Chapter 4), polynomial curve fitting, linear multiple regression analysis, and nonlinear least squares (Chapter 5), numerical integration (Chapter 6), box models (Chapter 7), and time series analysis (Chapter 8). The discussion of regression methods is much more thorough than the presentation typically found in introductory statistics texts, and includes, in particular, a rigorous analysis of the reasons for believing that certain methods produce unbiased parameter estimates.

My experience with the material in this book is that covering all eight chapters requires a full two semesters or three quarters. The first seven chapters require about two quarters, although they can be squeezed into one semester with some ingenuity and judicious selection of material. When teaching the course, I have strongly encouraged the students to make use of the software package MATLAB® It is a software package with which I am familiar, and its use greatly facilitates matrix operations. The presentation of material in several of the chapters relies heavily on the formulation of relationships in terms of matrices.

Understanding virtually any subject in mathematics requires doing problems, and that generalization applies to the material presented in this book. However, the exercises at the end of each chapter are more than just a means of reinforcing concepts presented in the chapter. The exercises in many cases expand on the material presented in the chapter, and the experience of many students has been that the concepts do not really sink in until the problems have been completed. The exercises are time consuming, but they are worth it. Persons who read the book without doing the exercises will derive far less benefit from the book than those who do the problems.

A course based on the first seven chapters of this book is required for all students in the geology and chemical oceanography programs in the graduate program in oceanography at the University of Hawaii, and the course is taken on an optional basis by many of the biological oceanography students. The first seven chapters have tended to interest fewer of the physical oceanography students, but Chapter 8 fits in rather well with a portion of their mathematical training. Questions will undoubtedly arise as to whether reading this book or taking a course based on this book is worthwhile for student X or indeed for faculty member or scientist Y, and the following litmus test may provide some help in answering such questions. A student wishes to measure the rate of a reaction which changes the concentration of Z. The student wants to know the average rate over a period of 24 hours and has enough resources to measure the concentration of Z at four points in time. At what times during a 24-hour period should the student measure the concentration of Z in order to minimize the uncertainty in the reaction rate estimated from the slope of a model I linear regression fit to the concentration versus time data? If your answer to this question is that the student should measure the concentration of Z at four equally spaced points in time, that is, at times 0, 8, 16, and 24 hours, then you probably need to read this book. More specifically, you need to read Chapter 2.

I am indebted to numerous persons for their help and encouragement during the preparation of this book. Much of the text was written during a sabbatical leave while I was the William McCurdy Fellow at the Duke University Marine Laboratory. I am indebted to Dr. Joseph Ramus, Director of the Duke Marine Laboratory, for making possible my sabbatical leave and to the faculty and staff of the lab for making my stay there both enjoyable and productive. I especially would like to thank Dr. Dan Rittschof for the use of his computer and Tim Boynton for his technical assistance. I am particularly indebted to Dr. Peter Howd, whose course on environmental data analysis I audited. It was through his course that I was introduced to MATLAB, and the organization of material in Chapter 8 owes much to his influence. I am also much indebted to Dr. Robert Berner for numerous discussions concerning his GEOCARB II model, one of the two case studies presented in Chapter 7. For the last two years I have used a draft of the manuscript of this book as the text for the mathematical methods course I teach at the University of Hawaii, and I am indebted to the students who took those courses and who pointed out errors and omissions in the text and/or who made suggestions for improving the presentation of material. I am also indebted to Dr. Philip Manor of John Wiley for his encouragement and advice concerning the preparation and publication of this book. Finally, I am most indebted to my wife Stephanie and my children Ryan and Jennifer, whose understanding and encouragement helped me through the many hours required for the writing of this book.

EDWARD LAWS

Honolulu, Hawaii

MATHEMATICAL METHODS FOR OCEANOGRAPHERS

1

CALCULUS REVIEW

We begin our study of mathematics applied to oceanography by reviewing some basic mathematical concepts, in particular, certain aspects of calculus. These concepts will be used throughout this book, and it is important that the reader feel comfortable with them.

RELATIONSHIPS AND FUNCTIONS

Let us first consider two variables X and Y. In some cases two variables are *independent* of each other, meaning that the value of one is unrelated to the value of the other. In other cases there is a *relationship* between X and Y, meaning that the value of one is in some sense *dependent* on the value of the other. A relationship between two variables is said to be a *functional relationship* if for each and every value of one variable there is only one value of the other variable. For example, the relationship described by the equation $X = 2Y$ is a functional relationship, because setting the value of X uniquely determines the value of Y, and vice versa.

Figure 1.1 illustrates in graphical form a relationship between X and Y. In this case the value of Y associated with a given value of X, say X_0, is the vertical distance from the X or horizontal axis (abscissa) at $X = X_0$ to the curve. Similarly, the value of X associated with a particular value of Y, say Y_0, is the horizontal distance from the Y or vertical axis (ordinate) at $Y = Y_0$ to the curve. The relationship as illustrated is a functional relationship, because for each value of X there is one and only one value of Y on the curve, and vice versa.

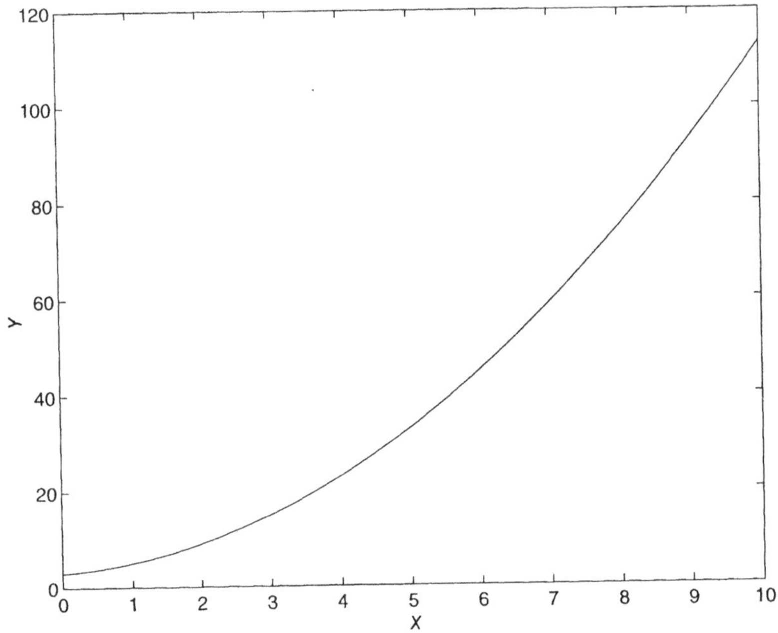

Figure 1.1 Simple example of a functional relationship between two variables X and Y.

Figure 1.2 illustrates another kind of relationship. In this case for each value of X there is one and only one value of Y on the curve, but for all but one value of Y there are two values of X on the curve. The single exception is the point at the top of the curve. In this case we would say that Y is a function of X, meaning that for each value of X there is one and only one value of Y. The converse, however, is not true. In this case X is not a function of Y, because it is not true that for each value of Y there is one and only one value of X.

When a variable Y is a function of a variable X, we commonly say that X is the *independent variable* and Y is the *dependent variable*, meaning that the value of Y is determined by the value of X, but the converse is not necessarily true. The functional relationship is often written $Y = F(X)$; that is, the dependent variable Y is a function of the independent variable X. In this case X is said to be the *argument* of the function F. The nature of the functional relationship is contained in the explicit description of or recipe for F, but for now suffice it to say that whatever F may be, it must satisfy the condition that for each value of X there is one and only one value of Y defined by or determined by $F(X)$. For example, we might say $Y = F(X) = X^2$, in which case we would have defined a function $F(X)$. For each value of X there is one and only one value of $F(X) = X^2$.

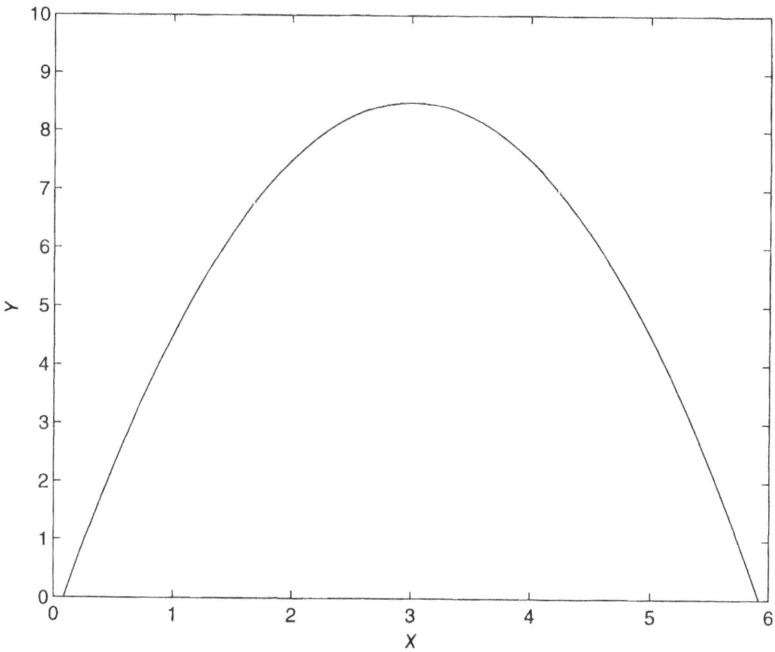

Figure 1.2 Graph of a relationship for which Y is a function of X but X is not a function of Y.

In defining a functional relationship, it is, strictly speaking, insufficient to just state the equation or recipe for calculating $F(X)$ from X. We must also specify the range of values of the independent variable over which the equation applies. The range of the independent variable is referred to as the *domain* of the function. Similarly, the range of values of $F(X)$ for the given domain is referred to as the *range* of the function. For example, if we defined $F(X) = X^2$ with the domain of the function being any real number between -2 and $+2$ inclusive, then the range of the function would be all real numbers between 0 and 4 inclusive.

DERIVATIVES OF FUNCTIONS

Representing a function in graphical form, as in Figure 1.1, is often a useful way to describe or depict a function. In this book, however, we will find that it is often desirable to have more than just a picture of the functional relationship between X and Y. In particular, it will often be desirable to have some measure of how much the dependent variable Y changes when the independent variable X changes. In other words, how sensitive is Y to X? For example, the respiration rate of an organism can be expected to increase as the temperature rises, at least until the temperature becomes too high and becomes stressful. A rule of thumb is that the respiration rate will rise by about a factor of 2 when the temperature rises by 10°C. In other words, the rate of change of respiration rate is a factor of 2 increase per 10°C rise in temperature. We could, of course, glean this information from a graph of respiration rate versus temperature, but it is sometimes more convenient to express or depict the rate of change explicitly rather than having to calculate it from a graph or other representation of the underlying functional relationship. In mathematical terms, how do we calculate rates of change of a function?

Figure 1.3 is a plot of respiration rate versus temperature for a particular organism. In this case temperatures below 10°C and above 35°C are considered to be lethal. The relevant domain of the respiration function is therefore 10 to 35°C. At temperatures between 17 and 27°C, there is a twofold increase in respiration rate, in accord with expectations. The change in respiration rate is the vertical distance between the value of the function at 27°C and the value of the function at 17°C. In this case that distance is 24 mg O_2 h^{-1} minus 12 mg O_2 h^{-1} = 12 mg O_2 h^{-1}. The change in temperature is, of course, the horizontal distance between 17 and 27°C, or 10°C. The mean rate of change of respiration over this temperature range is therefore 12 mg O_2 h^{-1}/10°C = 1.2 mg O_2 h^{-1} °C^{-1}.

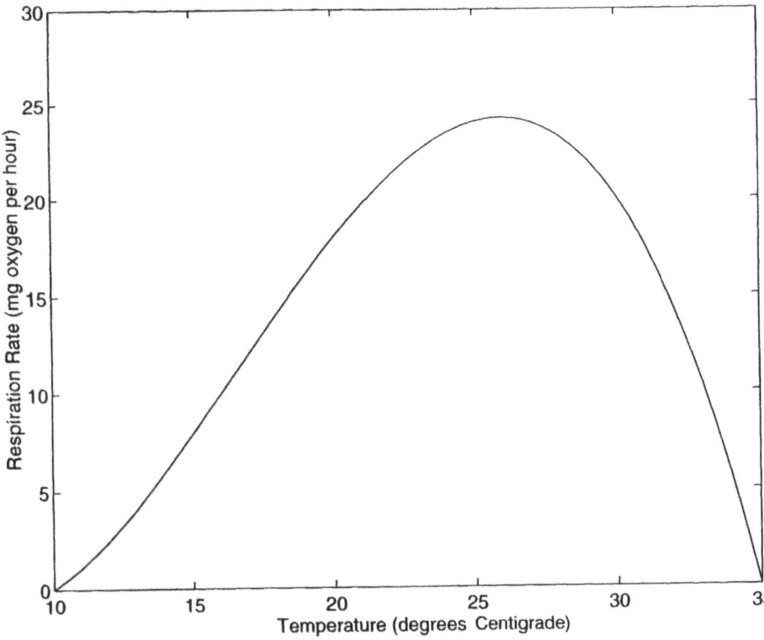

Figure 1.3 Hypothetical plot of respiration rate versus temperature for an organism.

Fair enough, but this is the average value over the temperature range 17 to 27°C. What if we want to know the rate of change over a narrower range of temperatures, or even at a particular temperature? Dealing with the first question is straightforward. For example, suppose we want to know the average rate of change of the respiration rate from 17 to 19°C. The respiration rate at 19°C is 14.7 mg O_2 h^{-1}, and the change in respiration rate between 17 and 19°C is therefore 14.7 − 12.0 = 2.7 mg O_2 h^{-1}. The average rate of change is therefore 2.7 mg O_2 h^{-1}/2°C = 1.35 mg O_2 h^{-1} $°C^{-1}$.

We can now continue the process of reducing the temperature range over which we calculate the average rate of change. For example, we can determine the average rate of change between 17.0 and 18.0°C, between 17.0 and 17.5°C, between 17.0 and 17.1°C, and so forth. As we reduce the range of temperature, the associated change in respiration rate will become smaller and smaller, and if the respiration rate has a well-defined rate of change at 17.0°C, the values we calculate for the average rate of change will approach the rate of change of respiration rate at 17.0°C. In mathematical terms, we could define the rate of change of the respiration rate at 17.0°C as follows:

$$\text{Rate of change at } 17.0°\text{C} = \lim_{\Delta T \to 0} \frac{F(17 + \Delta T) - F(17)}{\Delta T} \tag{1.1}$$

where $F(T)$ is the respiration rate at T°C, and the right-hand side of Equation 1.1 is the limit of the ratio $F(17 + \Delta T) - F(17)$ divided by ΔT as ΔT approaches 0. When F is given as a mathematical expression, it is important to evaluate the ratio for finite ΔT and then to take the limit as ΔT approaches 0. Obviously, it makes no sense to calculate separately the limits of the numerator and denominator as ΔT approaches 0, because in both cases that limit is 0, and 0 divided by 0 is undefined. In other words, the right-hand side of Equation 1.1 is meaningful only if we understand that it means to take the limit of the ratio, not the ratio of the limits.

We define the rate of change of a function at a particular value of its argument to be the derivative of the function at that value of the argument. The derivative of a function F is often designated F'. Thus, in general terms, we may say

$$F'(X) = \lim_{\Delta X \to 0} \frac{F(X + \Delta X) - F(X)}{\Delta X} \tag{1.2}$$

An important point about Equation 1.2 is that the limit should be independent of how ΔX approaches 0. In particular, ΔX can be positive or negative. If the right-hand side of Equation 1.2 depends on how ΔX approaches 0, then the derivative of the function is undefined at that point.

We will consider a few examples to see how to apply Equation 1.2. Suppose $F(X)$ is constant, that is, $F(X) = C$, where C is some number. In this case $F(X) = C$ and $F(X + \Delta X) = C$, so

$$\frac{F(X + \Delta X) - F(X)}{\Delta X} = \frac{C - C}{\Delta X} = \frac{0}{\Delta X} = 0$$

In this case the value of $[F(X + \Delta X) - F(X)]/\Delta X$ is independent of ΔX, and the process of taking the limit as ΔX approaches 0 is of no consequence. The answer makes good sense, since the rate of change of a constant function is, in fact, 0.

Now consider the case where $F(X) = X$. In this case $F(X + \Delta X) = X + \Delta X$, and we have

$$\frac{F(X + \Delta X) - F(X)}{\Delta X} = \frac{X + \Delta X - X}{\Delta X} = \frac{\Delta X}{\Delta X} = 1$$

Once again the value of $[F(X + \Delta X) - F(X)]/\Delta X$ is independent of ΔX, and the process of taking the limit as ΔX approaches 0 is of no consequence. The answer again makes good intuitive sense, since for the function $F(X) = X$ any change in the argument X produces exactly the same change in the value of $F(X)$.

Now consider the function $F(X) = X^2$. In this case $F(X + \Delta X) = (X + \Delta X)^2$ and

$$\frac{F(X + \Delta X) - F(X)}{\Delta X} = \frac{(X + \Delta X)^2 - X^2}{\Delta X}$$

$$= \frac{X^2 + 2X(\Delta X) + (\Delta X)^2 - X^2}{\Delta X} = 2X + \Delta X$$

Now

$$\lim_{\Delta X \to 0} \frac{F(X + \Delta X) - F(X)}{\Delta X} = \lim_{\Delta X \to 0} 2X + \Delta X = 2X$$

In this case the process of taking the limit as ΔX approaches 0 is important, and we conclude that the derivative of $F(X) = X^2$ with respect to X is $2X$. Note that in all cases we do not take the limit as $\Delta X \to 0$ until we have evaluated the expression $[F(X + \Delta X) - F(X)]/\Delta X$ for finite ΔX.

RULES FOR EVALUATING MORE COMPLEX DERIVATIVES

Now that we know in principle how to calculate the derivative of a function, let us consider some more complicated situations. Suppose we need to take the derivative of the sum of two functions. We will call the functions $F(X)$ and $G(X)$. We follow the same procedure as before:

$$[F(X) + G(X)]' = \lim_{\Delta X \to 0} \frac{F(X + \Delta X) + G(X + \Delta X) - F(X) - G(X)}{\Delta X} \tag{1.3}$$

Assuming that the limit of the sum is identical to the sum of the limits, we can write the right-hand side of Equation 1.3 as

$$\lim_{\Delta X \to 0} \frac{F(X + \Delta X) - F(X)}{\Delta X} + \lim_{\Delta X \to 0} \frac{G(X + \Delta X) - G(X)}{\Delta X}$$
$$= F'(X) + G'(X)$$

Thus the derivative of the sum of two functions is the sum of the derivatives of the two functions. This proof may seem rather simple, but it depends on the critical fact that the limit of a sum is the sum of the limits. This is not a trivial thing to say. We have already seen that the limit of a quotient is not necessarily the quotient of the limits, nor indeed is the limit of a product necessarily the product of the limits.

Suppose we want to calculate the derivative of the product of two functions, which we will call $F(X)$ and $G(X)$. We follow the same procedure as before:

$$[F(X)G(X)]' = \lim_{\Delta X \to 0} \frac{F(X + \Delta X)G(X + \Delta X) - F(X)G(X)}{\Delta X} \tag{1.4}$$

The answer in this case is not immediately obvious, but we can get to the answer by adding and subtracting $F(X)G(X + \Delta X)$ to the numerator. This has no effect on the value of the numerator, since we add and subtract the same expression. The result is that the right-hand side of Equation 1.4 becomes

$$\lim_{\Delta X \to 0} \frac{\begin{array}{r} F(X + \Delta X)G(X + \Delta X) + F(X)G(X + \Delta X) \\ -F(X)G(X + \Delta X) - F(X)G(X) \end{array}}{\Delta X}$$

Rearranging the numerator of this expression, we obtain

$$\lim_{\Delta X \to 0} \frac{[F(X + \Delta X) - F(X)]G(X + \Delta X) + F(X)[G(X + \Delta X) - G(X)]}{\Delta X}$$

Again assuming that the limit of the sum is the sum of the limits, we obtain

$$\lim_{\Delta X \to 0} \frac{[F(X+\Delta X) - F(X)]G(X+\Delta X)}{\Delta X} + \lim_{\Delta X \to 0} \frac{F(X)[G(X+\Delta X) - G(X)]}{\Delta X}$$

Since $F(X)$ is independent of ΔX, the limit of the second expression should be obvious. It is $F(X)G'(X)$. We must be a little more cautious in taking the first limit, because we are dealing with the limit of a product of two functions, both of which depend on ΔX. The first function is $[F(X + \Delta X) - F(X)]/\Delta X$, and its limit is obviously $F'(X)$. The second function is $G(X + \Delta X)$, and its limit is obviously $G(X)$. Can we assume, however, that the limit of the product equals the product of the limits in this case? The answer is yes if both $F'(X)$ and $G(X)$ are well defined and bounded at whatever value of X we wish to make the calculation. Assuming this to be the case, the limit of the first product is $F'(X)G(X)$. We conclude, therefore, that as long as $F(X)$ and $G(X)$ and their derivatives are well behaved, that is, they are well defined and bounded, then

$$[F(X)G(X)]' = F'(X)G(X) + F(X)G'(X) \tag{1.5}$$

Note, in particular, that although the derivative of the sum of two functions is the sum of the derivatives of the two functions, the derivative of the product of two functions is not the product of the derivatives of the two functions.

The last situation we wish to consider here is the derivative of a function of a function. We will call the first function F and the second function G. The argument of F is G, and the argument of G is X. We now ask, what is the derivative of F with respect to X? We proceed as before, with the understanding that a prime denotes differentiation with respect to X:

$$F'[G(X)] = \lim_{\Delta X \to 0} \frac{F[G(X+\Delta X)] - F[G(X)]}{\Delta X} \tag{1.6}$$

Here again the answer is not very obvious, but we can get to the answer by multiplying and dividing the quotient on the right-hand side of Equation 1.6 by $G(X + \Delta X) - G(X)$:

$$F'[G(X)] = \lim_{\Delta X \to 0} \frac{F[G(X+\Delta X)] - F[G(X)]}{G(X+\Delta X) - G(X)} \frac{G(X+\Delta X) - G(X)}{\Delta X}$$

Here again we are confronted with the problem of taking the limit of a product, and the rules are the same as before. If the functions in question are well defined and bounded at the point of interest, the limit of the product will be the product of the limits. The second function is $[G(X + \Delta X) - G(X)]/\Delta X$, and its limit is obviously $G'(X)$. Now let us define ΔG as $G(X + \Delta X) - G(X)$. Then $G(X + \Delta X) = G(X) + \Delta G$, and the first function in the product can be written as $[F(G + \Delta G) - F(G)]/\Delta G$. Since ΔG approaches 0 as ΔX approaches 0, the limit of this first function as ΔX approaches 0 is the derivative of F with respect to G. By analogy with our definition of ΔG, we could define ΔF to be $F(G + \Delta G) - F(G)$, and the derivative of F with respect to G would then be the limit as ΔG approaches 0 of $\Delta F/\Delta G$. We normally write this limit as dF/dG. Similarly, the derivative of G with respect to X can be considered to be the limit as ΔX approaches 0 of $\Delta G/\Delta X$ and is customarily written as dG/dX. We conclude therefore that

$$F'[G(X)] = \frac{dF}{dG}\frac{dG}{dX} \tag{1.7}$$

Equation 1.7 is generally referred to as the *chain rule*.

Now we will try a few problems to see how these theorems work in practice. Suppose we want to calculate the derivative of $F(X) = X^3$. We note that X^3 can be expressed as the product of two functions, X and X^2. Since we know how to calculate the derivatives of these two functions, we can apply our theorem for differentiating the product of two functions to get the answer. Let $f(X) = X$ and $g(X) = X^2$. Then $F(X) = X^3 = f(x)g(X)$ and

$$\begin{aligned} F'(X) &= f(X)g'(X) + f'(X)g(X) = (X)(2X) + (1)(X^2) \\ &= 2X^2 + X^2 = 3X^2 \end{aligned}$$

Now suppose we want to differentiate $F(X) = X^6$. We could treat this as the product of two functions, $f(X) = X^3$ and $g(X) = X^3$. We then have $F(X) = X^6 = f(X)g(X)$ and

$$\begin{aligned} F'(X) &= f(X)g'(X) + f'(X)g(X) = (X^3)(3X^2) + (3X^2)(X^3) \\ &= 3X^5 + 3X^5 = 6X^5 \end{aligned}$$

We could also use the chain rule to find the derivative of $F(X) = X^6$. To see how this would work, let $f(g) = g^2$ and $g(X) = X^3$. Since $(X^3)^2 = X^6$, it follows that $F(X) = f[g(X)]$ and

$$F'(X) = \frac{df}{dg}\frac{dg}{dX} = (2g)(3X^2) = (2X^3)(3X^2) = 6X^5$$

PROOF BY INDUCTION

By now we have learned to differentiate a constant function and $F(X) = X$, X^2, X^3, and X^6. We may have noticed a pattern to the results:

$$d(C)/dX = 0$$

$$d(X)/dX = 1$$

$$d(X^2)/dX = 2X$$

$$d(X^3)/dX = 3X^2$$

$$d(X^6)/dX = 6X^5$$

The pattern seems to be that $d(X^n)/dX = nX^{n-1}$. This is certainly true for all the examples we have worked out.[1] Can we show that it is true for all the positive integers? The answer is that we can, but it would take an infinite amount of time if we considered each integer separately. There must be a better way.

The better way is called proof by induction. Proof by induction is accomplished in two steps. In this case we first prove that the theorem is true for $n = 1$. Then we show that if the theorem is true for some value of n, say N, it must be true for $n = N + 1$. Basically, this is a bootstrap technique. Let us see how proof by induction works in this case.

We have already shown that the theorem is true for $n = 1$, so we can go directly to the second step. If the theorem is true for $n = N$, then

$$\frac{dX^N}{dX} = NX^{N-1}$$

Can we use this information to prove that the theorem must be true for $n = N + 1$; that is, can we show that $dX^{N+1}/dX = (N + 1)X^N$? We can consider X^{N+1} to be the product of two functions, X and X^N. Then

$$\frac{dX^{N+1}}{dX} = (X)\frac{dX^N}{dX} + \frac{dX}{dX}(X^N)$$

Now we have proven that $dX/dX = 1$ and we are assuming that $dX^N/dX = NX^{N-1}$. Hence

$$\frac{dX^{N+1}}{dX} = (X)(NX^{N-1}) + (1)(X^N) = NX^N + X^N = (N + 1)X^N$$

[1]Recall that $X^0 = 1$.

This completes the proof by induction. We have shown by induction that the theorem must be true for all positive integers.

Now we will try another proof by induction. Suppose we want to know the derivative of the sum of n functions, where $n > 1$. Call the functions $F_1, F_2, F_3, \ldots, F_n$. We already know the answer when $n = 2$: $(F_1 + F_2)' = F_1' + F_2'$. We hypothesize that the derivative of the sum of n functions is the sum of the derivatives of the n functions. We see that this is true for $n = 2$. Can we show that if the theorem is true for $n = N$, then it must be true for $n = N + 1$? If the theorem is true for $n = N$, then

$$(F_1 + F_2 + \cdots + F_N)' = F_1' + F_2' + F_3' + \cdots + F_N'$$

Now consider the function $G = F_1 + F_2 + F_3 + \cdots + F_N + F_{N+1}$. We can consider G to be the sum of two functions, G_1 and G_2, with $G_1 = F_1 + F_2 + F_3 + \cdots + F_N$ and $G_2 = F_{N+1}$. Now we have proven that the theorem is true for $n = 2$, so it must be true that

$$G' = G_1' + G_2'$$

Clearly, $G_2' = F_{N+1}'$. Furthermore, since G_1 is the sum of N functions and we are assuming the theorem is true for $n = N$, it follows that $G_1' = (F_1 + F_2 + F_3 + \cdots + F_N)' = F_1' + F_2' + F_3' + \cdots + F_N'$. It follows that $G' = G_1' + G_2' = F_1' + F_2' + F_3' + \cdots + F_N' + F_{N+1}'$. Hence we have shown by induction that the derivative of the sum of n functions is the sum of the derivatives of the n functions. We will use proof by induction from time to time in this book. There are some more proof by induction exercises at the end of this chapter.

MAXIMA AND MINIMA OF FUNCTIONS

One of the most common applications of derivatives in the natural sciences is the determination of the maxima and minima of functions. Figure 1.4 illustrates the nature of the problem. The function in this case is $F(X) = X^3/3 - X$. It is obvious that the function has extreme maxima and minima as X approaches $+\infty$ and $-\infty$, respectively. Of more interest are the local maximum and minimum just to the left and right, respectively, of the ordinate. How can we determine the values of X and $F(X)$ associated with these local extrema?

The trick is to take advantage of the fact that the derivative or *slope* of a function $F(X)$ at a local maximum or minimum will always be 0. Thus, by differentiating the function and setting its derivative equal to 0, we can determine the values of X associated with the local maximum or minimum. In this case $F'(X) = X^2 - 1 = (X + 1)(X - 1)$, and we can see that the values of X associated with $F'(X) = 0$ are $+1$ and -1. By plugging these

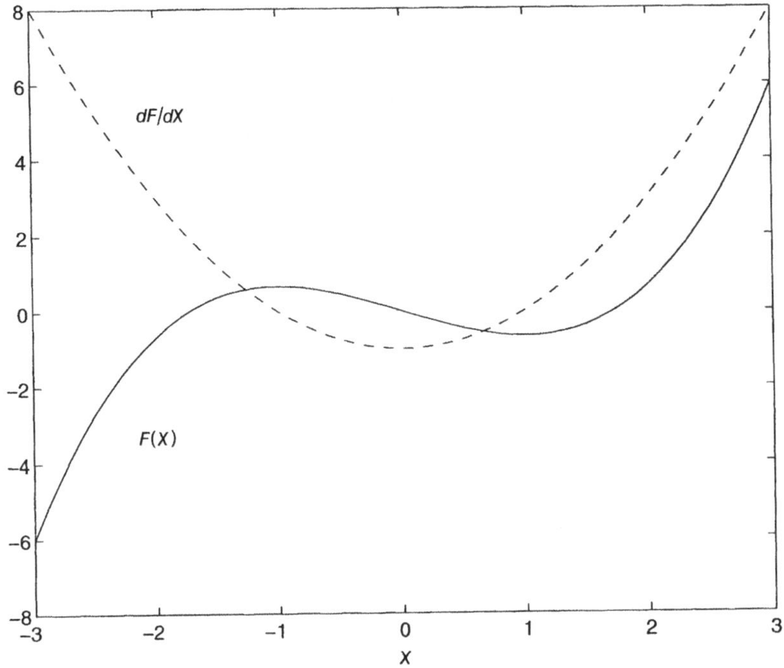

Figure 1.4 Graph of the function $F(X) = X^3/3 - X$ (solid curve) and dF/dX (dashed curve).

values of X back into the equation for $F(X)$, we find that the values of $F(X)$ at these points are $F(+1) = -2/3$ and $F(-1) = 2/3$.

In this case we can tell which point is a local maximum and which is a local minimum by just examining the graph of the function, but in other cases this might be impractical, or at least not easy to do. How can we tell whether the extrema we have identified are maxima or minima? The answer is to note that at a local maximum the derivative or slope of the function is positive to the left of the maximum and negative to the right of the maximum. Similarly, the slope of the function is negative to the left of a local minimum and positive to the right. This means that the slope of the function is decreasing as we pass through a local maximum and increasing as we pass through a local minimum. This is apparent in Figure 1.4. We can take advantage of this fact mathematically by examining the rate of change of the derivative of the function, that is, the derivative of the derivative, at the point where the function has a local maximum or minimum. In this case we have $F'(X) = X^2 - 1$, and if we let $F''(X)$ denote the derivative of the derivative, that is, the *second derivative*, then $F''(X) = 2X$. Now $2X$ is positive when $X = 1$ and negative when $X = -1$. Hence we conclude that the function has a local minimum at $X = 1$ and a local maximum at $X = -1$.

In the next chapter we will begin to work with functions of more than one independent variable, and at that time we will need to understand the concept of a *partial derivative*. Consider the function $F(X, Y) = 2X^2Y - 8XY - 6X^2 + 24X + Y^2$. If we wish to differentiate this function with respect to X, we consider Y to be a constant. If we wish to differentiate with respect to Y, we consider X to be a constant. In both cases the derivatives are called partial derivatives, because we are differentiating with respect to only one of the independent variables. The partial derivative with respect to X is denoted $\partial F/\partial X$, and the partial derivative with respect to Y is similarly denoted $\partial F/\partial Y$. In this case $\partial F/\partial X = 4XY - 8Y - 12X + 24$ and $\partial F/\partial Y = 2X^2 - 8X + 2Y$. Our job now is to determine the values of X and Y that make both of these partial derivatives 0. We note that

$$\frac{\partial F}{\partial X} = 4(X - 2)(Y - 3) \tag{1.8}$$

$$\frac{\partial F}{\partial Y} = 2[X(X - 4) + Y] \tag{1.9}$$

Equation 1.8 equals 0 if $X = 2$ or $Y = 3$. If $X = 2$, the second equation equals 0 when $Y = 4$. Hence both partial derivatives are 0 at $X = 2$ and $Y = 4$. If $Y = 3$, Equation 1.9 equals 0 when $X^2 - 4X + 3 = (X - 3)(X - 1) = 0$. This condition is satisfied when $X = 1$ or $X = 3$. Thus there are three points in the XY plane where both partial derivatives are 0, namely, (2, 4), (1, 3), and (3, 3), where the first number in each ordered pair indicates the X coordinate of the point and the second number indicates the Y coordinate.

We would now like to know whether any of these points are maxima or minima. When we are dealing with a function of two variables, the procedure for determining whether a function has a maximum or minimum at a point is more complicated than in the case of a function of only one variable. We must calculate the second partial derivative of F with respect to both X and Y. These partial derivatives are designated $\partial^2F/\partial X^2$ and $\partial^2F/\partial Y^2$, respectively. We must also calculate the mixed partial derivative, $\partial^2F/\partial X\,\partial Y$. The mixed partial derivative is calculated by first taking the partial derivative with respect to X and then taking the derivative of that function with respect to Y. The order of differentiation does not affect the final result; we could differentiate with respect to Y first and then differentiate with respect to X.

By differentiating Equation 1.8 with respect to X, we see that $\partial^2F/\partial X^2 = 4(Y - 3)$, and by differentiating Equation 1.9 with respect to Y, we see that $\partial^2F/\partial Y^2 = 2$. By differentiating Equation 1.8 with respect to Y, we see that $\partial^2F/\partial X\,\partial Y = 4(X - 2)$. We would get exactly the same answer by differentiating Equation 1.9 with respect to X. Check to convince yourself that this is true. Now let $P = \partial^2F/\partial X^2$, $Q = \partial^2F/\partial Y^2$, and $R = \partial^2F/\partial X\,\partial Y$. Then

TABLE 1.1 Calculation of P, Q, and R at designated points in (X,Y) Plane for $F(X,Y) = 2X^2Y - 8XY - 6X^2 + 24X + Y^2$

	(X,Y) Pair		
	(1, 3)	(2, 4)	(3, 3)
P	0	4	0
Q	2	2	2
R	−4	0	4

the function $F(X,Y)$ has a local maximum or minimum at the point (X_0, Y_0) if both $\partial F/\partial X$ and $\partial F/\partial Y$ are 0 at (X_0, Y_0) and $R^2 - PQ < 0$. The function has a local maximum if $P + Q < 0$ and a local minimum if $P + Q > 0$. The point (X_0, Y_0) is a saddle point if $R^2 - PQ > 0$, meaning the function decreases in at least one direction and increases in at least one direction as one moves away from the point (X_0, Y_0) in the XY plane. If $R^2 - PQ = 0$, then more information is needed to determine the behavior of the function near (X_0, Y_0).

Now we will determine the values of P, Q, and R at the three points where we have determined that both $\partial F/\partial X$ and $\partial F/\partial Y$ are 0. Table 1.1 summarizes the results. Clearly, $R^2 - PQ > 0$ at both (1, 3) and (3, 3). Hence both of these points are saddle points. At (2, 4), $R^2 - PQ = -8 < 0$, so the function has a maximum or minimum at that point. Since $P + Q > 0$ at (2, 4) the function has a local minimum at that point.

EXERCISES

1.1 The rates of certain chemical and biological processes are often assumed to be governed by an equation of the form

$$F(X) = \frac{X}{K + X} \tag{1.10}$$

where X is the concentration of a substrate or reactant and K is a constant. Use Equation 1.2 to show that

$$\frac{dF}{dX} = \frac{K}{(K + X)^2}$$

1.2 Use Equation 1.2 to show that the derivative with respect to X of $1/(K + X)$ equals $-1/(K + X)^2$. Use this result, the result of Exercise 1.1, and induction to show that the nth derivative with respect to X of $X/(K + X)$ equals $Kn!/(-K - X)^{n+1}$ for all integral values of n

greater than or equal to 1. In other words,

$$\left(\frac{d}{dX}\right)^n \frac{X}{K+X} = \frac{Kn!}{(-K-X)^{n+1}}$$

Hint: If the theorem is true for $n = N$, then

$$\left(\frac{d}{dX}\right)^N \frac{X}{K+X} = \frac{KN!}{(-K-X)^{N+1}}$$

Let $Y = 1/(K+X)$. We have already proven by induction that

$$\left(\frac{d}{dY}\right) Y^n = nY^{n-1}$$

for all positive integers n. Use the chain rule and the results of the first part of this exercise to show that

$$\left(\frac{d}{dX}\right) \frac{KN!}{(-K-X)^{N+1}} = \frac{K(N+1)!}{(-K-X)^{N+2}}$$

1.3 Find the values of X and Y for which $\partial F/\partial X = \partial F/\partial Y = 0$ for the following functions. Determine whether these values of X and Y correspond to maxima, minima, or saddle points of the functions.

a. $F(X,Y) = 1 + X^2 + Y^2$.

b. $F(X,Y) = 2X^2 - XY - 3Y^2 - 3X + 7Y$.

c. $F(X,Y) = X^2 - 5XY - Y^2$.

1.4 Let us define the operators $\Delta/\Delta X$ and $\Delta/\Delta Y$ as follows:

$$\frac{\Delta G}{\Delta X} = \frac{G(X+\Delta X, Y) - G(X,Y)}{\Delta X}$$

$$\frac{\Delta G}{\Delta Y} = \frac{G(X, Y+\Delta Y) - G(X,Y)}{\Delta Y}$$

Show that

$$\left(\frac{\Delta}{\Delta X}\right)\left(\frac{\Delta G}{\Delta Y}\right) = \left(\frac{\Delta}{\Delta Y}\right)\left(\frac{\Delta G}{\Delta X}\right)$$

that is, the operators $\Delta/\Delta X$ and $\Delta/\Delta Y$ commute with each other. Note that in the limit as ΔX and $\Delta Y \to 0$, these operators become $\partial/\partial X$ and $\partial/\partial Y$, respectively. Hence

$$\left(\frac{\partial}{\partial X}\right)\left(\frac{\partial G}{\partial Y}\right) = \left(\frac{\partial}{\partial Y}\right)\left(\frac{\partial G}{\partial X}\right)$$

that is, the answer is independent of the order of differentiation.

1.5 Assume that the photosynthetic rate P is a function of light intensity I and obeys the equation $P(I) = P_m(1 - e^{-\alpha I})$. Assume furthermore that I is a function of depth Z and obeys the equation $I = I_0 e^{-\beta Z}$. Use the chain rule (Equation 1.7) to calculate dP/dZ.

1.6 Let $G_N(X) = \Pi_{k=1}^{N} F_k(X)$, where $\Pi_{m=1}^{N} F_m = F_1 F_2 F_3 \cdots F_N$. Use induction to show that

$$\frac{dG_N(X)}{dX} = \sum_{k=1}^{N} \frac{dF_k(X)}{dX} \frac{1}{F_k(X)} \prod_{m=1}^{N} F_m(X)$$

1.7 Assume that light intensity I in a water column is described by the equation $I = \exp(-Z/20)$, where Z is the depth in meters. The chlorophyll concentration (chl) in the same water column is described by the equation

$$\text{chl} = 0.1\exp(-Z/20) + 0.2\exp\left(-(Z - 100)^2/700\right)$$

where the units of chl are mg m^{-3}. The productivity index (PI), in units of mg C mg^{-1} chl *a* d^{-1}, follows the equation PI $= 200I/(0.1 + I)$. The photosynthetic rate, with units of mg C m^{-3} d^{-1}, is the product of chl and PI. Use the chain rule, the result of Exercise 1.1, and the result of Exercise 1.6 to find an equation for dP/dZ. Make graphs of chl, PI, P, and dP/dZ versus depth from the surface to 150 m (e.g., plot the functions at 1-m intervals from the surface to 150 m). Based on your graph of P versus Z, at approximately what depths does P have a local maximum and minimum? Is this conclusion consistent with your graph of dP/dZ?

1.8 Evaluate the function $F(X, Y) = 2X^2Y - 8XY - 6X^2 + 24X + Y^2$ at the following values of X and Y:

a.

X:	1.1	1.1	1.1	1.0	1.0	1.0	0.9	0.9	0.9
Y:	3.1	3.0	2.9	3.1	3.0	2.9	3.1	3.0	2.9

b.

X:	2.1	2.1	2.1	2.0	2.0	2.0	1.9	1.9	1.9
Y:	4.1	4.0	3.9	4.1	4.0	3.9	4.1	4.0	3.9

c.

X:	3.1	3.1	3.1	3.0	3.0	3.0	2.9	2.9	2.9
Y:	3.1	3.0	2.9	3.1	3.0	2.9	3.1	3.0	2.9

Are your answers to parts a, b, and c, consistent with the conclusion that (1, 3) and (3, 3) are saddle points, and that (2, 4) is a local minimum of the function?

2

MODEL I LINEAR REGRESSION

We will now apply what we learned in Chapter 1 to a problem that frequently arises in data analysis, namely, fitting a straight line to some experimental data. Figure 2.1 illustrates the nature of the problem. A student has collected some experimental data concerning the photosynthetic rate of a plant as a function of irradiance. The data are plotted in Figure 2.1. The relationship between irradiance and photosynthetic rate appears to be linear, but the data do not lie exactly on a straight line, presumably due to errors associated with the values of irradiance and/or photosynthesis. If the underlying relationship is indeed linear, what is the best way to estimate the equation of the line which describes that relationship? The line in this case would be described by an equation of the form $Y(X) = A + BX$, where Y is the photosynthetic rate and X is the irradiance. The student is faced with the job of choosing the values of A and B. One could, of course, lay a ruler on the graph and move the ruler around until the data all lie close to the edge of the ruler. The student could then draw a line along the edge of the ruler and estimate the values of A and B from the line. Clearly, this would be a rather subjective procedure. We want something more objective.

LEAST SQUARES

The solution in this case is to choose the values of A and B so that the sum of the squares of the deviations of the experimental photosynthetic rates from the straight line is a minimum. Mathematically, this means that we want

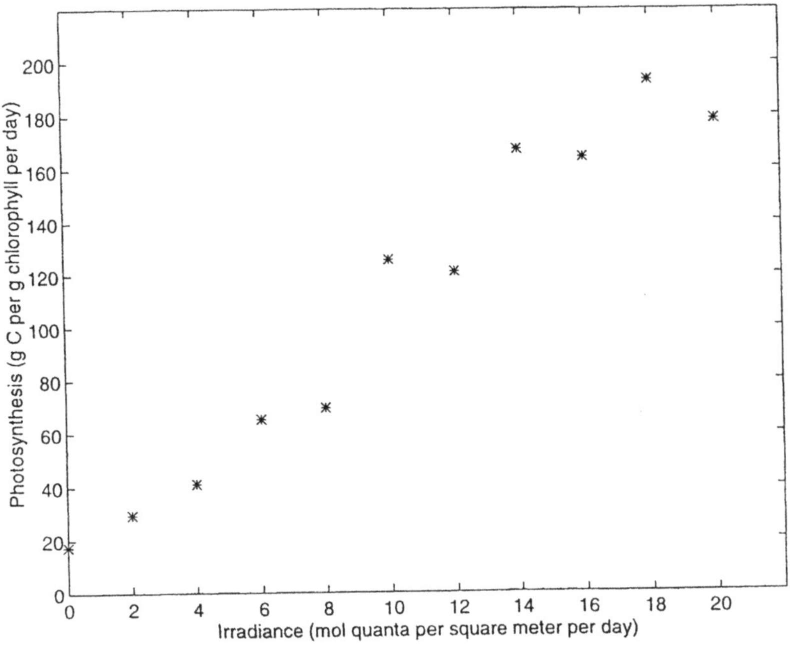

Figure 2.1 Hypothetical experimental data for photosynthetic rates measured at a series of light intensities.

to minimize the following function:

$$F(A, B) = \sum_{k=1}^{N} (Y_k - A - BX_k)^2 \tag{2.1}$$

where N is the number of (X, Y) pairs in the data set. Why do we use this approach to determine A and B?

There are several assumptions that underlie the use of this approach. First, we assume that the underlying relationship between X and Y is linear. In other words we assume that the true relationship between X and Y is described by the equation $Y = \alpha + \beta X$. The values of A and B that we choose will be estimates of α and β, respectively. We do not know the values of α and β, but we want our procedure for estimating α and β to give the best possible estimates. When we say best in the context of parameter estimation, we normally mean that we would like to satisfy the following three conditions:

1. The estimator is *unbiased.*
2. The estimator is *efficient.*
3. The estimator is *consistent.*

We will consider these conditions one at a time. What do we mean by an unbiased estimator? We mean that if we repeated the experiment M times and each time used the same procedure for estimating α and β, then the average of the M values of A and B would converge to α and β, respectively, as $M \to \infty$. Mathematically, we want the following conditions to be satisfied:

$$\lim_{M\to\infty} \frac{1}{M} \sum_{k=1}^{M} A_k = \alpha \tag{2.2}$$

$$\lim_{M\to\infty} \frac{1}{M} \sum_{k=1}^{M} B_k = \beta \tag{2.3}$$

The left-hand sides of Equations 2.2 and 2.3 are called the *expectation values* of A and B, respectively, and are generally written as $E(A)$ and $E(B)$. Hence we want our procedure to assure that $E(A)$ and $E(B)$ equal α and β, respectively. If this condition is satisfied, then our procedure is giving us unbiased estimates of α and β.

By efficient, we mean that

$$E[A - \alpha]^2 < E[\tilde{A} - \alpha]^2 \tag{2.4}$$

$$E[B - \beta]^2 < E[\tilde{B} - \beta]^2 \tag{2.5}$$

where $\tilde{A}$ and $\tilde{B}$ are any other estimators of the parameters α and β. To appreciate the meaning of an efficient estimator, consider Figure 2.2. Here we imagine the results of some time spent on the rifle range by two soldiers. The first soldier has a good rifle but a rather unsteady hand. His shots, indicated by the $+$'s in the figure, are badly scattered, although their average position is almost exactly on the center of the target. The second soldier has a steady hand, but his gun has a faulty sight, which causes his shots to tend to lie to the left and below the center of the target. If by efficient in this case we mean getting one's shots inside the innermost circle on the target, then the second soldier was clearly more efficient. Only 2 of his 50 shots lie outside the inner circle. Alternatively, we could gage the efficiency of the shots by calculating $(1/N)\Sigma_{k=1}^{N} d_k^2$, where d_k is the distance of the kth shot from the center of the target. This, of course, is similar to Equations 2.4 and 2.5. The second soldier would again be judged more efficient. If we estimated the position of the bull's-eye by calculating the average position of the second soldier's shots, however, the estimate would be biased due to the faulty sight on his gun. The average position of the first soldier's shots actually give an unbiased estimate of the position of the bull's-eye. Thus an unbiased estimator is not necessarily efficient, and vice versa.

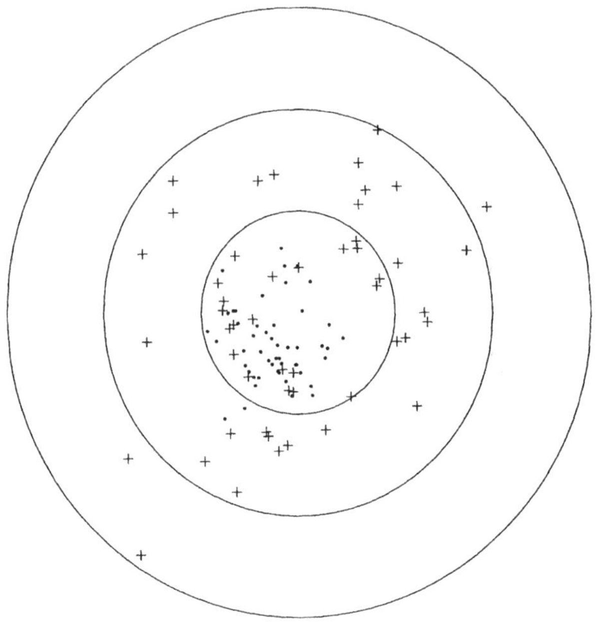

Figure 2.2 Hypothetical results of a target practice session. The +'s are the positions of unbiased shots with a large noise component. The dots are the positions of biased shots with a small noise component.

An estimator is said to be consistent if the value of the estimator approaches the value of the parameter being estimated with a probability approaching unity as the sample size becomes large. Mathematically, A and B are consistent estimators if, for any $\varepsilon > 0$,

$$\lim_{N\to\infty} P(|A - \alpha| \geq \varepsilon) = 0 \tag{2.6}$$

$$\lim_{N\to\infty} P(|B - \beta| \geq \varepsilon) = 0 \tag{2.7}$$

where $P(X)$ is the probability of X and A and B are the estimators of α and β based on a single sampling of N (X, Y) pairs. The requirements that the estimators be unbiased and consistent might at first seem equivalent, but they are not. In the case of an unbiased estimator, we require that if we calculate A and B from a finite number of (X, Y) pairs and repeat this exercise a theoretically infinite number of times, the average of all the A and B values will equal α and β, respectively. In the case of a consistent estimator, we carry out the exercise only once, but with a theoretically infinite number of (X, Y) pairs, and require that the values of A and B calculated from this one exercise equal α and β, respectively.

The condition of greatest interest and concern is usually that estimators be unbiased, and we will focus on that issue in this discussion. Under certain conditions it can be shown that choosing A and B so as to minimize the right-hand side of Equation 2.1 does yield unbiased estimates of α and β. The conditions most frequently cited are the following:

1. There is no error associated with the values of X.
2. The errors in Y are random numbers with a mean of 0 and are independent of X.

If these two conditions are satisfied, then we use what statisticians call model I linear regression to estimate α and β. We choose A and B so as to minimize the right-hand side of Equation 2.1.

Whether or not there are errors associated with the values of X can be a subtle point. It is sometimes mistakenly assumed that if the measurements of X are made with great accuracy, then the errors in X are negligible. There are, however, two sources of error associated with both X and Y. One is measurement error. The other, as noted by Ricker (1973, p. 410) is the error "inherent in the material being measured." For example, suppose we were interested in knowing the relationship between the height and weight of school children. We could measure the height and weight of each child with great accuracy, but the fact is that not all children of a given height have the same weight, nor do all children of a given weight have the same height. In this case, regardless of whether we consider height or weight to be X, the fact is that there will be some error in X caused by natural variability, and this error will exist no matter how accurately we make our measurements. It is therefore very important to keep in mind that errors in X can be caused by both measurement errors and natural variability. We can often reduce the first source of error to a negligible level by making accurate measurements. Minimizing the impact of the second source of error on our estimates of α and β may be much more difficult, although there are certainly ways to try to deal with this problem (see the Exercises).

There is another very important circumstance under which it can be shown that minimizing the right-hand side of Equation 2.1 gives unbiased estimates of α and β. Even if there are errors associated with the values of X, model I regression methods will give unbiased estimates of α and β if condition 2 is satisfied, that is, if the errors in Y are random numbers with a mean of 0 and are independent of X, and if the values of X are *controlled observations*. Since all observations are associated with some error, the distinction between controlled and uncontrolled observations is very important. "An 'uncontrolled' observation is made when, wishing to ascertain the value of some unknown quantity u_i, we measure it with an instrument... A 'controlled' observation is one made when, instead of wanting to know the value of some unknown u_i, we wish to bring the quantity to a value u_i" (Berkson, 1950). For example, suppose we want to know the rate of a chemical reaction as a function of temperature. We decide to measure the rate of the reaction at five temperatures, 5, 10, 15, 20, and 25°C. We have some device available to us for controlling the temperature in the reaction mixture. For example, we may run the reaction in a double-walled reaction flask. We circulate water from a thermoregulated water bath through the jacket of the flask, thereby controlling the temperature. The temperature of each observation becomes a controlled observation, because, to the best of our ability, we set the temperature to exactly the value we want. A very important point is that if we repeated the experiment many times, we would use the same set of X values, that is, the same five temperatures, every time. Although errors would cause the true temperatures to deviate from the values we had selected, we would think that the set of five temperatures were the same every time. When this condition is satisfied, it can be shown (see the following discussion) that model I regression methods will give unbiased estimates of α and β as long as the errors in X and Y are random and independent of each other.

Now that we have some feeling for the usefulness and limitations of model I regression, let us see how to choose A and B so as to minimize the right-hand side of Equation 2.1. Obviously, we want $\partial F/\partial A = \partial F/\partial B = 0$, and we want to ensure that we have indeed identified a minimum in the value of the function. We note that the right-hand side of Equation 2.1 is a sum of terms, and since the derivative of a sum is the sum of the derivatives, we can concentrate on one term at a time. The kth term is $(Y_k - A - BX_k)^2$. This is a function of a function, the first function being $G = Y_k - A - BX_k$ and the second function being $H = G^2$. Now $\partial G/\partial A = -1$, and $\partial G/\partial B = -X_k$. Since $dH/dG = 2G$, we see that

$$\frac{\partial}{\partial A}(Y_k - A - BX_k)^2 = \left(\frac{dH}{dG}\right)\left(\frac{\partial G}{\partial A}\right) = (2G)(-1) = 2(A + BX_k - Y_k) \tag{2.8}$$

$$\frac{\partial}{\partial B}(Y_k - A - BX_k)^2 = \left(\frac{dH}{dG}\right)\left(\frac{\partial G}{\partial B}\right)$$
$$= (2G)(-X_k) = 2\left(AX_k + BX_k^2 - X_kY_k\right) \tag{2.9}$$

Summing over all the terms in Equation 2.1 and setting the partial derivatives equal to 0 gives the following equations:

$$\sum_{k=1}^{N} Y_k = AN + B\sum_{k=1}^{N} X_k \tag{2.10}$$

$$\sum_{k=1}^{N} X_kY_k = A\sum_{k=1}^{N} X_k + B\sum_{k=1}^{N} X_k^2 \tag{2.11}$$

The average values of Y and X, which we will denote $\bar{Y}$ and $\bar{X}$, respectively, are defined by

$$\bar{X} = \frac{1}{N}\sum_{k=1}^{N} X_k \quad \text{and} \quad \bar{Y} = \frac{1}{N}\sum_{k=1}^{N} Y_k$$

Therefore Equation 2.10 can be written as

$$\bar{Y} = A + B\bar{X} \tag{2.12}$$

We can solve Equations 2.10 and 2.11 for B by first multiplying Equation 2.10 by $\Sigma_{k=1}^{N} X_k$ and Equation 2.11 by N:

$$\sum_{k=1}^{N} X_k \sum_{m=1}^{N} Y_m = AN\sum_{k=1}^{N} X_k + B\left(\sum_{k=1}^{N} X_k\right)^2 \tag{2.13}$$

$$N\sum_{k=1}^{N} X_k Y_k = AN\sum_{k=1}^{N} X_k + BN\sum_{k=1}^{N} X_k^2 \tag{2.14}$$

By subtracting Equation 2.13 from Equation 2.14, we eliminate A and can solve for B:

$$B = \frac{N\sum X_k Y_k - \sum X_n \sum Y_m}{N\sum X_j^2 - \left(\sum X_i\right)^2} \tag{2.15}$$

where the index in each summation extends from 1 to N. Equation 2.15 can be rewritten as

$$B = \frac{\sum X_k Y_k - N\bar{X}\bar{Y}}{\sum X_j^2 - N\bar{X}^2} \tag{2.16}$$

Now consider the expression $\Sigma_{k=1}^{N}(X_k - \bar{X})(Y_k - \bar{Y})$. Expanding this expression gives us

$$\begin{aligned}\sum(X_k - \bar{X})(Y_k - \bar{Y}) &= \sum X_k Y_k - \bar{X}\sum Y_k - \bar{Y}\sum X_k + N\bar{X}\bar{Y} \\ &= \sum X_k Y_k - N\bar{X}\bar{Y}\end{aligned} \tag{2.17}$$

Thus the numerator of the right-hand side of Equation 2.16 is $\Sigma(X_k - \bar{X})(Y_k - \bar{Y})$. Similar reasoning shows that the denominator is equal to $\Sigma(X_k - \bar{X})^2$. Now define $x_k = X_k - \bar{X}$ and $y_k = Y_k - \bar{Y}$. Then $\Sigma(X_k - \bar{X})(Y_k - \bar{Y}) = \Sigma x_k y_k$ and $\Sigma(X_k - \bar{X})^2 = \Sigma x_k^2$. Hence

$$B = \frac{\sum x_k y_k}{\sum x_m^2} \tag{2.18}$$

Now let us see whether the values of A and B defined by Equations 2.12 and 2.18 do minimize $F(A, B)$. From Equations 2.8 and 2.9 it follows that

$$\left(\frac{\partial}{\partial A}\right)^2 (Y_k - A - BX_k)^2 = 2 \tag{2.19}$$

$$\left(\frac{\partial}{\partial B}\right)^2 (Y_k - A - BX_k)^2 = 2X_k^2 \tag{2.20}$$

$$\left(\frac{\partial}{\partial A}\right)\left(\frac{\partial}{\partial B}\right)(Y_k - A - BX_k)^2 = 2X_k \tag{2.21}$$

Hence

$$P = \left(\frac{\partial}{\partial A}\right)^2 F(A, B) = 2N$$

$$Q = \left(\frac{\partial}{\partial B}\right)^2 F(A, B) = 2\sum_{k=1}^{N} X_k^2,$$

and

$$R = \left(\frac{\partial}{\partial A}\right)\left(\frac{\partial}{\partial B}\right)F(A, B) = 2\sum_{k=1}^{N} X_k$$

Clearly, both P and Q are positive, so $P + Q > 0$. Furthermore,

$$R^2 - PQ = 4\left[\left(\sum_{k=1}^{N} X_k\right)^2 - N\sum_{k=1}^{N} X_k^2\right] = -4N\sum_{k=1}^{N} \left(X_k - \overline{X}\right)^2 \tag{2.22}$$

Clearly, $R^2 - PQ < 0$, and based on the discussion in Chapter 1, we have identified a minimum of the function $F(A, B)$.

EXPECTATION VALUES AND ERROR BOUNDS

Equations 2.12 and 2.18 define the estimates of α and β, respectively, obtained using model I linear regression. Here A is the estimated *intercept* of the least-squares line, since $Y = A$ when $X = 0$. Similarly, B is the estimated *slope* of the least-squares line, since $dY/dX = B$. In model I we assume that there are no errors associated with X and that any scatter in the data points about the regression line is due to errors in Y. How do the errors in Y affect our estimates of α and β?

To address this question, we first review some facts about statistics and error analysis. We begin by considering the function Z defined by

$$Z = a_1 X + a_2 Y \tag{2.23}$$

We want to know the expectation value of $[Z - E(Z)]^2$, where $E(Z)$ is the expectation value of Z and equals $a_1 E(X) + a_2 E(Y)$. The expectation value of $[Z - E(Z)]^2$ is called the variance of Z and is denoted by $V(Z)$. It follows that

$$\begin{aligned} V(Z) &= E[Z - E(Z)]^2 \\ &= E[a_1 X + a_2 Y - a_1 E(X) - a_2 E(Y)]^2 \\ &= E\{a_1[X - E(X)] + a_2[Y - E(Y)]\}^2 \\ &= a_1^2 E[X - E(X)]^2 + a_2^2 E[Y - E(Y)]^2 \\ &\quad + 2a_1 a_2 E\{[X - E(X)][Y - E(Y)]\} \\ &= a_1^2 V(X) + a_2^2 V(Y) + 2a_1 a_2 E\{[X - E(X)][Y - E(Y)]\} \end{aligned} \tag{2.24}$$

$E\{[X - E(X)][Y - E(Y)]\}$ is the *covariance* of X and Y and is denoted $\mathrm{Cov}(X, Y)$. If X and Y are *independent* of each other,[1] then $\mathrm{Cov}(X, Y) = 0$, and we say that X and Y are *uncorrelated.* Two variables are *correlated* if their covariance is nonzero. Hence if $Z = a_1 X + a_2 Y$ and X and Y are independent variables, then

$$V(Z) = a_1^2 V(X) + a_2^2 V(Y) \tag{2.25}$$

This theorem can be generalized to any function Z which is a linear combination of M independent variables. In other words, if $Z = \sum_{k=1}^{M} a_k X_k$, where the X_k are independent variables, then

$$V(Z) = \sum_{k=1}^{M} a_k^2 V(X_k) \tag{2.26}$$

[1]Let X and Y be continuous variables and let $p(X_0, Y_0)dX\,dY$ be the probability that X lies between X_0 and $X_0 + dX$ and that Y lies between Y_0 and $Y_0 + dY$ in the limit as dX and dY approach 0. X and Y are independent of each other if $p(X, Y) = p_1(X)p_2(Y)$. $p(X, Y)$ is called the joint probability density function of X and Y, and in the case of independent variables, it equals the product of the probability density functions of each variable. This last statement is also true if X and/or Y is a discrete variable. In plain English, X and Y are independent of each other if the probability that X has a certain value or lies within a certain range of values is independent of Y, and vice versa.

Now let us examine the implications of the assumption that for two independent variables $\mathrm{Cov}(X, Y) = 0$. We note that $\mathrm{Cov}(X, Y) = E\{[X - E(X)][Y - E(Y)]\} = E(XY) - E(X)E(Y)$. Hence if $\mathrm{Cov}(X, Y) = 0$, $E(XY) = E(X)E(Y)$. In other words, if two variables are independent of each other, then their covariance is 0 and the expectation value of their product is equal to the product of their expectation values. More generally, if X and Y are independent variables and $F(X)$ and $G(Y)$ are functions of X and Y, respectively, then $E[F(X)G(Y)] = E[F(X)]E[G(Y)]$; that is, the expectation value of the product of the functions equals the product of the expectation values of the functions.

Now we will determine whether or not A and B are unbiased estimators of α and β, respectively. We begin by noting that $\Sigma_{k=1}^{N} x_k y_k = \Sigma_{k=1}^{N} x_k Y_k$, since $\Sigma_{k=1}^{N} x_k \overline{Y} = \overline{Y} \Sigma_{k=1}^{N} x_k = 0$, because $\Sigma_{k=1}^{N} x_k = 0$. We conclude that $B = \Sigma x_k Y_k / \Sigma x_m^2$ and hence that $B = \Sigma_{k=1}^{N} c_k Y_k$, where $c_k = x_k / \Sigma x_m^2$. Each Y_k can be written as $\alpha + \beta X_k + \varepsilon_k$, where $\alpha + \beta X_k$ is the true value of Y at $X = X_k$ and ε_k is a random error term having a mean of 0. Since we assume no error in the X's, it follows that

$$B = \sum_{k=1}^{N} c_k(\alpha + \beta X_k) + \sum_{k=1}^{N} c_k \varepsilon_k \tag{2.27}$$

It is easy to show (see Exercise 2.1) that the first summation on the right-hand side of Equation 2.27 equals β. The error in B is therefore due entirely to the second term. The expectation value of B is given by

$$E(B) = \beta + E\left(\sum_{k=1}^{N} c_k \varepsilon_k \right) \tag{2.28}$$

We are assuming, however, that the errors in Y are independent of X, which means that ε_k is independent of c_k. Hence $E(c_k \varepsilon_k) = E(c_k)E(\varepsilon_k)$. However, since $E(\varepsilon_k) = 0$ for all k, it follows that $E(B) = \beta$.

Now let us consider the intercept A. Recall that $A = \overline{Y} - B\overline{X}$. Assuming that there is an error ε_k associated with Y_k, we can write

$$A = \frac{1}{N} \sum_{k=1}^{N} (\alpha + \beta X_k + \varepsilon_k) - \sum_{m=1}^{N} [c_m(\alpha + \beta X_m) + c_m \varepsilon_m] \overline{X} \tag{2.29}$$

It is easy to show that Equation 2.29 reduces to

$$A = \alpha + \frac{1}{N} \sum_{k=1}^{N} \varepsilon_k - \overline{X} \sum_{m=1}^{N} c_m \varepsilon_m \tag{2.30}$$

It is obvious that the expectation value of $(1/N)\Sigma_{k=1}^{N} \varepsilon_k$ is 0, and the expectation value of $\overline{X} \Sigma_{m=1}^{N} c_m \varepsilon_m$ is 0 because the ε_k are independent of the X_k and $E(\varepsilon_k) = 0$. Thus $E(A) = \alpha$; that is, A is an unbiased estimator of α.

Now consider the case where there are errors associated with X, but the observations are controlled observations. Let δ_k be the error associated with X_k. We then have $Y_k = \alpha + \beta(X_k + \delta_k) + \varepsilon_k$. The deviation of Y_k from the true regression line is therefore $\beta\delta_k + \varepsilon_k$. It is important to realize that in the controlled case, the X_k are the target values which, in the words of Berkson, "we wish to bring the quantity to." These target values never change from one repetition of the experiment to the next. In the example given, where we wanted to measure the rate of a chemical reaction at 5, 10, 15, 20, and 25°C, the five target values of X_k would always be 5, 10, 15, 20, and 25°C. The actual temperatures in any particular repetition of the experiment might deviate from these values due to errors, but we would think that we were running the experiments at 5, 10, 15, 20, and 25°C and hence would use exactly those values in calculating A and B on each repetition of the experiment. The value of B calculated from a particular experimental run would therefore be

$$B = \sum_{k=1}^{N} c_k(\alpha + \beta X_k) + \sum_{k=1}^{N} c_k(\beta\delta_k + \varepsilon_k) \tag{2.31}$$

The first summation again equals β, and the error is therefore confined to the second summation. The expectation value of B is given by

$$E(B) = \beta + E\left[\sum_{k=1}^{N} c_k(\beta\delta_k + \varepsilon_k)\right] \tag{2.32}$$

However, since the c_k are based on the target values of the X's and hence do not change from one repetition of the experiment to the next, it follows that $E(c_k\,\beta\delta_k) = \beta c_k E(\delta_k) = 0$ and $E(c_k\,\varepsilon_k) = c_k E(\varepsilon_k) = 0$. Hence $E(B) = \beta$.

It is similarly straightforward to show, by analogy with Equation 2.30, that

$$A = \alpha + \frac{1}{N}\sum_{k=1}^{N}(\beta\delta_k + \varepsilon_k) - \bar{X}\sum_{m=1}^{N} c_m(\beta\delta_m + \varepsilon_m) \tag{2.33}$$

The expectation value of the first error term is clearly 0, and the expectation value of the second error term is 0 for the reasons we have just discussed. Hence, even when there are errors associated with X, model I regression will still give us an unbiased estimate of α and β as long as the observations are controlled and the errors in X and Y are random and have an expectation value of 0. If the observations are not controlled, then c_k will not be the same from one repetition of the experiment to the next, and, as we will see in Chapter 4, the c_k under these conditions will tend to systematically underestimate the magnitude of the true c_k. The result will be biased estimates of α and β.

If there are errors associated with either X or Y, the estimations of β obtained by applying Equation 2.18 to successive data sets will not all be the same. It is obviously of interest to have some measure of the scatter of B values about the true slope, β. In particular, we would like to have an unbiased estimate of the variance of B, or, in other words, an unbiased estimate of $E[B - E(B)]^2 = E[(B - \beta)^2] = V(B)$.

In the case where there are no errors associated with X, it follows from Equation 2.27 that

$$V(B) = E\left(\sum_{k=1}^{N} c_k \varepsilon_k\right)^2 \tag{2.34}$$

When we square $\sum_{k=1}^{N} c_k \varepsilon_k$, we obtain squared terms like $c_k^2 \varepsilon_k^2$ as well as cross terms like $c_k c_j \varepsilon_k \varepsilon_j$, where $k \neq j$. Now $E(c_k c_j \varepsilon_k \varepsilon_j) = E(c_k c_j) E(\varepsilon_k \varepsilon_j)$, since the errors in Y are independent of the X's and hence of c_k and c_j. Furthermore, since the errors in Y are independent of each other, $E(\varepsilon_k \varepsilon_j) = E(\varepsilon_k)E(\varepsilon_j)$. However, $E(\varepsilon_k) = E(\varepsilon_j) = 0$. Hence the expectation value of all the cross terms is 0.

For the squared terms, we again take advantage of the fact that the errors in Y are independent of the X's. Hence $E(c_k^2 \varepsilon_k^2) = E(c_k^2)E(\varepsilon_k^2)$. Now since $Y_k - E(Y_k) = \varepsilon_k$, it follows that $V(Y_k) = V(\varepsilon_k)$. We now make the assumption that $V(Y_k)$ is the same for all Y_k. In other words, not only are the errors in Y independent of the X's and of each other, but they all have the same variance. We will call this common variance $V(Y|X)$, meaning the variance of Y, given X. We conclude that

$$V(B) = \sum_{k=1}^{N} E(c_k^2)E(\varepsilon_k^2) = V(Y|X) \sum_{k=1}^{N} E(c_k^2) = V(Y|X) E\left[\sum_{k=1}^{N} (c_k^2)\right] \tag{2.35}$$

However, since $c_k = x_k / \sum x_m^2$, $\sum_{k=1}^{N} c_k^2 = 1/\sum x_m^2$ (see Exercise 2.1b). Hence

$$V(B) = V(Y|X) E\left(\frac{1}{\sum x_m^2}\right) \tag{2.36}$$

Thus an unbiased estimate of $V(B)$ is

$$V(B) = \frac{V(Y|X)}{\sum x_m^2} \tag{2.37}$$

In the case where the observations are controlled but there are errors associated with both X and Y, we use Equation 2.31 to deduce that

$$V(B) = E\left[\sum_{k=1}^{N} c_k(\beta\delta_k + \varepsilon_k)\right]^2 \tag{2.38}$$

The logic follows as in the previous case, but now we note that $Y_k - E(Y_k) = \beta\delta_k + \varepsilon_k$, and hence $V(Y_k) = E(\beta\delta_k + \varepsilon_k)^2$. We have already assumed that δ_k and ε_k are independent, but we now also assume that $E(\delta_k^2)$ is the same for all k and that $E(\varepsilon_k^2)$ is the same for all k. Given these assumptions, we again arrive at Equation 2.37.

Now what about the variance of A? In the case where no error is associated with the X's, it follows from Equation 2.30 that

$$V(A) = E(A - \alpha)^2 = E\left\{\frac{1}{N}\sum_{k=1}^{N} \varepsilon_k - \left(\sum_{k=1}^{N} c_k\varepsilon_k\right)\overline{X}\right\}^2 = E\left(\sum_{k=1}^{N} d_k\varepsilon_k\right)^2 \tag{2.39}$$

where $d_k = ((1/N) - c_k\overline{X})$. We again take advantage of the fact that the ε_k are independent of each other as well as of the X's. Hence the expectation value of all the cross terms in the expansion of the square on the right-hand side of Equation 2.39 will be 0, and

$$V(A) = E\left[\sum_{k=1}^{N} (d_k\varepsilon_k)^2\right] \tag{2.40}$$

However, since the ε_k are independent of the X's, Equation 2.40 becomes

$$V(A) = \sum_{k=1}^{N} E(d_k^2)E(\varepsilon_k^2) \tag{2.41}$$

Here again, we assume that $E(\varepsilon_k^2)$ is the same for all k and equal to $V(Y|X)$. Hence Equation 2.41 becomes

$$V(A) = V(Y|X)\sum_{k=1}^{N} E(d_k^2) = V(Y|X)E\left(\sum_{k=1}^{N} d_k^2\right) \tag{2.42}$$

Now with a little manipulation (see Exercise 2.1d), one can show that

$$\sum_{k=1}^{N} d_k^2 = \frac{\sum X_k^2}{N\sum x_m^2}$$

Hence an unbiased estimate of $V(A)$ is

$$V(A) = V(A) = V(Y|X)\left[\frac{\sum X_k^2}{N\sum x_m^2}\right] = V(B)\frac{\sum X_k^2}{N} \tag{2.43}$$

In the case where the observations are controlled but there are errors associated with both X and Y, we use Equation 2.33 to deduce that

$$V(A) = E\left[\sum_{k=1}^{N} d_k(\beta\delta_k + \varepsilon_k)\right]^2 \tag{2.44}$$

The logic follows as before, again noting that in this case $V(Y_k) = E(\beta\delta_k + \varepsilon_k)^2$. We assume this number to be the same for all k, and are again led to Equation 2.43.

We have now derived equations for both $V(B)$ and $V(A)$, but both depend on $V(Y|X)$, which we do not know. We assume that $V(Y|X) = E(Y_k - \alpha - \beta X_k)^2$ for all k, and, based on this assumption, it follows that $(1/N)\Sigma_{k=1}^{N}(Y_k - \alpha - \beta X_k)^2$ is an unbiased estimator of $V(Y|X)$. The problem is that we do not know α and β. We could estimate $V(Y|X)$ by calculating $(1/N)\Sigma_{k=1}^{N}(Y_k - A - BX_k)^2$, but this estimate would clearly be biased, because both A and B were chosen to minimize $\Sigma_{k=1}^{N}(Y_k - A - BX_k)^2$. How do we calculate an unbiased estimate of $V(Y|X)$?

Unbiased Estimates and Degrees of Freedom

To answer this question, we will first consider a simpler problem. Consider the sum $\Sigma_{n=1}^{N}(X_n - \mu)^2$, where μ is the true mean of X. We define ε_n to be $X_n - \mu$. Clearly, $(1/N)\Sigma\varepsilon_n^2$ is an unbiased estimate of $V(X)$. Now let $\bar{X} = (1/N)\Sigma X_n$ be the mean of the N values of X in our sample. We can now write

$$\sum_{n=1}^{N}(X_n - \mu)^2 = \sum_{n=1}^{N}(X_n - \bar{X} + \bar{X} - \mu)^2 \tag{2.45}$$

$$= \sum_{n=1}^{N}(X_n - \bar{X})^2 + 2(\bar{X} - \mu)\sum_{n=1}^{N}(X_n - \bar{X}) + N(\bar{X} - \mu)^2 \tag{2.46}$$

The second summation is clearly 0. If we now take the expectation values of both sides of Equation 2.46, we obtain

$$E\left[\sum_{n=1}^{N}(X_n-\mu)^2\right]=NV(X)=E\left[\sum_{n=1}^{N}(X_n-\bar{X})^2\right]+NE\left[(\bar{X}-\mu)^2\right] \tag{2.47}$$

However, $E[(\bar{X}-\mu)^2]=V(\bar{X})=V(X)/N$. Hence

$$E\left[\sum_{n=1}^{N}(X_n-\bar{X})^2\right]=(N-1)V(X) \tag{2.48}$$

We conclude that $[1/(N-1)]\Sigma(X_n-\bar{X})^2$ is an unbiased estimator of $V(X)$.

We will use a similar strategy to derive the relationship between $\Sigma_{n=1}^{N}(Y_n-A-BX_n)^2$ and $V(Y|X)$. We begin with the sum $\Sigma_{n=1}^{N}(Y_n-\alpha-\beta X_n)^2$. We define ε_n to be $Y_n-\alpha-\beta X_n$ and assume that $E(\varepsilon_n^2)=V(Y|X)$ is the same for all n. We then have

$$\sum_{n=1}^{N}\varepsilon_n^2=\sum_{n=1}^{N}\left[Y_n-\bar{Y}-B(X_n-\bar{X})+(\bar{Y}-\alpha-\beta\bar{X})+(B-\beta)(X_n-\bar{X})\right]^2 \tag{2.49}$$

$$=\sum_{n=1}^{N}\left[(Y_n-A-BX_n)+(\bar{Y}-\alpha-\beta\bar{X})+(B-\beta)(X_n-\bar{X})\right]^2 \tag{2.50}$$

When we expand the expression on the right-hand side of Equation 2.50, we find that all the cross terms cancel out. The reason can be traced directly to our least-squares equations, Equations 2.10 and 2.11. From Equation 2.10 we can see that

$$\sum_{n=1}^{N}(Y_n-A-BX_n)=0 \tag{2.51}$$

and from Equation 2.11 it follows that

$$\sum_{n=1}^{N}(Y_n-A-BX_n)X_n=0 \tag{2.52}$$

We therefore have

$$\sum_{n=1}^{N}\varepsilon_n^2=\sum_{n=1}^{N}(Y_n-A-BX_n)^2+(B-\beta)^2\sum_{n=1}^{N}(X_n-\bar{X})^2+N(\bar{Y}-\alpha-\beta\bar{X})^2 \tag{2.53}$$

Taking the expectation value of both sides of Equation 2.53, we note that $(B-\beta)^2 = \Sigma c_k^2 \varepsilon_k^2$ and hence

$$E\left[(B-\beta)^2 \sum_{n=1}^{N} (X_n - \bar{X})^2\right] = V(Y|X) E\left[\sum_{k=1}^{N} c_k^2 \sum_{n=1}^{N} (X_n - \bar{X})^2\right] = V(Y|X)$$

(see Exercise 2.1b). Similarly,

$$E\left[(\bar{Y} - \alpha - \beta\bar{X})^2\right] = \frac{1}{N^2} E\left[\left(\sum_{n=1}^{N} \varepsilon_n\right)^2\right] = \frac{1}{N} V(Y|X) \qquad (2.54)$$

We conclude that

$$E\left(\sum_{n=1}^{N} \varepsilon_n^2\right) = NV(Y|X) = E\left[\sum_{n=1}^{N} (Y_n - A - BX_n)^2\right] + 2V(Y|X) \quad (2.55)$$

Hence

$$E\left[\sum_{n=1}^{N} (Y_n - A - BX_n)^2\right] = (N-2)V(Y|X)$$

and we conclude that an unbiased estimate of $V(Y|X)$ is

$$V(Y|X) = \frac{1}{N-2}\left[\sum_{n=1}^{N} (Y_n - A - BX_n)^2\right] \qquad (2.56)$$

We reach exactly the same conclusion in the case where the observations are controlled, the only difference being that $V(Y|X)$ in that case equals the expectation value of $(\beta\delta_k + \varepsilon_k)^2$.

We have now seen two examples in which a sum of squared deviations based on sample statistics can yield an unbiased estimate of a variance. In both cases the statistics were chosen so as to minimize the sum of squared deviations. In the first case, involving $\Sigma_{k=1}^{N}(X_k - \bar{X})^2$, one sample statistic, $\bar{X}$, appeared in the expression, and we divided the sum of squared deviations by $N-1$ to obtain an unbiased estimate of $V(X)$. In the second case, involving $\Sigma_{k=1}^{N}(Y_k - A - BX_k)^2$, two sample statistics, A and B, appeared in the expression, and we divided the sum of squared deviations by $N-2$ to obtain an unbiased estimate of $V(Y|X)$. In these two calculations $N-1$ and $N-2$ are called the number of *degrees of freedom* associated with the two variance estimates. The term degrees of freedom is often abbreviated df and may be defined as the number of observations in a sample minus the number of parameters estimated from the sample. In other words, the df is the number of observations in excess of those necessary to estimate the parameters. In the case of $\Sigma(X_k - \bar{X})^2$, we obviously need at least one observation to estimate μ, and in the case of $\Sigma(Y_k - A - BX_k)^2$, we need at least two points to estimate α and β, since two points determine a line. Looking at the problem another way, we could not estimate $V(X)$ in the first case with only one value of X, and we could not estimate $V(Y|X)$ in the second case with only two (X, Y) pairs. Thus the df associated with a particular statistic must be a positive number and is a measure of the number of excess observations which were available for calculating the statistic.

Now that we know how to calculate an unbiased estimate of $V(Y|X)$, we can also calculate unbiased estimates of $V(A)$ and $V(B)$ using Equations 2.37 and 2.43. There are two other variances that we may also wish to calculate. For example, suppose we measure the respiration rate of herring as a function of temperature under conditions where we control the temperature. Over the range of temperatures studied, the relationship between temperature and respiration rate appears to be linear, and we fit a straight line to the data using model I techniques, since temperature was under our control. We now ask the question, with what confidence can we predict the respiration rate of herring at temperature T_0? Obviously, we need to know the variance of the predicted respiration rates about the population regression line. We could also ask, how much is the average respiration rate at temperature T_0 of M herring chosen at random likely to deviate from the value predicted from the regression line? The second question is different from the first and will have a different answer. We will consider the first question first.

Call $\hat{Y}_0$ our estimate of herring respiration rate at temperature T_0, and let μ be the true respiration rate. In other words, $\hat{Y}_0 = A + BT_0$, and $\mu = \alpha + \beta T_0$. Then

$$\hat{Y}_0 - \mu = A + BT_0 - \alpha - \beta T_0 = (A - \alpha) + T_0(B - \beta)$$

Hence $V(\hat{Y}_0) = E[(A - \alpha) + T_0(B - \beta)]^2$. However, $A - \alpha = \Sigma_{n=1}^{N} d_n \varepsilon_n$ and $B - \beta = \Sigma_{m=1}^{N} c_m \varepsilon_m$. Taking into account the independence of the error terms, we conclude that

$$V(\hat{Y}_0) = V(Y|T) \sum_{n=1}^{N} (d_n^2 + 2T_0 c_n d_n + T_0^2 c_n^2) \tag{2.57}$$

Given the definitions of the c_n and d_n, one can show that Equation 2.57 becomes

$$V(\hat{Y}_0) = V(Y|T)\left[\frac{1}{N} + \frac{(T_0 - \bar{T})^2}{\sum t_k^2}\right] \tag{2.58}$$

Now we will consider the second case, where we measure the respiration rates of M herring at temperature T_0 after having calculated a regression line based on a sampling of N herring. We want to know how far the average of the M respiration rates is likely to differ from the value predicted by the regression line. The average of the M respiration rates, $\bar{Y}_0$, is equal to $\alpha + \beta T_0 + (1/M)\Sigma_{m=1}^{M} \varepsilon_m$, and the value predicted by the regression line, $\hat{Y}_0$, is $A + BT_0$. In both cases the expectation value is $\alpha + \beta T_0$. Hence the expectation value of the difference is 0, and $E(\hat{Y}_0 - \bar{Y}_0)^2 = V(\hat{Y}_0 - \bar{Y}_0)$. Since the sample of M herring is taken after the sampling of N herring, the errors in $\hat{Y}_0$ and $\bar{Y}_0$ are independent, and $V(\hat{Y}_0 - \bar{Y}_0) = V(\hat{Y}_0) + V(\bar{Y}_0)$. Now

$$V(\bar{Y}_0) = V\left(\frac{1}{M}\sum_{m=1}^{M} \varepsilon_m\right) = \frac{1}{M}V(Y|T) \tag{2.59}$$

and we already know $V(\hat{Y}_0)$ from Equation 2.58. We conclude that

$$E(\bar{Y}_0 - \hat{Y}_0)^2 = V(\hat{Y}_0 - \bar{Y}_0) = V(Y|T)\left[\frac{1}{N} + \frac{1}{M} + \frac{(T_0 - \bar{T})^2}{\sum_{k=1}^{N} t_k^2}\right] \tag{2.60}$$

Equations identical to Equations 2.58 and 2.60 are obtained in cases where there are errors associated with the X's, but the observations are controlled. The flow of logic is exactly the same, with $\beta\delta_k + \varepsilon_k$ replacing ε_k.

The Normal and *t* Distributions

At this point we have made no assumption about the distribution of the errors in X and Y, other than the fact that they are random and have a mean of 0. In order to assign confidence limits to our estimates, however, we need to assume something about the distribution of these errors. The assumption commonly made is that the errors are normally distributed, meaning that their distribution follows the normal distribution function first described by De Moivre in 1733. If a variable is normally distributed, the probability in the limit as dX approaches 0 of the variable's lying between X and $X + dX$ is given by $p(X)\,dX$, where

$$p(X) = \frac{1}{\sigma\sqrt{2\pi}}\exp\left[-\frac{(X-\mu)^2}{2\sigma^2}\right]$$

where μ is the average value of X and σ^2 is the variance of X. The square root of the variance, or σ, is called the *standard deviation* of X. It is common

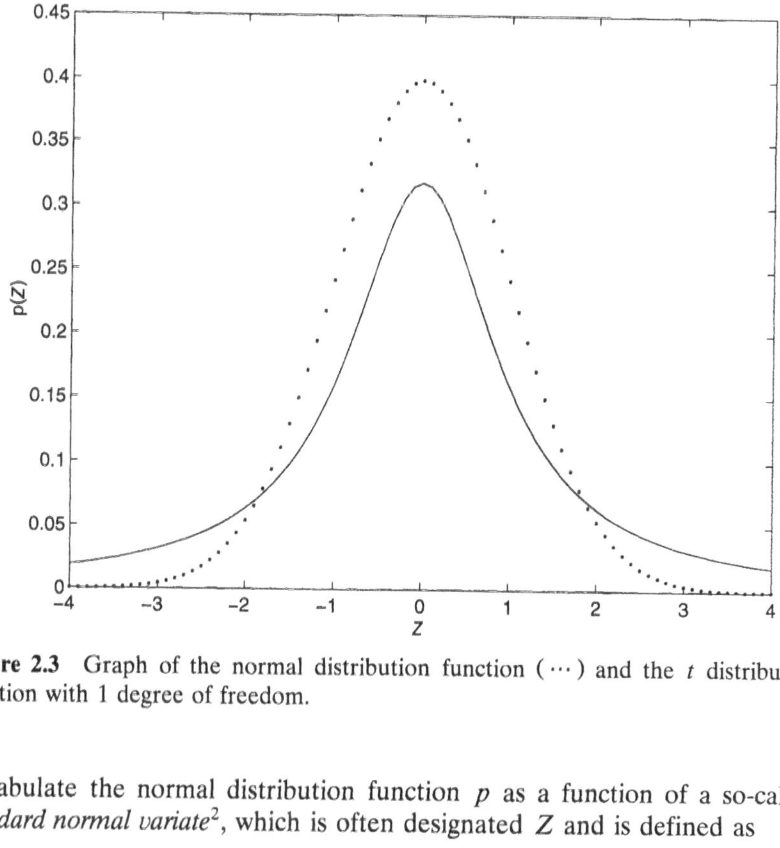

Figure 2.3 Graph of the normal distribution function (···) and the t distribution function with 1 degree of freedom.

to tabulate the normal distribution function p as a function of a so-called *standard normal variate*[2], which is often designated Z and is defined as

$$Z = \frac{X - \mu}{\sigma}$$

Here Z has a mean value of 0 and a variance of 1. Figure 2.3 shows a standard normal curve. If a variable is normally distributed, it will lie within $\pm\sigma$ of μ 68.3% of the time, within $\pm 2\sigma$ of μ 95.4% of the time, and within $\pm 3\sigma$ of μ 99.7% of the time.

In statistics it is common to ask for a confidence interval about the statistic which would include the true parameter a given percentage of the time if the experiment were repeated many times. For example, if we knew that estimates of β calculated from many repetitions of an experiment followed a normal distribution function, and if we knew σ, we could say that in 95.4% of the repetitions of the experiment β would lie somewhere between $B - 2\sigma$ and $B + 2\sigma$. We could say this because 95.4% of the B values would lie between $\beta - 2\sigma$ and $\beta + 2\sigma$.

[2]Also referred to as a *standard normal deviate, normal variate in standard measure,* or *standard score*.

Unfortunately, we do not know σ, although we do have an unbiased estimate of σ^2 for the parameters of interest. For example, Equations 2.37 and 2.56 give us an unbiased estimate of $V(B)$. However, the distribution of $(X - \mu)/s$ is not the same as the distribution of $(X - \mu)/\sigma$. It is true that our expressions for $s(A)$ and $s(B)$ are consistent estimators of $\sigma(A)$ and $\sigma(B)$, respectively. Hence, in the limit of a large number of (X, Y) data points, we can expect that $s \to \sigma$. However, if our data set is not large, we must somehow deal with the fact that $(X - \mu)/s$ is not a standard normal variate.

This problem had been recognized for many years and was resolved in 1908 by W. S. Gosset. Gosset worked out the distribution function for $Z/(y/n)^{1/2}$, where Z and y are independent random variables, Z following a normal distribution with a mean of 0 and a variance of 1, and y following a chi-square (χ^2) distribution[3] with n degrees of freedom. For our purposes, $Z = (X - \mu)/\sigma$ and $y/n = s^2/\sigma^2$. Hence $Z/(y/n)^{1/2} = (X - \mu)/s$. The distribution function worked out by Gosset is called *Student's t distribution function*[4] or just the t distribution function. It is a function of both $(X - \mu)/s$ and n and becomes identical to the normal distribution function as $n \to \infty$. The distribution function $t[(X - \mu)/s, n]$ is defined by the equation

$$t\left[\frac{X-\mu}{s}, n\right] = \frac{\Gamma[(n+1)/2]}{\Gamma(n/2)}\left[1 + \frac{(X-\mu)^2}{ns^2}\right]^{-(n+1)/2}(n\pi)^{-1/2}$$

where $\Gamma(n + 1) = n\Gamma(n)$, $\Gamma(1/2) = \pi^{1/2}$, and $\Gamma(n) = (n - 1)!$ for integer n. Here $(n - 1)!$ means $(n - 1)$ factorial, that is, the product of the first $n - 1$ integers. Figure 2.3 shows the t distribution function with 1 degree of freedom. The difference between the t and normal distribution functions is very dramatic for small values of n. In the case of the normal distribution function, for example, 95% of the area under the curve lies within $\mu \pm 1.96\sigma$. In the case of the t distribution function with 1 degree of freedom, 95% of the area under the curve lies within $\mu \pm 12.706s$. The much wider spread of the t function for small values of n arises primarily because statistics become much more variable from one repetition of an experiment to the next when there are few degrees of freedom associated with their estimation.

We can now establish confidence intervals for our statistics, realizing that they follow the t distribution function. One note of caution, however, is in order. Suppose we are interested in calculating a 95% confidence interval. The meaning behind a 95% confidence interval is that if we repeated an experiment many times and calculated an unbiased estimate of a parameter and a 95% confidence interval each time, then 95% of the confidence intervals would include the true value of the parameter. This does not mean

[3]The χ^2 distribution is defined by $p(\chi^2) = [2^{n/2}\Gamma(n/2)]^{-1}(\chi^2)^{((n/2)-1)}\exp(-\chi^2/2)$.
[4]Gosset published his results for the distribution function under the pseudonym Student.

that after doing the experiment once we can say with 95% confidence that the true value lies within the interval we have calculated from that particular experiment.

To perhaps see the distinction more clearly, consider rolling a six-sided die. Before we roll the die, we can say that the probability of rolling a 5 is one in six. However, once the die is cast, there is no uncertainty about the result. Either we have a 5 or we do not have a 5. It makes no sense to say that what we are looking at after we roll the die has one chance in six of being a 5. Similarly, before doing an experiment, we can say that the 95% confidence interval we are going to estimate has a 95% probability of including the true value of the parameter we are going to estimate. However, once the experiment is done and the confidence interval calculated, the true parameter value either does or does not lie within the confidence interval. The distinction is subtle, but it is real and worth remembering.

ASSUMPTIONS ABOUT ERRORS, PROPAGATION OF ERRORS, AND MONTE CARLO METHODS

At this point it is appropriate to review some of the assumptions we have made about errors and to think a bit about ways of testing our propagation of error analysis.

In model I regression we always assume that the errors in Y are random and independent of each other and of X. Given this assumption, we were able to show that Equations 2.12 and 2.18 gave unbiased estimates of α and β if

1. there were no errors associated with X, or
2. there were errors associated with X, but the errors were random and independent of each other and of X and the observations were controlled.

We then went on to derive expressions for $V(Y|X)$, $V(A)$, $V(B)$, $V(\hat{Y}_0)$, and $E(\hat{Y}_0 - \bar{Y}_0)^2$. We were able to derive these expressions and to show that they were unbiased estimates because we assumed that the variances of the errors in Y were the same for all values of Y and that the variances of the errors in X were the same for all values of X. The fact that the errors are independent of each other and of X does not require that they have the same variance, so this is an additional assumption imposed to allow us to derive simple unbiased estimates for the parameters of interest. Finally, we assumed that the errors in X and Y were normally distributed. This allowed us to derive expressions for confidence intervals for the statistics we had calculated. It is important to keep in mind the hierarchy of assumptions here and the implications of the various assumptions. It is not, for example, necessary to assume that the errors in Y and X are normally distributed in order to conclude that Equations 2.12 and 2.18 give unbiased estimates of α and β.

Here is a summary of the equations we have derived for calculating unbiased estimates of parameters:

$$A = \overline{Y} - B\overline{X} \tag{2.61}$$

$$B = \frac{\sum_{k=1}^{N} x_k y_k}{\sum_{n=1}^{N} x_n^2} \tag{2.62}$$

$$V(Y|X) = \frac{1}{N-2} \sum_{n=1}^{N} (Y_k - A - BX_k)^2 \tag{2.63}$$

$$V(A) = V(Y|X) \frac{\sum_{k=1}^{N} X_k^2}{N \sum_{n=1}^{N} x_n^2} \tag{2.64}$$

$$V(B) = \frac{V(Y|X)}{\sum_{n=1}^{N} x_n^2} \tag{2.65}$$

$$V(\hat{Y}_0) = V(Y|X)\left[\frac{1}{N} + \frac{(X_0 - \overline{X})^2}{\sum x_k^2}\right] \tag{2.66}$$

$$V(\overline{Y}_0 - \hat{Y}_0) = V(Y|X)\left[\frac{1}{N} + \frac{1}{M} + \frac{(X_0 - \overline{X})^2}{\sum_{k=1}^{N} x_k^2}\right] \tag{2.67}$$

Finally, we learned that if Z is a normally distributed variable, $\overline{Z}$ is an unbiased estimate of $E(Z)$, and s^2 is an unbiased estimate of $V(\overline{Z})$ with n degrees of freedom, then $[\overline{Z} - E(Z)]/s$ follows Student's t distribution with n degrees of freedom.

Deriving expressions such as Equations 2.63 to 2.67 falls under the general category of propagation of error analysis. The idea is to determine in a quantitative way the impact of observational errors on statistics calculated from those observations. Propagation of error is relatively straightforward in the case of model I regression if one is willing to assume that $V(Y|X)$ is the same for all X. Obviously, life would become much more difficult if we had to drop this assumption, and life would certainly become more difficult if we were not dealing with such simple equations. Is there some way to perform a propagation of error analysis in more complex situations?

The answer is yes, at least if we have some idea of the distribution of observational errors. The idea is to generate what are called noise-corrupted data sets, in which we deliberately create errors in our data and determine the impact of those errors on our calculations. For example, let us go back to the case where we measure the rate of a chemical reaction at 5, 10, 15, 20, and 25°C. Assume for the sake of argument that the underlying relationship between temperature and reaction rate is $Y = 10 + 2X$, where Y is the reaction rate and X is the temperature. We will assume the observations are controlled, which means we use the same five values of X in each repetition of the experiment. Now assume that the errors in Y are normally distributed with a standard deviation of 3 and a mean of 0. A noise-corrupted data set would consist of five (X, Y) pairs, with the X values being the five chosen temperatures and the Y values determined from $Y_k = 10 + 2X_k + \varepsilon_k$, where ε_k is a normally distributed variable with a mean of 0 and a standard deviation of 3. We fit a straight line through the data, using Equations 2.61 and 2.62 to calculate A and B. We use Equations 2.63 to 2.65 to estimate $V(Y|X)$, $V(A)$, and $V(B)$. We then repeat this procedure many times, for example, 1000 times. We keep track of $\Sigma_{k=1}^{N} A_k$, $\Sigma_{k=1}^{N} A_k^2$, $\Sigma_{k=1}^{N} B_k$, and $\Sigma_{k=1}^{N} B_k^2$, and with this information we can calculate unbiased estimates of $V(A)$ and $V(B)$. These estimates should agree well with the average of the estimates of $V(A)$ and $V(B)$ that we made with each noise-corrupted data set, at least if the latter are unbiased estimates. The average of the $V(Y|X)$ estimates should be approximately $3^2 = 9$. We cannot expect the agreement to be perfect, because we simulated running the experiment 1000 times, not an infinite number of times. However, the more simulated repetitions we do, the closer the statistics should come to the true parameters.

In this case $\Sigma_{k=1}^{N} x_k^2 = 250$, and $V(Y|X) = 9$, so $V(B) = 9/250$, and $\sigma(B) = 3\sqrt{10}/50 = 0.190$. We expect that $(B - 2)/s(B)$ will follow the t distribution function with $5 - 2 = 3$ degrees of freedom. Reference to Appendix A shows that 95% of the B values should therefore lie between $2 - (0.19)(1.96)$ and $2 + (0.19)(1.96)$, or 1.6 and 2.4.

The procedure of using large numbers of simulated repetitions of an experiment to estimate statistics is called a Monte Carlo procedure. In the case of model I linear regression, it may seem like using an axe to cut butter, but it does provide a check on our theoretical analysis of propagation of error. The power of Monte Carlo methods, however, lies in their ability to propagate errors in the numerous cases where there is no simple theoretical solution. We will try a few Monte Carlo problems in the Exercises.

If we are going to use Monte Carlo methods, how do we generate random numbers that follow a normal distribution? Some computer software packages will do it for us, but we can do it ourselves if we can generate uniformly distributed random numbers between 0 and 1. Virtually any programming language or software package that deals with mathematical calculations includes a random number generator that produces uniformly distributed random numbers on the interval [0, 1]. If U_1 and U_2 are two such random numbers, then

$$Z = \left[-2\ln(U_1)\right]^{1/2}\cos(2\pi U_2)$$

is a normally distributed random number with a mean of 0 and a standard deviation of 1. Here $\ln(U_1)$ is the natural logarithm, that is, logarithm to the base e, of U_1, and π is the ratio of the circumference of a circle to its diameter, that is, 3.14159 . . . . Then $V = \mu + \sigma Z$ is a normally distributed random number with a mean of μ and a standard deviation of σ.

EXERCISES

2.1 Given that $c_k = x_k/\Sigma x_m^2$ and that $d_k = (1/N) - c_k\overline{X}$, show that

a.
$$\sum_{k=1}^{N} c_k = 0$$

b.
$$\sum_{k=1}^{N} c_k^2 = \frac{1}{\Sigma x_m^2}$$

c.
$$\sum_{k=1}^{N} c_k(\alpha + \beta X_k) = \beta$$

d.
$$N\sum_{k=1}^{N} d_k^2 = \frac{\Sigma X_m^2}{\Sigma x_n^2}$$

e.
$$\sum_{k=1}^{N} d_k = 1$$

2.2 Show that Equation 2.29 reduces to Equation 2.30.

2.3 Show that $V(\hat{Y}_0)$ in Equation 2.66 is a minimum when $X_0 = \overline{X}$. Make a graph of the 95% confidence intervals to $\hat{Y}_0$ for the example given in the discussion of Monte Carlo methods. Plot the confidence intervals for temperatures (X) between 5 and 25°C. The 95% confidence intervals for $\hat{Y}_0$ will be $\hat{Y}_0 \pm 3.182s(\hat{Y}_0)$, where $\hat{Y}_0 = A + BX_0$, A and B having been determined from a least-squares fit to a noise-corrupted data set of five (X, Y) pairs, $s(\hat{Y}_0)$ is the square root of your estimate of $V(\hat{Y}_0)$, and 3.182 is the appropriate t statistic corresponding to 3 degrees of freedom (Appendix B).

2.4 Suppose you want to measure the rate of nutrient uptake by phytoplankton in a body of water. To determine the rate of uptake, you propose to measure nutrient concentrations in the water at 10 points in time over a 24-hour period. You will then fit a regression line to the nutrient concentration versus time data. The slope of the line will be the uptake rate. At what times should you take your 10 samples in order to make the variance of the slope a minimum? *Hint*: You can sample more than once at the same time.

2.5 Look at Equations 2.64 and 2.65. Assume that we have no way to reduce $V(Y|X)$. For example, our measurement errors may be very small compared to natural variability, and we have no way to control natural variability.

a. Assume that you can sample X at both positive and negative values of X and that there will be 10 (X, Y) data points in your data set. The domain of possible X values is $[-20, 10]$. Under what conditions is $\Sigma_{k=1}^{10} X_k^2/(10\Sigma_{m=1}^{10} x_m^2)$ a minimum, and what is its minimum value? Assume that your first priority is to minimize $V(A)$ and your second priority is to reduce $V(B)$ as much as possible subject to the constraints imposed by minimizing $V(A)$. What should be the 10 values of X in your data set? Assume again that repeated measurements may be made at the same value of X. *Hint*: Note that $X_k = x_k + \bar{X}$. Since $\Sigma_{k=1}^{N} x_k = 0$, it follows that

$$\frac{\sum_{k=1}^{N} X_k^2}{N \sum_{m=1}^{N} x_m^2} = \frac{1}{N} + \frac{\bar{X}^2}{\sum_{m=1}^{N} x_m^2}$$

For what value of $\bar{X}$ is this expression a minimum?

b. Now assume that the domain of possible X values is $[0, 20]$. Assume that you have enough resources to measure N (X, Y) pairs. M of the measurements are made at $X = 0$, and $N - M$ are made at some other value of X, where $1 \leq M \leq N - 1$. Show that under these conditions

$$\frac{\sum_{k=1}^{N} X_k^2}{N \sum_{m=1}^{N} x_m^2} = \frac{1}{M}$$

What value of M minimizes $V(A)$? What value of X minimizes $V(B)$?

2.6 Equation 2.63 requires us to first calculate A and B and then to go back and calculate the sum of the squared differences of Y_k from the regression line. We can estimate $V(Y|X)$, however, without knowing A and B. Show that Equation 2.63 is equivalent to

$$V(Y|X) = \frac{1}{N-2}\Sigma(Y_k - A - BX_k)^2$$

$$= \frac{1}{N-2}\left[\Sigma y_k^2 - \frac{\left(\Sigma x_m y_m\right)^2}{\Sigma x_n^2}\right] \qquad (2.68)$$

It follows that $(\Sigma x_m y_m)^2 \leq \Sigma x_n^2 \Sigma y_k^2$, since $V(Y|X) \geq 0$, and $V(Y|X) = 0$ when $(\Sigma x_m y_m)^2/\Sigma x_n^2 \Sigma y_k^2 = 1$. The statistic $(\Sigma x_m y_m)^2/\Sigma x_n^2 \Sigma y_k^2$ is a measure of the degree of *correlation* between X and Y and is designated r^2. The statistic $r = \Sigma x_m y_m / \sqrt{\Sigma x_n^2 \Sigma y_k^2}$. Clearly, $-1 \leq r \leq 1$. X and Y are said to be negatively correlated if $r < 0$ and positively correlated if $r > 0$. More about this in Chapter 3.

2.7 Let us do a Monte Carlo calculation to see if the equations we have derived are correct and give us unbiased estimates of population parameters. We will use the example described under Monte Carlo calculations. Generate 1000 noise-corrupted data sets, assuming the X values are controlled observations and the Y values follow the equation $Y_k = 10 + 2X_k + \varepsilon_k$, where ε_k is a normally distributed random number with a mean of zero and a standard deviation of 3. The five X values will always be 5, 10, 15, 20, and 25. For each data set calculate the following:

a. A and B from Equations 2.61 and 2.62.

b. $V(Y|X)$, $V(A)$, and $V(B)$ from Equations 2.63 and 2.64 and 2.65.

c. 95% confidence intervals for A and B. Since there are $5 - 2 = 3$ degrees of freedom, these intervals will be $A \pm 3.182[V(A)]^{1/2}$ and $B \pm 3.182[V(B)]^{1/2}$. Keep track of how many times these confidence intervals do not include the true values of α and β.

Calculate

$$\frac{1}{999}\left[\sum_{k=1}^{1000} A_k^2 - \frac{\left(\sum_{k=1}^{1000} A_k\right)^2}{1000}\right] \quad \text{and} \quad \frac{1}{999}\left[\sum_{k=1}^{1000} B_k^2 - \frac{\left(\sum_{k=1}^{1000} B_k\right)^2}{1000}\right]$$

These should be unbiased estimates of $V(A)$ and $V(B)$. Similarly, calculate

$$\frac{1}{1000}\sum_{k=1}^{1000} V_k(Y|X) \qquad \frac{1}{1000}\sum_{k=1}^{1000} V_k(A) \qquad \frac{1}{1000}\sum_{k=1}^{1000} V_k(B)$$

These should be unbiased estimates of $V(Y|X)$, $V(A)$, and $V(B)$. We know that $V(Y|X) = 9$, that $\Sigma_{m=1}^{5} x_m^2 = 250$, and that $\Sigma_{m=1}^{5} X_m^2 = 1375$. What then are the true values of $V(A)$ and $V(B)$? How do your two estimates of $V(A)$ and $V(B)$ compare to the true values? How does your estimate of $V(Y|X)$ compare to the true value of 9? Calculate the average of the 1000 values of A and B. How do these compare to 10 and 2, respectively? Do you find that the true values of α and β fall outside the 95% confidence intervals only about 5% of the time?

2.8 Repeat Exercise 2.7, but now assume that the X values are not under your control. Assume furthermore that there is no error associated with the X values. In each experiment the five X values will be chosen at random from a normally distributed population having a mean of 15 and a variance of 100/3. In this case the expectation value of

$$\sum_{m=1}^{5} x_m^2 = (N-1)V(X) = (4)\left(\frac{100}{3}\right) = \frac{400}{3}$$

and the expectation value of

$$\sum_{m=1}^{5} X_m^2 = N\left(15^2 + \frac{100}{3}\right) = 5\left(\frac{775}{3}\right) = \frac{3875}{3}$$

If you plug these values into Equations 2.64 and 2.65, taking $V(Y|X) = 9$, you will find out that this time your estimates of $V(A)$ and $V(B)$ do not agree with the values you have calculated. Why? The reason is that, in general, the average value of a quotient is not equal to the average value of the numerator divided by the average value of the denominator. The average value of the quotient will equal the quotient of the average values if the denominator is constant, and that was the case in Exercise 2.7. In this case the denominator is not constant, because X follows a normal distribution. The estimates you have made of $V(A)$ and $V(B)$ are good unbiased estimates, and they should not agree with the numbers you calculate by plugging the expectation values of $\Sigma_{m=1}^{5} x_m^2$ and $\Sigma_{m=1}^{5} X_m^2$ into Equations 2.64 and 2.65.

2.9 Repeat Exercise 2.8, but now choose the five X values randomly from a uniformly distributed population on the interval [5, 25]. You can generate uniformly distributed random numbers X on the interval [5, 25] through the transformation $X = 5 + 20\eta$, where η is a uniformly distributed number on the interval [0, 1]. The variance of a uniformly distributed random variable on the interval [5, 25] is 100/3, and $E(X^2) = 775/3$. You should again get good unbiased estimates of all parameters, but $V(A)$ and $V(B)$ will be different than in Exercise 2.8. Why? Although the expectation values of X, X^2, and x^2 are the same in both problems, the expectation values of $\Sigma X_m^2 / \Sigma x_k^2$ and $1/\Sigma x_k^2$ are different.

2.10 Show that the values of A and B calculated from Equations 2.30 and 2.27 approach α and β, respectively, in the limit as $N \to \infty$. Hence they are consistent estimators of α and β. *Hint*: This has proven to be a difficult problem for many students. The difficulty lies in determining $\lim_{N\to\infty} \sum_{k=1}^{N} c_k \varepsilon_k$. In order to calculate this limit, it is useful to write

$$c_k = \frac{x_k}{\sum x_n^2} = \frac{N x_k}{N \sum x_n^2}$$

It then follows that

$$\lim_{N\to\infty} \sum_{k=1}^{N} c_k \varepsilon_k = \lim_{N\to\infty} \sum_{k=1}^{N} \frac{N}{\sum x_n^2} \frac{x_k \varepsilon_k}{N} = \lim_{N\to\infty} \frac{N}{\sum x_n^2} \sum_{k=1}^{N} \frac{x_k \varepsilon_k}{N}$$

Assuming the limit of the product equals the product of the limits, this expression becomes

$$\left(\lim_{N\to\infty} \frac{N}{\sum x_n^2} \right) \left(\lim_{N\to\infty} \frac{1}{N} \sum_{k=1}^{N} x_k \varepsilon_k \right)$$

The assumption that the limit of the product equals the product of the limits is valid if both limits in the product are well defined, which is the case here. From this point on the proof is straightforward.

REFERENCES

Berksen, J. B. 1950. Are there two regressions? *J. Amer. Statist. Assoc.* **45** 164–180.

Ricker, W. E. 1973. Linear regressions in fishery research. *J. Fish. Res. Board Canada* **30** 409–434.

3

CORRELATION

Correlation is a measure of the extent to which two variables covary. We say, for example, that X and Y are positively (negatively) correlated if X tends to be large (small) when Y is large and small (large) when Y is small. Correlation analysis is related to regression analysis, but the two are not strictly equivalent, although they are sometimes treated as if they were.

The most common measure of correlation is the product–moment correlation coefficient ρ, defined by

$$\rho = \mathrm{Cov}(X, Y)/[V(X)V(Y)]^{1/2} \tag{3.1}$$

For finite data sets which are subsets of the population of (X, Y) pairs, an estimate, r, of ρ may be calculated from the equation

$$r = \frac{\sum x_m y_m}{\sqrt{\sum x_n^2 \sum y_k^2}} \tag{3.2}$$

Both ρ and r are associated with a bivariate population known as the bivariate normal distribution. When two variables X and Y follow a *bivariate normal distribution* function, their joint probability density function[1] $p(X, Y)$

[1]See Chapter 2, footnote 1.

is equal to

$$\frac{1}{2\pi\sigma_x\sigma_y\sqrt{1-\rho^2}}\exp\left\{-\frac{\left[\dfrac{(X-\mu_x)^2}{\sigma_x^2}-\dfrac{2\rho(X-\mu_x)(Y-\mu_y)}{\sigma_x\sigma_y}+\dfrac{(Y-\mu_y)^2}{\sigma_y^2}\right]}{[2(1-\rho^2)]}\right\}$$

where μ_x and μ_y are the expectation values of X and Y, respectively, and σ_x and σ_y are the corresponding standard deviations. This is a rather formidable looking expression, but it has a straightforward interpretation. Given X, Y follows a normal distribution function with an expectation value and variance given by

$$E(Y|X_0) = \mu_y + \rho\frac{(X_0-\mu_x)}{\sigma_x}\sigma_y \tag{3.3}$$

$$V(Y|X_0) = (1-\rho^2)\sigma_y^2 \tag{3.4}$$

Here $E(Y|X_0)$ and $V(Y|X_0)$ are the expectation value and variance, respectively, of Y at $X = X_0$. Equations 3.3 and 3.4 may not seem familiar, but recall that in model I linear regression

$$\alpha = \mu_y - \beta\mu_x \tag{3.5}$$

$$\beta = \frac{\text{Cov}(X,Y)}{\sigma_x^2} = \frac{\rho\sigma_y}{\sigma_x} \tag{3.6}$$

Plugging Equations 3.5 and 3.6 into Equation 3.3 gives

$$E(Y|X_0) = \alpha + \beta X_0 \tag{3.7}$$

an equation which should certainly look familiar. Furthermore, Equation 3.4 is easily shown to be identical to Equation 2.68 in the limit as $N \to \infty$. The connection between correlation and model I regression should now be obvious, but there are some important differences in the purposes and assumptions that underlie the two methods of data analysis. In correlation analysis we wish to establish and estimate the degree of association between two variables. Neither variable is considered to be independent and the other dependent, and the probability distribution function for r (Eq. 3.2) is based on the assumption that X and Y follow a bivariate normal distribution. There is no requirement that either X or Y be measured without error. In regression analysis we wish to establish and estimate the dependence of one variable on another; that is, we wish to describe the functional relationship between X and Y and/or predict one in terms of the other. We do not have to assume that X and Y follow a bivariate normal distribution, but in model I regression we require either that no error be associated with the independent variable X or that the observations be controlled. If the observations are uncontrolled and there are errors associated with both variables, model II regression methods are required (Chapter 4). In cases where observations are controlled, X and Y will not follow a bivariate normal distribution; in cases where the observations are uncontrolled, X and Y may or may not follow a bivariate normal distribution. In model I regression we do require that the errors in Y be normally distributed for purposes of calculating confidence intervals for our statistics, but otherwise it is sufficient for the errors in Y to be random and independent of X.

Thus the purposes and assumptions of correlation and regression analysis are quite different. An important relaxation to the bivariate normal distribution assumption may be made in correlation analysis, however, when $\rho = 0$. In that case, if Y is normally distributed, the probability distribution function for r is independent of whether X is normally distributed or not. Similarly, if X is normally distributed, the probability distribution function for r is independent of whether Y is normally distributed or not. Thus a test of the null hypothesis that $\rho = 0$ requires only that either X or Y be normally distributed.

DISTRIBUTION OF *r*

It should be obvious from Equation 3.4 that $-1 \leq \rho \leq 1$, and we already noted from Equation 2.68 that $-1 \leq r \leq 1$. Thus both ρ and r are confined to the interval $[-1, 1]$. Now what about the probability distribution function for r? Perhaps surprisingly, Equation 3.2 does not, in general, give an unbiased estimate of ρ. There is no bias when $\rho^2 = 1$, but for finite N and $\rho^2 < 1$, $|E(r)| < |\rho|$. Figure 3.1 shows the extent of the bias for several values of N. The bias is never greater than 25% and is greatest for small N and small ρ. For most purposes the bias is negligible when $N > 10$. It may seem surprising that there is bias as $\rho \to 0$, since $E(r) = 0$ when $\rho = 0$. Consider, however, that $f(X) = X$ and $g(X) = X^2$ both equal 0 when $X = 0$, yet $|g(X)| < |f(X)|$ for $|X| < 1$, and indeed $|g(X)|/|f(X)| \to 0$ as $|X| \to 0$. Hence the fact that $E(r) = 0$ when $\rho = 0$ does not imply that $|E(r)|/|\rho| \to 1$ as $\rho \to 0$, and indeed this is not the case.

Appendix C gives critical values of r for testing the hypothesis $\rho = 0$. When $\rho \neq 0$, it can be shown that the function

$$z = \tfrac{1}{2}\ln\left(\frac{1+r}{1-r}\right) \tag{3.8}$$

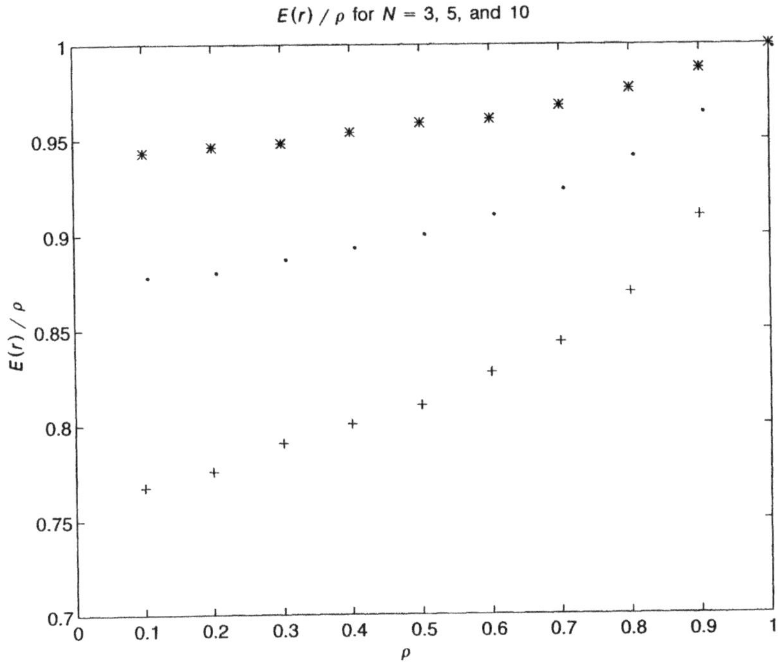

Figure 3.1 $E(r)/\rho$ versus ρ for $N = 3$ (+), 5 (·), and 10 (∗). $E(r)$ in each case was calculated by generating 10,000 (X, Y) data sets in which X and Y followed a bivariate normal distribution.

closely follows a normal distribution with mean and variance given by

$$\mu_z = \tfrac{1}{2}\ln\left(\frac{1+\rho}{1-\rho}\right) \tag{3.9}$$

$$V(z) = \frac{1}{N-3} \tag{3.10}$$

As an example of the application of Equations 3.9 and 3.10, suppose that 10 (X, Y) pairs are drawn from what we believe to be a bivariate normal population, and the value of r calculated from the 10 pairs is 0.7. From Equation 3.8 we calculate $z = 0.867$, and, according to Equation 3.10, $V(z) = 1/7$. From Appendix A, the 95% confidence interval for z is therefore $0.867 - 1.96/\sqrt{7}$ to $0.867 + 1.96/\sqrt{7} = [0.126, 1.608]$. From Equation 3.8 the corresponding 95% confidence interval for ρ is $[0.125, 0.923]$.

If we are content to know merely whether there is a significant correlation between X and Y, there is a simple correspondence between the critical values of r and the critical values of the t distribution used to test for the significance of the model I regression slope. From Equations 2.65, 2.68, and 3.2 it follows that

$$V(B) = \frac{1}{N-2}\left[\frac{\sum y_k^2}{\sum x_m^2} - \left(\frac{\sum x_n y_n}{\sum x_j^2}\right)^2\right]$$

$$= \left(\frac{1-r^2}{N-2}\right)\frac{\sum y_k^2}{\sum x_m^2} \tag{3.11}$$

Since $B = \Sigma x_n y_n / \Sigma x_m^2$, it follows that

$$\frac{B}{\sqrt{V(B)}} = r\sqrt{\frac{N-2}{1-r^2}} \tag{3.12}$$

The left-hand side of Equation 3.12 is the statistic to be compared to the critical value of the argument of the t distribution function in Appendix B for the purposes of testing the hypothesis $\beta = 0$. We therefore expect that the critical values of the t distribution function will equal the right-hand side of Equation 3.12 when the corresponding critical values of r are substituted in the equation. Let us see if this is, in fact, the case.

Assume that the null hypothesis is that $\rho = 0$. Assume that we draw 10 $[X, Y]$ pairs at random from a population for which either X or Y is normally distributed. We calculate $\bar{X}$ and $\bar{Y}$ and then proceed to calculate r. Since we estimated two population parameters, namely, μ_x and μ_y, from our random sample, there are $10 - 2 = 8$ degrees of freedom associated with our estimate of ρ. From Appendix C, there is only a 5% chance with df $= 8$ that $|r|$ will exceed 0.632 if $\rho = 0$. Plugging this value of r into the right-hand side of Equation 3.12, we calculate a critical t value of 2.306. Reference to Appendix B shows that with df $= 8$ there is indeed only a 5% chance that the magnitude of $B/\sqrt{V(B)}$ will exceed 2.306 if $\beta = 0$. In other words, if either X or Y is normally distributed and $\rho = 0$, then values of r calculated from a random sampling of 10 $[X, Y]$ pairs would exceed 0.632 in magnitude, that is, would be less than -0.632 or greater than 0.632, only 5% of the time. Similarly, if $\beta = 0$, the statistic $B/\sqrt{V(B)}$ would exceed 2.306 in magnitude only 5% of the time. Thus testing the hypothesis that $\rho = 0$ from the table of critical r values is equivalent to testing the hypothesis that $\beta = 0$ from the table of critical t values. It must be stressed, however, that the purpose and design of an experiment often make one or the other of these tests meaningless. The fact that one can flip back and forth from the table of critical t values to the table of critical r values does not imply that regression and correlation analyses are interchangeable.

ANALYSIS OF VARIANCE OF Y

By definition,

$$V(Y) = \sigma_y^2 = E\left[(Y - \mu_y)^2\right] = \lim_{N\to\infty} \frac{1}{N} \sum_{k=1}^{N} (Y_k - \mu_y)^2$$

However,

$$\begin{aligned} Y_k - \mu_y &= Y_k - \alpha - \beta X_k + \alpha + \beta X_k - \mu_y \\ &= \varepsilon_k + \alpha + \beta X_k - \mu_y = \varepsilon_k + \beta(X_k - \mu_x) \end{aligned}$$

Hence

$$V(Y) = \lim_{N\to\infty} \frac{1}{N} \sum_{k=1}^{N} [\varepsilon_k + \beta(X_k - \mu_x)]^2 = V(Y|X) + \beta^2\sigma_x^2$$

since the cross terms average to 0. However, from Equation 3.6 we see that $\beta^2\sigma_x^2 = \rho^2\sigma_y^2 = \rho^2 V(Y)$. Hence the variance of Y can be expressed as the sum of two terms, one of which, $V(Y|X)$, is the variance of Y about the regression line $\alpha + \beta X$, and the second of which, $\rho^2 V(Y)$, is the portion of the variance accounted for by the regression line. It is for this reason that ρ^2 is often referred to as the fraction of the variance accounted for by the regression line.

The statistic r^2 is commonly called the *coefficient of simple determination* and is in some sense a measure of the fraction of the variance of Y accounted for by the regression line $A + BX$. From Equation 2.68 it follows that

$$\frac{\sum (Y_k - A - BX_k)^2}{\sum (Y_n - \bar{Y})^2} = 1 - r^2 \qquad (3.13)$$

but the left-hand side of Equation 3.13 is not an unbiased estimator of $V(Y|X)/V(Y)$. Based on the discussion in Chapter 2, one might suspect that the bias could be removed by multiplying the left-hand side of Equation 3.13 by $(N-1)/(N-2)$, in which case the numerator $\Sigma(Y_k - A - BX_k)^2/(N-2)$ would become an unbiased estimator of $V(Y|X)$, and the denominator $\Sigma(Y_n - \bar{Y})^2/(N-1)$ would become an unbiased estimator of $V(Y)$. However, as we learned in Exercise 2.8, the expectation value of a quotient is, in general, not equal to the quotient of the expectation values, and in this case $[(1-r^2)(N-1)]/(N-2)$ is an unbiased estimator of $1-\rho^2$ only when $\rho = 0$ or $\rho^2 = 1$, even when X and Y follow a bivariate normal distribution.

When $\rho = 0$, $E(r^2) = 1/(N - 1)$. Hence, for $N = 3$ and $\rho = 0$, $E(r^2) = 0.5$. The difference between saying that a regression accounts for none of the variance of Y and saying that it accounts for 50% of the variance of Y is considerable. Thus for small values of N it may be desirable to estimate ρ^2 from the equation

$$\rho^2 = 1 - \frac{(1 - r^2)(N - 1)}{N - 2} = \frac{(N - 1)r^2 - 1}{N - 2} \tag{3.14}$$

Figure 3.2 shows the function $[(N - 1)E(r^2) - 1]/[(N - 2)\rho^2]$ for several values of N when X and Y follow a bivariate normal distribution. The right-hand side of Equation 3.14 does eliminate much of the bias in $E(r^2)$, but as in the case of Figure 3.1, some bias remains. In fact $[(N - 1)E(r^2) - 1]/(N - 2)$ underestimates ρ^2, the degree of bias again being greatest for small N and small ρ. The bias is negligible for most purposes when $N > 20$.

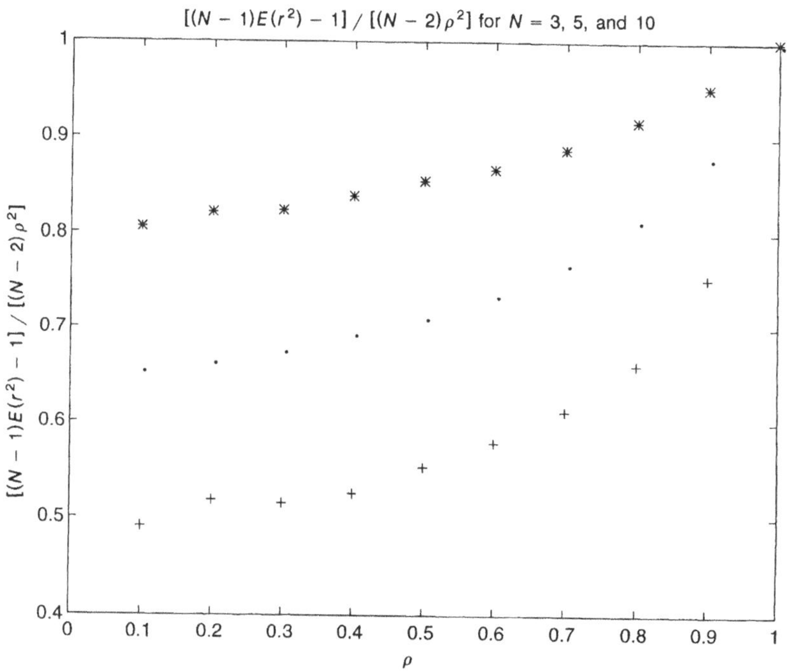

Figure 3.2 $\{(N - 1)[1 - E\{r^2\}]\}/[(N - 2)\rho^2]$ versus ρ for $N = 3$ (+), 5 (·), and 10 (∗). $E(r^2)$ was calculated by generating 20,000 (X, Y) data sets in which X and Y followed a bivariate normal distribution.

RANK CORRELATION

The assumption that X and Y follow a bivariate normal distribution is rather restrictive, and it would be desirable to have some other measure of correlation that is not so restricted by assumptions regarding the distributions of the variates. One of the most common solutions to this problem is to calculate correlation coefficients based on rank rather than using the actual numerical values of the variates. This approach falls under the general category of *nonparametric statistics*. Two of the most frequently used rank correlation coefficients are the *Spearman* and *Kendall* correlation coefficients, generally designated r_s and τ, respectively. Both vary between -1 and $+1$, with positive values implying a positive correlation and negative values a negative correlation.

In both cases we begin by ranking all the X values and all the Y values from smallest to largest. The original set of $[X_k, Y_k]$ pairs then becomes a set of ranks $[L_k, M_k]$. The value of r_s is calculated in the same way as the product–moment correlation coefficient r, but using the set of ranks $[L_k, M_k]$ rather than the original set of $[X_k, Y_k]$ values. If there are no ties in the rankings, the same result is obtained by calculating the sum of the squares of the differences of the ranks for all the $[L_k, M_k]$ pairs. We will call this sum of squared differences $\Sigma_{k=1}^{N} d_k^2$. If N is the number of $[L_k, M_k]$ pairs, then

$$r_s = 1 - \frac{6}{N(N^2 - 1)} \sum_{k=1}^{N} d_k^2 \tag{3.15}$$

As an example, consider the data in Table 3.1, which shows concentrations of particulate carbon (PC) and particulate nitrogen (PN) in 11 different subarctic lakes. In this case $\Sigma_{k=1}^{N} d_k^2 = 8$, and $r_s = 1 - (6)(8)/[(11)(120)] = 0.964$.

TABLE 3.1 Concentrations of Particulate Carbon (PC) and Particulate Nitrogen (PN) in 11 Subarctic Lakes Reported by Hecky et al. (1993)[a]

Lake	PC	PN	*C* Rank	*N* Rank	d^2
Chitty	25.7	2.7	9	8	1
Baptiste	18.0	2.0	5	5	0
Drygeese	16.3	1.7	3	2	1
Alexis	20.0	2.1	6	6	0
Wood	39.7	5.6	11	11	0
Prosperous	20.3	2.5	7	7	0
Gordon	17.5	1.9	4	4	0
S. Indian	29.6	3.0	10	9	1
McLeod Bay	10.5	1.2	1	1	0
Hearne Channel	14.5	1.8	2	3	1
West basin	23.0	4.2	8	10	4
Σd^2					8

[a]Units are mmol/m^3.

Appendix D lists the critical values of r_s. They are larger than the corresponding values of r, but for $N > 30$ the differences are insignificant. In this case, with $N = 11$, the calculated r_s value of 0.964 is judged to be significant at 99% confidence. In other words, if there were no correlation between particulate carbon and particulate nitrogen concentrations in subarctic lakes, there is less than a 1% chance that a random sample of 11 such lakes would give an r_s value greater than 0.964 in magnitude.

Calculation of Kendall's τ is a bit more complicated, because we must first order the ranks of either X or Y from smallest to largest. Table 3.2 shows this ordering for the 11 lakes in Table 3.1, with the particulate carbon concentrations ranked from smallest to largest. We now look at the PN ranks and for each rank count the number of smaller ranks to the right. These counts are shown in the last row of Table 3.2. The total count, Q, in this case is 3. The Kendall τ statistic is then calculated from the equation

$$\tau = 1 - \frac{4Q}{N(N-1)} \tag{3.16}$$

where N is the number of ranks. In this case we get $\tau = 1 - 12/110 = 0.891$. The critical values of Kendall's τ are shown in Appendix E. For large N they approach a value equal to $2/(3\sqrt{N})$ times the corresponding critical value of the t statistic. In this case, with $\tau = 0.891$, we again conclude with better than 99% confidence that the correlation is significant.

All this may seem straightforward enough, but what about the case where there are ties in the ranks? Table 3.3 is one such example. In this case, again involving concentrations of PC and PN in lakes, there is one tie in the ranks of particulate carbon and three ties in the ranks of particulate nitrogen. It is reasonable to assume in such a data set that the ties are not real, but merely reflect imprecision in the numbers. The probability distribution functions for r_s and τ, however, have been calculated on the assumption that there are no ties. The best we can do in this case is to assign an average rank to the tied variates, as shown in Table 3.3. The value of r_s is then calculated by plugging the PC and PN ranks into Equation 3.2. The result in this case is $r_s = 0.929$, which is significant at 99% confidence (Appendix D). In the case of τ we

TABLE 3.2 Rankings of Particulate Carbon (PC) and Particulate Nitrogen (PN) for the 11 Lakes in Table 3.1, Ordered by Particulate Carbon.

PC rank	1	2	3	4	5	6	7	8	9	10	11
PN rank	1	3	2	4	5	6	7	10	8	9	11
Number smaller PN ranks to right	0	1	0	0	0	0	0	2	0	0	

TABLE 3.3 Concentrations of Particulate Carbon (PC) and Particulate Nitrogen (PN) in 26 Temperate Lakes Reported by Hecky et al. (1993).[a]

Lake	PC	PN	*C* Rank	*N* Rank	Number of Smaller *N* Ranks Below
Superior	16.1	1.5	1	1	0
Mirror	17.2	2.0	2	3	1
ELA 375	24.4	1.7	3	2	0
Nipigon	25.6	2.7	4	5.5	1.5
Erie	33.3	4.4	5	14.5	9.5
ELA 373 epi	34.8	2.4	6	4	0
Trout	36.4	3.0	7	8	2
ELA 239	37.2	2.8	8	7	1
ELA 240	37.8	2.7	9	5.5	0
Sydney	41.0	3.7	10	12	3
ELA 442 epi	41.5	3.3	11	11	2
ELA 377 epi	43.2	3.1	12	9	0
ELA 221	43.6	3.2	13	10	0
ELA 302N	44.4	4.7	14	16.5	2.5
Winnipeg	46.4	5.9	15	20	5
ELA 222	53.8	4.2	16.5	13	0
ELA 304	53.8	4.7	16.5	16.5	1
Green	56.9	4.4	18	14.5	0
ELA 302S	57.3	7.0	19	22	3
Orange	59.6	5.2	20	18	0
Linge	60.5	5.5	21	19	0
Musclow	63.7	6.1	22	21	0
ELA 149	92.8	7.8	23	23	0
ELA 303	146.0	11.2	24	24	0
ELA 227	239.7	23.0	25	25	0
Prairie Lake 885	369.3	53.9	26	26	0

[a]Units are mmol/m^3.

proceed as in the previous example, adding 0.5 to the last column whenever an identical rank appears below a given rank in the table. In this case we obtain $Q = 31.5$ and $\tau = 0.806$, which is also significant at 99% confidence (Appendix E). Note, however, that there is ambiguity in the value of τ. Had we reversed the order of lakes ELA 222 and ELA 304, Q would have increased by 1, and the calculated value of τ would have been 0.800. Because lakes ELA 222 and ELA 304 have the same C rank, there is no reason, based on their C ranks, to list one before the other. In order to minimize this ambiguity, it is best when calculating τ to rank the data according to the variate which is associated with the fewest number of ties, which in this case is the PC concentration.

EXERCISES

3.1 Show by induction that $\Sigma_{k=1}^{N-1} k = N(N-1)/2$.

3.2 Show, using the result of Exercise 3.1, that the maximum value of Q in Equation 3.16 is $N(N-1)/2$. What is the value of τ corresponding to this value of Q?

3.3 Assume that the X and Y variates in a set of N $[X, Y]$ pairs are ranked from smallest to largest. Assume that there are no ties in the ranks. If the X ranks are designated L and the Y ranks are designated M, show, using the result of Exercise 3.1, that

$$\bar{L} = \frac{1}{N} \sum_{k=1}^{N} L_k = \bar{M} = \frac{1}{N} \sum_{k=1}^{N} M_k = \frac{N+1}{2}$$

Show by induction that

$$\sum_{k=1}^{N} L_k^2 = \sum_{k=1}^{N} M_k^2 = \frac{N(N+1)(2N+1)}{6}$$

Using the previous results, show that

$$\sum_{k=1}^{N} L_k^2 - NL^2 = \frac{N}{12}(N^2 - 1)$$

Using the preceding results, show that

$$r_s = \sum_{k=1}^{N} (L_k - \bar{L})(M_k - \bar{M}) \bigg/ \left[\sum_{k=1}^{N} (L_k - \bar{L})^2 \sum_{m=1}^{N} (M_m - \bar{M})^2 \right]^{1/2}$$

$$= \left[\sum_{k=1}^{N} L_k M_k - N(\bar{L})^2 \right] \bigg/ \left[\sum_{m=1}^{N} L_m^2 - N\bar{L}^2 \right]$$

$$= 1 - \frac{6}{N(N^2 - 1)} \sum_{k=1}^{N} (L_k - M_k)^2$$

3.4 Given the data in Table 3.3, assume that you did not assign tied ranks, but instead assigned each lake an integer rank, choosing the ranks randomly in cases where either the X or the Y variates were identical. What is the largest possible value of τ that could arise from such assignments? What is the smallest possible value of τ? What is the significance level of these two values of τ (Appendix E)?

3.5 Assume that you draw five $[X, Y]$ pairs at random from a bivariate normal population with $\rho = 0.5$. Calculate r. Repeat this process 10,000 times (Monte Carlo calculation). Calculate $\bar{r}$ and $s(r)$. What is the ratio $\bar{r}/\rho$? Is this result consistent with Figure 3.1? Does your answer depend on the values you assign to μ_x, μ_y, σ_x, and σ_y?

3.6 Draw 10 $[X, Y]$ pairs at random from a bivariate normal population with $\rho = 0.5$ and calculate r. Repeat this process 10,000 times. Calculate z for each value of r using Equation 3.8. Use Equation 3.10 to calculate a 95% confidence interval for each value of z. Plug these upper and lower bounds for z into Equation 3.8 and invert to solve for the corresponding 95% confidence interval for ρ. What percentage of your 10,000 confidence intervals did not include the true value of ρ, 0.5? The answer should be close to 5%.

REFERENCE

Hecky, R. E., P. Campbell, and L. L. Hendzel 1993. The stoichiometry of carbon, nitrogen, and phosphorus in particular matter of lakes and oceans. *Limnol. Oceanogr.* **38** 709–724.

4

MODEL II LINEAR REGRESSION

We now enter the rather murky realm of model II linear regression analysis. The motivation for using something other than model I under certain conditions is straightforward enough. Suppose we are faced with a situation in which random errors are associated with both X and Y and the observations are not controlled. The assumptions of model I regression are not satisfied in this case, but how will this affect our results?

The answer depends on our intended use of the regression equation. There are two basic goals of regression analysis, prediction and estimation of functional relationships. In model I regression the distinction between these two goals is often ignored, because the same equation serves both purposes. However, as noted by Sokal and Rohlf (1981, p. 549), "Unbiased estimates of Y given X, and minimal width confidence belts around the estimates of Y, are not generally given by functional equations that best describe the joint variation of two random variables." If the goal of the regression analysis is to be able to predict Y given X or X given Y, then in many cases the best approach is to use model I regression, even when both X and Y are subject to error and the observations are uncontrolled (Ricker, 1973). The model I regression equations are therefore often referred to as the predictive equations.

However, when the goal of the regression analysis is to estimate the functional relationship between X and Y, model I regression methods are no longer appropriate if both X and Y are subject to error and the observations are uncontrolled. Under these conditions we can expect a bias in our estimates of the slope and intercept of the underlying relationship between X and Y. To understand why this is true, we will look at the effect of errors on the expectation values of the numerator and denominator of the right-hand side of Equation 2.18. We assume that $X_k = X_{kt} + \delta_k$ and that $Y_k = \alpha + \beta X_{kt} + \varepsilon_k$, where X_{kt} is the true value of the kth value of X and δ_k and ε_k are random errors with a mean of 0. As before, we assume that the δ_k and ε_k have common variances, which we designate σ_δ^2 and σ_ε^2, respectively. We can now see that

$$\begin{aligned} E\left(\sum X_k Y_k\right) &= E\left[\sum (X_{kt} + \delta_k)(\alpha + \beta X_{kt} + \varepsilon_k)\right] \\ &= E\left[\sum X_{kt}(\alpha + \beta X_{kt})\right] \\ &= E\left(\sum X_{kt} Y_{kt}\right) \end{aligned} \tag{4.1}$$

since the expectation values of the error terms and the product of the error terms are 0. Similarly,

$$\begin{aligned} E(\overline{XY}) &= \frac{1}{N^2} E\left[\sum X_k \sum Y_m\right] \\ &= \frac{1}{N^2} E\left[\sum (X_{kt} + \delta_k) \sum (\alpha + \beta X_{mt} + \varepsilon_m)\right] \\ &= \frac{1}{N^2} E\left[\sum X_{kt} \sum (\alpha + \beta X_{mt})\right] \\ &= \frac{1}{N^2} E\left[\sum X_{kt} \sum Y_{mt}\right] \\ &= E\left(\overline{X}_t \overline{Y}_t\right) \end{aligned} \tag{4.2}$$

Thus there is no reason to expect that $\sum_{k=1}^{N} x_k y_k$ is biased in any way by the errors in X and Y. However,

$$\begin{aligned} E\left(\sum X_k^2\right) &= E\left[\sum (X_{kt} + \delta_k)^2\right] \\ &= E\left(\sum X_{kt}^2\right) + N\sigma_\delta^2 \end{aligned} \tag{4.3}$$

$$\begin{aligned} E(\overline{X})^2 &= \frac{1}{N^2} \sum X_k \sum Y_m \\ &= \frac{1}{N^2} \sum (X_k + \delta_k) \sum (Y_m + \delta_m) \\ &= E(\overline{X})^2 + \frac{1}{N}\sigma_\delta^2 \end{aligned} \tag{4.4}$$

Hence the expectation value of $\Sigma x_k^2 (= \Sigma X_k^2 - N\bar{X}^2)$ is biased upward by an amount equal to $(N-1)\sigma_\delta^2$. We might thus expect that Equation 2.18 would tend to underestimate B when there are errors in X and the observations are uncontrolled, and indeed this is the case. As we have seen before, however, the expectation value of a quotient is not, in general, equal to the expectation value of the numerator divided by the expectation value of the denominator, and therefore in this case the extent of bias cannot, in general, be accurately estimated from the ratio of the expectation values. An exception to this statement occurs when the δ_k, ε_k, and X_{mt}, are all normally and independently distributed and it is known that the Y_k and X_k follow a bivariate normal distribution. Under this special set of circumstances, one can show that

$$E(B) = \frac{\beta V(X)}{V(X) + \sigma_\delta^2} = \frac{\beta}{1 + \dfrac{\sigma_\delta^2}{V(X)}} = \frac{\beta}{1 + \lambda^2} \tag{4.5}$$

where $\lambda = \sigma_\delta / \sqrt{V(X)}$. The right-hand side of Equation 4.5 is, in fact, $E(xy)/E(x^2)$ in this particular case.

In the more general model II case, can we still expect Equation 2.18 to give us a biased estimate of β, and if so can we predict the magnitude of the bias? The answer to the first question is yes. If N is large the right-hand side of Equation 4.5 will closely approximate $E(B)$, even when X and Y do not follow a bivariate normal distribution. The reason is that for large N the numerator and denominator of Equation 2.18 will closely approximate their expectation values each time the experiment is repeated. An important implication of this realization is that collecting more and more data will not reduce the bias caused by the errors in X.

For small samples Equation 4.5 is approximately correct when X is not normally distributed as long as λ is small. What happens, however, when λ is not small? To get some feeling for the problem, look at Figure 4.1. Here we have a plot of noise-corrupted data, the underlying functional relationship being $X = Y$. There are 10 $[X, Y]$ pairs. Each value of X and each value of Y was noise corrupted with a normally distributed error signal with a mean of 0 and a standard deviation of 1.5. The true values of X and Y were the first 10 integers. If we ignore the fact that this is not a model I situation, we can do a least-squares fit of Y to X and obtain the solid line in Figure 4.1. The slope of this line is 0.82. If we are interested in the underlying relationship between X and Y, however, there is in this case no more reason to call Y the dependent variable than to call X the dependent variable. If we do the latter and fit X to Y using model I regression methods, we obtain the dotted line in the figure. The slope of this line is 1.31. Obviously both lines cannot be giving us the best estimate of the true slope. Why is our estimate of the slope so sensitive to which variable we call dependent?

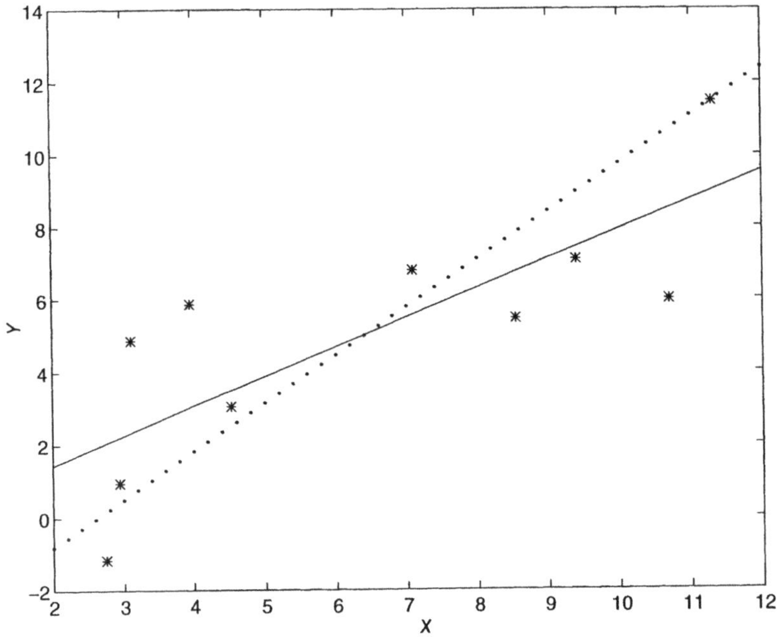

Figure 4.1 Model I linear regression lines fit to a data set for which the underlying functional relationship was $X = Y$. Both X and Y were noise corrupted using normally distributed random numbers with a standard deviation of 1.5. The slopes of the two regression lines are 0.82 and 1.31. $r = 0.79$.

To understand what is happening, consider the fact that when we fit X to Y by model I least squares, the slope of the regression line, which we will call D, is given by

$$D = \frac{\sum x_k y_k}{\sum y_m^2} \tag{4.6}$$

This is logical enough. We simply exchange x and y in Equation 2.18. Now note that

$$BD = \left(\frac{\sum x_k y_k}{\sum x_m^2}\right)\left(\frac{\sum x_n y_n}{\sum y_j^2}\right) = r^2 \tag{4.7}$$

and we know that $r^2 \leq 1$. It should be obvious from Equations 2.18 and 4.6 that the signs of B and D are the same and equal to the sign of $\sum_{k=1}^{N} x_k y_k$. It follows from the foregoing that $BD \leq 1$ and

$$|B| \leq \frac{1}{|D|} \tag{4.8}$$

Now if we regressed X on Y and obtained an Equation of the form $X = C + DY$, the inverse Equation would be $Y = -C/D + X/D$. Hence the slope of the dotted line in Figure 4.1 must be $1/D$, and indeed $1.31 > 0.82$. From Equation 4.7 it follows that B and $1/D$ will be identical only when $r^2 = 1$, which could only happen if all the data lie on a straight line. We can also see that the ratio of B to $1/D$, that is, BD, equals r^2. In other words, the ratio of the two model I slopes equals r^2, and the poorer the correlation between X and Y, the greater the discrepancy between the two slopes. The best estimate of the slope of the functional relationship between X and Y lies somewhere between B and $1/D$, although we have no guarantee that β lies within that range. By best estimate we again mean that the estimate is unbiased, and we can say with 100% confidence that β lies somewhere between $E(B)$ and $E(1/D)$. If $r^2 \cong 1$ we may be able to live with this uncertainty, but when r^2 is small, there is a considerable difference between B and $1/D$. Is there some way to get an unbiased estimate of β under these conditions?

In general and indeed in most cases, the answer to this question is, unfortunately, no. Several different approaches to the problem have been worked out, but none can be said to give an unbiased estimate of β. The debate over which method to use is ongoing. None of the methods is without critics. The three most commonly used approaches to model II regression are the *major axis method*, the *reduced major axis* or *geometric mean* (*GM*) *method*, and methods based on arithmetic averages (*AM methods*). We consider each of these separately in the following discussion.

MAJOR AXIS METHOD

If X and Y follow a bivariate normal distribution, a plot of random (X, Y) pairs will appear elliptical, as illustrated in Figure 4.2. The ellipse can be described by its two principal axes, the major axis and the minor axis, which are at right angles to each other. The major axis is the longest possible axis of the ellipse, and in the case of a bivariate normal distribution passes through the point (μ_x, μ_y) and has a slope equal to $\rho\sigma_y/\sigma_x$. The goal of the major axis model II regression method is to find the equation of the major axis.

In model I regression, when we assumed no errors to be associated with the X variate, we set out to minimize the sum of the squares of the deviations of our data from the regression line in the direction perpendicular to the X axis. In the major axis method, we minimize the sum of the squares of the perpendicular distances of our data from the regression line. In other

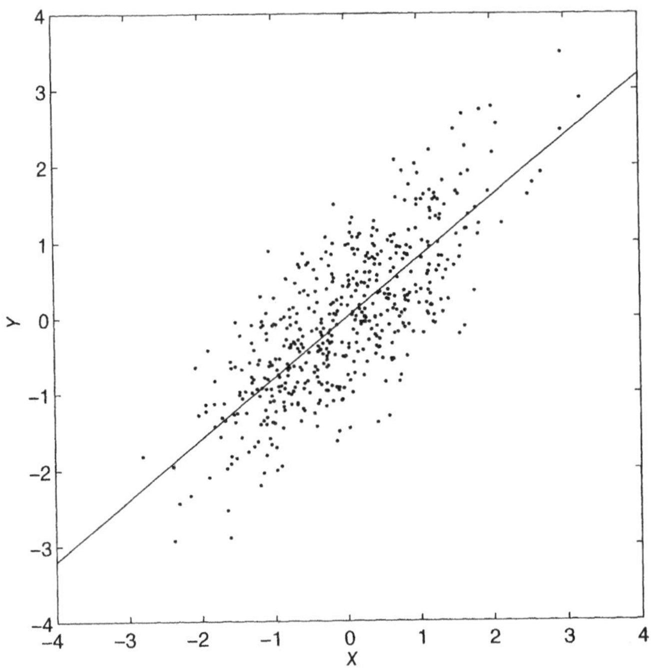

Figure 4.2 Random data points from a bivariate normal population with $\rho = 0.8$ and $\sigma_y = \sigma_x$.

words, we envision drawing lines from our data points to the regression line in a direction perpendicular to the regression line (not perpendicular to the X axis) and minimizing the sum of the squares of the lengths of these lines.

In order to derive the major axis regression equations, we begin by considering a point (X_k, Y_k) which lies somewhere off the line $Y = A_{ma} + B_{ma}X$, where A_{ma} and B_{ma} are the intercept and slope, respectively, of the major axis regression line. We know that a line perpendicular to $Y = A_{ma} + B_{ma}X$ will have a slope equal to $-1/B_{ma}$, and hence

$$\frac{Y_k - Y}{X_k - X} = -\frac{1}{B_{ma}} \Rightarrow B_{ma}^2(Y - Y_k)^2 = (X - X_k)^2 \tag{4.9}$$

Now let r_k^2 be the square of the perpendicular distance from (X_k, Y_k) to the regression line. It follows that

$$r_k^2 = (X - X_k)^2 + (Y - Y_k)^2 = (B_{ma}^2 + 1)(Y - Y_k)^2 \tag{4.10}$$

From Equation 4.9 it follows that $X = X_k - B_{ma}(Y - Y_k)$ and hence

$$Y = A_{ma} + B_{ma}X = A_{ma} + B_{ma}[X_k - B_{ma}(Y - Y_k)] \tag{4.11}$$

Rearranging Equation 4.11 gives

$$Y - Y_k = \frac{A_{ma} + B_{ma} X_k - Y_k}{1 + B_{ma}^2} \tag{4.12}$$

Substituting the right-hand side of Equation 4.12 for $Y - Y_k$ in the right-hand side of Equation 4.10 gives

$$r_k^2 = \frac{(Y_k - A_{ma} - B_{ma} X_k)^2}{1 + B_{ma}^2} \tag{4.13}$$

Hence our job is to choose A_{ma} and B_{ma} so as to minimize

$$\sum r_k^2 = \frac{1}{1 + B_{ma}^2} \sum (Y_k - A_{ma} - B_{ma} X_k)^2$$

When we differentiate this expression with respect to A_{ma} and set the derivative equal to zero, we obtain the familiar equation

$$A_{ma} = \bar{Y} - B_{ma} \bar{X} \tag{4.14}$$

Setting the partial derivative of Σr_k^2 with respect to B_{ma} equal to 0 and substituting the right-hand side of Equation 4.14 for A_{ma} leads to the quadratic equation

$$B_{ma}^2 \sum x_k y_k + B_{ma}\left(\sum x_m^2 - \sum y_n^2\right) - \sum x_j y_j = 0 \tag{4.15}$$

Equation 4.15 has two solutions, but only one satisfies the condition that B_{ma} have the same sign as $\Sigma_{k=1}^{N} x_k y_k$. The solution is

$$B_{ma} = \frac{\sum y_n^2 - \sum x_m^2 + \sqrt{\left(\sum x_j^2 - \sum y_j^2\right)^2 + 4\left(\sum x_h y_h\right)^2}}{2 \sum x_k y_k} \tag{4.16}$$

Fair enough, so what is the problem? The problem is in Equation 4.10. We are minimizing a sum of terms of the form $(X - X_k)^2 + (Y - Y_k)^2$. If X and Y are not measured in the same units, for example, if X is a respiration rate and Y is a weight, it makes no sense to sum these two quantities. We are guilty of mixing apples and oranges. Suppose, for example, that we report the weight (X) of a striped bass in metric tons and its respiration rate (Y) in microliters of oxygen consumed per year. Obviously, the respiration rates so reported will be numerically enormous and the weights numerically miniscule. By scaling Y so that it is numerically large compared to X, we increase the slope of the major axis, as shown in Figure 4.3, and in the limit of very large Y perpendiculars dropped from our data to the regression line will have a slope of virtually 0; that is, they will be parallel to the X axis. This is exactly the model I situation when we treat X as the dependent variable. Hence when Y is numerically large compared to X, we can expect B_{ma} to approach $1/D = \Sigma y_k^2 / \Sigma x_m y_m$. On the other hand, if we reported the weight (X) of the striped bass in micrograms and their respiration rates (Y) in cubic kilometers of oxygen consumed per second, the former would be numerically enormous and the latter numerically miniscule. In this case, as shown in Figure 4.4, the slope of the major axis would be small, and perpendiculars dropped from our data to the regression line would be almost parallel to the Y axis. This is, of course, the model I situation when we treat Y as the dependent variable. Hence when X is numerically large compared to Y, we can expect B_{ma} to approach $\Sigma x_m y_m / \Sigma x_k^2$. As noted by Ricker (1973, p. 415), "Thus the position of the major axis can be made to lie anywhere between the two predictive regressions [i.e., the two model I regression lines] simply by changing the units of measurement." Sokal and Rohlf (1981, pp. 549–550) comment, "When the two variables are in different units of measurement the slope of the major axis is generally meaningless and another technique for Model II regression should be employed."

When X and Y are measured in the same units, it may be appropriate to use the major axis method, but only with caution. The truth is that even when X and Y are measured in the same units, if $(X - X_k)^2$ is numerically much larger than $(Y - Y_k)^2$ or vice versa, then for all intents and purposes we will

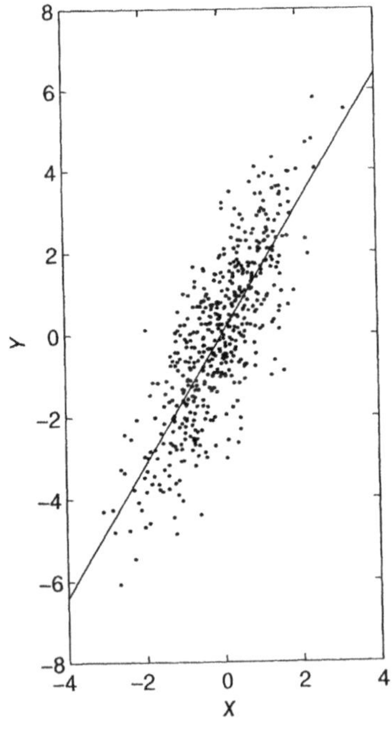

Figure 4.3 Random data points from a bivariate normal population with $\rho = 0.8$ and $\sigma_y = 2\sigma_x$.

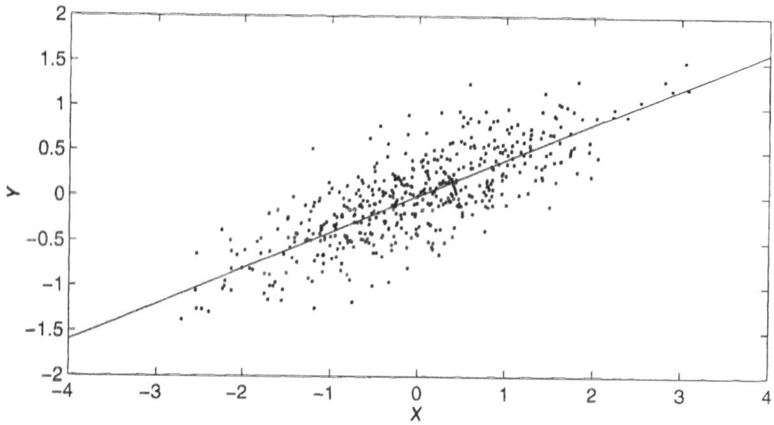

Figure 4.4 Random data points from a bivariate normal population with $\rho = 0.8$ and $\sigma_y = 0.5\sigma_x$.

be back to a model I regression. For example, if we compare the height, X, of a person with the length of his/her little toe, Y, and report both results in millimeters, it is safe to assume that X will be much larger than Y. Under these conditions the major axis regression equation will be virtually identical to the model I regression equation obtained by treating height as the independent variable and little toe length as the dependent variable.

One case which circumvents this issue is the situation in which both variates are logarithms, since logarithms are dimensionless. One should not, however, immediately transform to logarithms in order to avoid the issue. A logarithmic transformation may be appropriate if the underlying functional relationship is believed to be of the form $Y = aX^b$, as is the case in allometric work. In other cases, however, we may have to live with the fact that the units of measurement confound the use of the major axis method and turn to another form of model II regression.

REDUCED MAJOR AXIS OR GEOMETRIC MEAN METHOD

Since a major criticism of the major axis method is its dependence on the units of the X and Y variates, it might seem reasonable to try to circumvent the problem by scaling X and Y in some logical way. The reduced major axis method uses this approach. The original variates X and Y are transformed into dimensionless variates $\ddot{X}$ and $\ddot{Y}$ as follows:

$$\ddot{X} = \frac{X - \bar{X}}{S_x} \tag{4.17}$$

$$\ddot{Y} = \frac{Y - \bar{Y}}{S_y} \tag{4.18}$$

It is easy to show that

$$\sum_{k=1}^{N} \ddot{x}_k^2 = \sum_{k=1}^{N} \ddot{y}_k^2 \quad \text{and} \quad \sum_{k=1}^{N} \ddot{x}_k \ddot{y}_k = (N-1)r$$

Hence B_{ma} in Equation 4.16 equals ± 1 when calculated using $\ddot{X}$ and $\ddot{Y}$, the sign of the slope being equal to the sign of r. When we back-transform to get the slope of the reduced major axis regression line, B_{rma}, in terms of X and Y, we get

$$B_{\text{rma}} = \frac{dY}{dX} = \left(\frac{dY}{d\ddot{Y}}\right)\left(\frac{d\ddot{Y}}{d\ddot{X}}\right)\left(\frac{d\ddot{X}}{dX}\right)$$

$$= (S_y)(\pm 1)\left(\frac{1}{S_x}\right) = \pm\sqrt{\frac{\sum y_k^2}{\sum x_m^2}} \tag{4.19}$$

the sign of the slope again being the same as the sign of r. The reduced major axis method is often referred to as the geometric mean method because the right-hand side of Equation 4.19 is, in fact, the square root of the product, that is, the geometric mean, of the two model I regression line slopes:

$$\sqrt{\frac{B}{D}} = \sqrt{\left(\frac{\sum x_k y_k}{\sum x_m^2}\right)\left(\frac{\sum y_j^2}{\sum x_n y_n}\right)} = \sqrt{\frac{\sum y_j}{\sum x_m^2}} \tag{4.20}$$

Thus the position of the geometric mean regression line is completely insensitive to the choice of units for X and Y. It will always equal the geometric mean of the two model I slopes and hence lie midway between them.

The equation for the geometric mean regression line may be derived directly from least squares without resort to transformations. We require that the slope and intercept of the regression line be chosen so as to minimize the magnitude of the sum of the products of the horizontal and vertical deviations of our data from the regression line. For the kth point the vertical deviation is $Y_k - A_{\text{rma}} - B_{\text{rma}} X_k$, and the horizontal deviation is $X_k - A_{\text{rma}}/B_{\text{rma}} + Y_k/B_{\text{rma}}$. We therefore seek to minimize the magnitude of

$$\sum_{k=1}^{N} (Y_k - A_{\text{rma}} - B_{\text{rma}} X_k)\left(X_k + \frac{A_{\text{rma}}}{B_{\text{rma}}} - \frac{Y_k}{B_{\text{rma}}}\right)$$

$$= -\frac{1}{B_{\text{rma}}} \sum_{k=1}^{N} (Y_k - A_{\text{rma}} - B_{\text{rma}} X_k)^2 \tag{4.21}$$

Note that the product of the horizontal and vertical deviations is always negative, so it is the magnitude of the sum of products that we seek to minimize. When we set the partial derivative with respect to A_{rma} of the right-hand side of Equation 4.21 equal to 0, we obtain the familiar equation

$$A_{\text{rma}} = \overline{Y} - B_{\text{rma}}\overline{X} \tag{4.22}$$

so our regression line will again pass through $(\overline{X}, \overline{Y})$. Differentiating with respect to B_{rma} and setting the derivative equal to 0 leads directly to Equation 4.19.

The geometric mean (GM) method has its advocates, notably Ricker (1973), and its critics (e.g., Jolicoeur, 1975; Kuhry and Marcus, 1977). Since the GM slope always splits the difference between the two model I predictive slopes, it should be obvious that the GM method will not give an unbiased estimate of the functional slope in all cases. An unbiased estimate of the functional slope approaches B as $\sigma_\delta \to 0$ and $1/D$ as $\sigma_\varepsilon \to 0$. In the absence of information on the magnitude of σ_δ and σ_ε, the GM method does in a sense hedge one's bets by placing the slope midway between the two predictive model I slopes. As noted before, an unbiased estimate of the true slope must lie somewhere between the expectation values of the two predictive slopes.

ARITHMETIC MEAN METHODS

There are two commonly used arithmetic mean (AM) methods, one developed by Wald (1940) and the other by Nair and Bartlett (Nair and Shrivastava, 1942; Nair and Banerjee, 1942; Bartlett, 1949). In the Wald method, the set of (X, Y) pairs is ranked according to its X values and then divided into two groups, the first group having the smaller X values and the second group the larger X values. If there are an odd number of pairs, the pair with the median X value is dropped. The slope of the relationship, B_w, is then estimated from the equation

$$B_w = \frac{\left(\sum_{k_2=1}^{N_2} Y_{k_2} - \sum_{k_1=1}^{N_1} Y_{k_1}\right)}{\left(\sum_{k_2=1}^{N_2} X_{k_2} - \sum_{k_1=1}^{N_1} X_{k_1}\right)} \tag{4.23}$$

where the subscripts 1 and 2 refer to the two groups. In the Nair–Bartlett method the set of (X, Y) pairs is again ranked according to its X values, but is divided into three rather than two groups. The first and third groups, containing the smallest and largest X values, respectively, are made of equal size, and the middle group contains as close as possible to $N/3$ pairs, where N is the total number of (X, Y) pairs. The slope, B_{nb}, is then calculated from the equation

$$B_{\text{nb}} = \frac{\left(\sum_{k_3=1}^{N_3} Y_{k_3} - \sum_{k_1=1}^{N_1} Y_{k_1}\right)}{\left(\sum_{k_3=1}^{N_3} X_{k_3} - \sum_{k_1=1}^{N_1} X_{k_1}\right)} \tag{4.24}$$

In both AM methods, the intercept A is set equal to $\bar{Y} - B\bar{X}$. One criticism of AM methods is that they do not yield a conventional least-squares regression line, since the equations for the slope are not derived from any minimization procedure. Given B, however, it is clear that A minimizes $\sum_{k=1}^{N}(Y_k - A - BX_k)^2$.

BIAS AND CONSISTENCY

If any of the model II methods gave unbiased estimates of the functional relationship between X and Y, there would be little debate about which method to use. Without information about the magnitude of the errors associated with X and Y, however, it is really impossible to devise a method which will always give an unbiased estimate of the true relationship. The best one can say in the general case is that the true relationship lies somewhere between the expectation values of the two model I prediction lines. It is possible, however, to determine under what conditions the major axis, GM, or AM methods would give consistent estimates of the true functional relationship. For example, Sprent (1969) has shown that when N is large the GM slope follows the equation

$$B_{\text{rma}}^2 = \frac{\beta^2 V(X) + \sigma_\varepsilon^2}{V(X) + \sigma_\delta^2} \tag{4.25}$$

Hence the GM slope is an unbiased estimate of β for large N when $\sigma_\varepsilon^2/\sigma_\delta^2 = \beta^2$. A similar analysis of the equations for the major axis slope and AM slopes leads to the following conclusions as $N \to \infty$:

1. The major axis slope, B_{ma}, gives a consistent estimate of β when $\sigma_\varepsilon = \sigma_\delta$. The magnitude of B_{ma} tends to overestimate $|\beta|$ when $\sigma_\varepsilon > \sigma_\delta$ and to underestimate $|\beta|$ when $\sigma_\varepsilon < \sigma_\delta$.

2. The GM slope gives a consistent estimate of β when $\sigma_\varepsilon^2/\sigma_\delta^2 = \beta^2$. The magnitude of the GM slope tends to overestimate $|\beta|$ when $\sigma_\varepsilon^2/\sigma_\delta^2 > \beta^2$ and to underestimate $|\beta|$ when $\sigma_\varepsilon^2/\sigma_\delta^2 < \beta^2$.
3. The AM slope gives a consistent estimate of β when the groupings of variates is independent of whether the groupings are based on ordering X or Y. The AM slope will also give a consistent and unbiased estimate of β in the limit as $\sigma_\delta \to 0$.

ERROR BOUNDS

A number of papers have been published concerning the error bounds for model II regression slopes (B_{II}) and intercepts (A_{II}) (Wald, 1940; Bartlett, 1949; Kermack and Haldane, 1950; Anderson, 1963; Jolicoeur and Mosimann, 1968; Clarke, 1980). The published techniques, however, are not concerned with $E(B_{\text{II}} - \beta)^2$ or $E(A_{\text{II}} - \alpha)^2$. Rather, they are concerned with $E(B_{\text{II}} - E(B_{\text{II}}))^2$ and $E(A_{\text{II}} - E(A_{\text{II}}))^2$. As we have seen, in general we cannot expect $E(B_{\text{II}}) = \beta$ or $E(A_{\text{II}}) = \alpha$. Is there any way to get a measure of $E(B_{\text{II}} - \beta)^2$ or $E(A_{\text{II}} - \alpha)^2$?

The answer is yes, but only in a conservative way. We know, for example, that the true slope of the functional relationship lies somewhere between $E(B)$ and $1/E(D)$. Assume, for example, that the 95% confidence intervals for B and D are $B \pm C_B$ and $D \pm C_D$. If the correlation between X and Y is positive, we can say that at least 95% of the time the true functional slope will lie between $B - C_B$ and $1/(D - C_D)$. If the correlation is negative, then at least 95% of the time the true slope should lie between $1/(D + C_D)$ and $B + C_B$. These confidence intervals will be large when r^2 is small, since $1/D = B/r^2$, and a small value of r^2 means that C_B and C_D will be large. This may seem a bit discouraging, but it is a true reflection of our uncertainty about the functional relationship between X and Y in model II situations when r^2 is small.

EXERCISES

4.1 Show that only one root of Equation 4.15 has the same sign as $\sum_{k=1}^{N} x_k y_k$.

4.2 Show that if all values of y in Equation 4.16 are multiplied by a number k, then, as $k \to 0$, $B_{\text{ma}} \to \Sigma(ky_m)^2/\Sigma x_n(ky_n)$. In other words, the major axis slope becomes identical to $1/D$ when y is numerically large. In doing this problem you will need to use the fact that, as $z \to 0$, $(1 + z)^{1/2} \to 1 + z/2$. *Hint*: In doing this problem and Exercise 4.3, you may find it convenient to write Equation 4.16 in the form,

$$B_{\text{ma}} = \frac{A - B + \sqrt{A^2 - 2AB + B^2 + 4C^2}}{2C}$$

where $A = \Sigma y_n^2$, $B = \Sigma x_n^2$, and $C = \Sigma x_n y_n$. Then, for example, if $y \to ky$, $A \to k^2 A$ and $C \to kC$. Be careful not to discard any terms until you have expanded the square root, particularly in Exercise 4.3.

4.3 Show that if all values of x in Equation 4.16 are multiplied by a number k, then, as $k \to \infty$, $B_{\text{ma}} \to \Sigma(kx_m)y_m/\Sigma(kx_n)^2$. In other words, the major axis slope becomes identical to B when x is numerically large.

4.4 Show that

$$\sum_{k=1}^{N} \ddot{x}_k \ddot{Y}_k = \sum_{k=1}^{N} \ddot{x}_k \ddot{y}_k = (N-1)r,$$

where $\ddot{X}$ and $\ddot{Y}$ are as defined in Equations 4.17 and 4.18.

4.5 Show that differentiating Equation 4.21 with respect to B_{rma} and setting the derivative equal to zero does lead to Equation 4.19 if $A_{\text{rma}} = \overline{Y} - B_{\text{rma}}\overline{X}$.

4.6 Show that exchanging x and y in the right-hand side of Equation 4.16 gives an expression equal to $1/B_{\text{ma}}$. This means that with the major axis method we get the same regression line regardless of whether we consider X or Y to be the dependent variable. Note that this is also true of the GM regression line.

4.7 Assume that the underlying functional relationship between X and Y is $Y = 10X$. Assume that the true values of X in a data set lie in the interval [0, 10]. Assume that X and Y are measured with normally distributed errors that have means of 0 and standard deviations of σ_δ and σ_ε, respectively.

a. For $N = 9$, 99, 999, and 9999, generate N noise-corrupted (X, Y) pairs and fit each of the four data sets with a major axis, reduced major axis, and Bartlett arithmetic mean model II regression. The true X values are to be evenly spaced on the interval [0, 10]. Let $\sigma_\delta = \sigma_\varepsilon = 1$. Which methods appear to be giving a consistent estimate of the true slope (10)?

b. Repeat part a, but let $\sigma_\delta = 1$ and $\sigma_\varepsilon = 10$. Now which methods appear to be giving a consistent estimate of the true slope?

c. Repeat part a, but let the true relationship be $Y = X$. Now which methods appear to be giving a consistent estimate of the true slope?

d. Repeat parts a–c for the Bartlett AM method, but base the groups on the ranks of the true X values rather than on the ranks of the noise-corrupted X values.

REFERENCES

Anderson, T. W. 1963. Asymptotic theory for principal component analysis. *Ann. Math. Statist.* **34**: 122–148.

Bartlett, M. S. 1949. Fitting a straight line when both variables are subject to error. *Biometrics* **5**: 207–212.

Clarke, M. R. B. 1980. The reduced major axis of a bivariate sample. *Biometrika* **67**: 441–446.

Jolicoeur, P. 1975. Linear regressions in fishery research: Some comments. *J. Fish. Res. Board Canada* **32**: 1491–1494.

Jolicoeur, P., and J. E. Mosimann. 1968. Intervalles de confiance pour la pente de l'axe majeur d'une distribution normale bidimensionnelle. *Biom.-Praxim* **9**: 121–140.

Kermack, K. A., and J. B. S. Haldane. 1950. Organic correlation and allometry. *Biometrika* **37**: 30–41.

Kuhry, B., and L. F. Marcus. 1977. Bivariate linear models in biometry. *Syst. Zool.* **26**: 201–209.

Nair, K. R., and K. S. Banerjee. 1942. A note on fitting of straight lines when both variables are subject to error. *Sankhyā* **6**: 331.

Nair, K. R., and M. P. Shrivastava. 1942. On a simple method of curve fitting. *Sankhyā* **6**(2): 121–132.

Ricker, W. E. 1973. Linear regressions in fishery research. *J. Fish. Res. Board Canada* **30**: 409–434.

Sokal, R. R., and F. J. Rohlf. 1981. *Biometry*, 2nd ed. W. H. Freeman, New York.

Sprent, P. 1969. *Models in Regression and Related Topics*. Methuen, London.

Wald, A. 1940. The fitting of straight lines if both variables are subject to error. *Ann. Math. Statist.* **11**: 284.

5

POLYNOMIAL CURVE FITTING, LINEAR MULTIPLE REGRESSION ANALYSIS, AND NONLINEAR LEAST SQUARES

Although in many cases the relationship between two variables can be adequately described by a straight line, it is sometimes the case that an adequate description of the data require something more sophisticated than a straight line and/or the underlying relationship between X and Y is believed to be nonlinear. Our purposes in fitting an analytical function to such data are the same as in linear regression analysis. We wish to predict one variable from another or we wish to estimate the functional relationship between the variables. We again encounter model I and model II situations. If our aim is prediction, we can apply more-or-less straightforward least-squares methods, minimizing the errors in the variable we wish to predict. Our goal is not to estimate parameters, and hence we are not concerned about whether the statistics we calculate are biased or not. If we are trying to estimate functional parameters, life becomes complicated and imperfect. In a model I situation we can obtain unbiased parameter estimates when the dependent variable is a linear function of the parameters. When the dependent variable is a nonlinear function of the parameters, there is, in general, no way to obtain unbiased estimates of the parameters, even in a model I situation. In some cases a transformation may remove the nonlinearity, but transformations can create other types of problems, as we will see.

Figure 5.1 illustrates a simple case where some sort of nonlinear curve fitting is needed. The experimental data are the transmission coefficients of visible light in type IB seawater measured at 25-nm intervals from 400 to 700 nm. It seems clear that the relationship over this wavelength range is nonlinear, and if we were interested in interpolating the results to

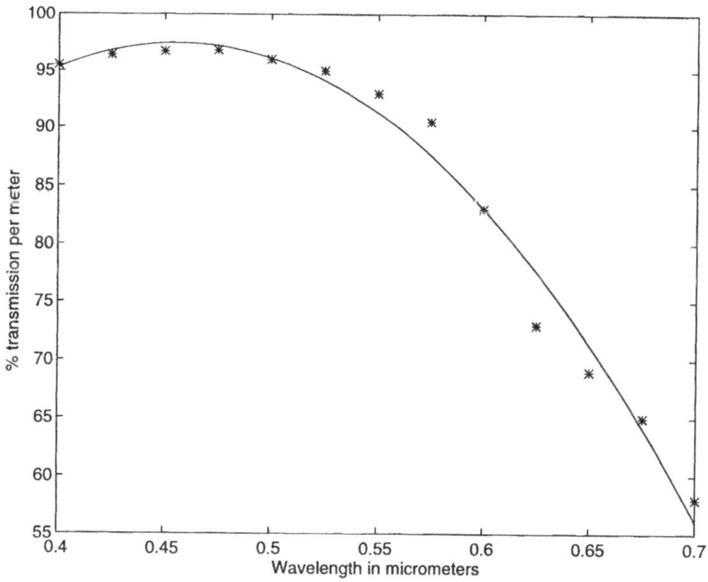

Figure 5.1 Light transmission characteristics of type IB seawater. From Jerlov (1968).

wavelengths other than those for which there are experimental results, one approach would be to draw a smooth curve through the data.

There are two basic approaches to this problem. One is to fit a simple polynomial function of the form

$$Y(X) = \sum_{k=0}^{M} c_k X^k \tag{5.1}$$

The function $Y(X)$ in Equation 5.1 is a linear function of the c_k, and, as we will see, this fact greatly simplifies the determination of the c_k. The fact that $Y(X)$ is a nonlinear function of X does not create any serious problems unless we feel that model I regression analysis is inappropriate. A second approach is to assume that $Y(X)$ is a nonlinear function of both X and at least some of the c_k. We might use this approach if we wanted to constrain the behavior of the function in some way or if we felt we knew the exact form of the underlying functional relationship. In the case of the data in Figure 5.1, for example, we might try fitting a function of the form

$$Y(X) = A + B\sin(KX) + C\cos(KX) \tag{5.2}$$

where A, B, C, and K are constants to be determined by some kind of curve-fitting procedure. In this case $Y(X)$ is a linear function of A, B, and

C, but a nonlinear function of K. We could determine A, B, and C by straightforward means, but determining K would require something more sophisticated.

MATRIX ALGEBRA

In deriving the procedures for solving such problems, we will find it convenient to use matrix algebra to formulate the relevant equations. It is assumed here that the reader has been introduced to matrices before, and the following discussion is therefore intended to constitute a review rather than an introduction to the topic.

A matrix is a rectangular array of elements arranged in rows and columns. We will here designate matrices as boldface letters to distinguish them from conventional scalar numbers. An example of a matrix would be

$$\mathbf{A} = \begin{bmatrix} 1 & 6 & 3 \\ 9 & 2 & 8 \end{bmatrix}$$

This matrix contains two rows and three columns. The element in row i and column j is designated a_{ij}. Thus in this case $a_{12} = 6$ and $a_{23} = 8$. A matrix with only one row or one column is referred to as a *vector*, a *row vector* in the former case and a *column vector* in the latter case.

Matrices may be manipulated like ordinary numbers, subject to certain constraints. Two matrices may be added or subtracted if they contain the same number of rows and columns. The sum or difference of two matrices is a third matrix whose elements are the sum or difference, respectively, of the corresponding elements of the original two matrices. Thus if

$$\mathbf{A} = \begin{bmatrix} 1 & 6 & 3 \\ 9 & 2 & 8 \end{bmatrix} \quad \text{and} \quad \mathbf{B} = \begin{bmatrix} 7 & 5 & 2 \\ 6 & 4 & 0 \end{bmatrix}$$

then

$$\mathbf{A} + \mathbf{B} = \begin{bmatrix} 8 & 11 & 5 \\ 15 & 6 & 8 \end{bmatrix} \quad \text{and} \quad \mathbf{A} - \mathbf{B} = \begin{bmatrix} -6 & 1 & 1 \\ 3 & -2 & 8 \end{bmatrix}$$

Multiplying or dividing a matrix by a pure number (scalar) is straightforward. One simply multiplies or divides each element in the matrix by the number. Thus if

$$\mathbf{A} = \begin{bmatrix} 1 & 6 & 3 \\ 9 & 2 & 8 \end{bmatrix}$$

then

$$2\mathbf{A} = \begin{bmatrix} 2 & 12 & 6 \\ 18 & 4 & 16 \end{bmatrix}$$

The product or quotient of two matrices is not so straightforward. The product of two matrices is defined if the number of columns of the first matrix equals the number of rows of the second matrix. The product of **A** and **B** is not defined for the two matrices given previously. However, if

$$\mathbf{C} = \begin{bmatrix} 1 & 2 \\ 3 & 4 \end{bmatrix} \quad \text{and} \quad \mathbf{D} = \begin{bmatrix} 5 & 6 \\ 7 & 8 \end{bmatrix}$$

then the product **CD** and the product **DC** are both defined. The rule in matrix multiplication is to multiply element by element the rows of the left-hand matrix by the columns of the right-hand matrix. The product of the ith row of the left-hand matrix times the jth column of the right-hand matrix becomes the element in the ith row and jth column of the product matrix. For example, if $\mathbf{E} = \mathbf{CD}$, then $e_{11} = (1)(5) + (2)(7) = 19$, $e_{12} = (1)(6) + (2)(8) = 22$, $e_{21} = (3)(5) + (4)(7) = 43$, and $e_{22} = (3)(6) + (4)(8) = 50$. Hence

$$\mathbf{E} = \begin{bmatrix} 19 & 22 \\ 43 & 50 \end{bmatrix}$$

Following the same rules, one can show that

$$\mathbf{F} = \mathbf{DC} = \begin{bmatrix} 23 & 34 \\ 31 & 46 \end{bmatrix}$$

It is clear from this example that $\mathbf{CD} \neq \mathbf{DC}$. Thus whether one multiplies another matrix on the left or right (i.e., premultiplies or postmultiplies) is an important consideration in matrix multiplication. Indeed the products **CD** and **DC** are both defined only if **C** and **D** are square matrices with the same number of rows. In general, matrix multiplication of two square matrices is not a *commutative* operation. On the other hand, it is always true that $\mathbf{A} + \mathbf{B} = \mathbf{B} + \mathbf{A}$. Hence addition of matrices is commutative.

The *identity matrix*, which is designated **I**, is defined to be the square matrix such that for any other square matrix **A** with the same number of rows and columns, $\mathbf{AI} = \mathbf{IA} = \mathbf{A}$. It is straightforward to show that $i_{jj} = 1$ and $i_{jk} = 0$ for $j \neq k$. Hence the identity matrix for any 2×2 matrix is $\begin{bmatrix} 1 & 0 \\ 0 & 1 \end{bmatrix}$.

Matrix division is defined only for square matrices and is accomplished by multiplying by the inverse of the matrix divisor. Since matrix multiplication is not, in general, commutative, neither is matrix division. Therefore division on the right does not, in general, give the same result as division on the left. The *inverse* of a square matrix **A** is designated $\mathbf{A}^{-1}$ and is defined to be the square matrix such that $\mathbf{A}\mathbf{A}^{-1} = \mathbf{A}^{-1}\mathbf{A} = \mathbf{I}$. For example, if

$$\mathbf{A} = \begin{bmatrix} 1 & 2 \\ 3 & 4 \end{bmatrix}$$

then one can show that

$$\mathbf{A}^{-1} = \begin{bmatrix} -2 & 1 \\ 1.5 & -0.5 \end{bmatrix}$$

The inverse of a matrix **A** exists only if the *determinant* of **A**, which we designate $|\mathbf{A}|$, is nonzero. In the example just given, $|\mathbf{A}| = (1)(4) - (3)(2) = -2$. Hence $\mathbf{A}^{-1}$ is defined. If the inverse of a matrix does not exist, the matrix is said to be *singular*.

The *transpose* of a matrix is the matrix obtained by interchanging the rows and columns of the original matrix. We will call $\mathbf{A}'$ the transpose of the matrix **A**. Thus if

$$\mathbf{A} = \begin{bmatrix} 1 & 2 \\ 3 & 4 \end{bmatrix}$$

then

$$\mathbf{A}' = \begin{bmatrix} 1 & 3 \\ 2 & 4 \end{bmatrix}$$

A *symmetric* matrix is a matrix for which $a_{jk} = a_{kj}$. If **A** is a symmetric matrix, then $\mathbf{A}' = \mathbf{A}$. The *trace* of a square matrix is the sum of the diagonal elements of the matrix. The trace of the matrix **A** will be designated tr(**A**). We can therefore say that $\mathrm{tr}(\mathbf{A}) = \Sigma_{k=1}^{N} a_{kk}$. In the example just cited, $\mathrm{tr}(\mathbf{A}) = 1 + 4 = 5$.

We can define a matrix of expectation values, similar to the expectation value of a scalar quantity. We will call **E(A)** the *expectation value* of **A**. The elements of **E(A)** are the expectation values of the elements of **A**. If the elements of **A** are fixed, we say that **A** is a constant matrix. If the elements of **A** are random numbers, then we say that **A** is a random matrix. If a vector **Y** is a random matrix, then we can define the *variance–covariance matrix* of **Y**, which we designate $\boldsymbol{\sigma}^2(\mathbf{Y})$. $\boldsymbol{\sigma}_{jj}^2 = \mathrm{Var}(Y_j)$ and $\boldsymbol{\sigma}_{jk}^2 = \mathrm{Cov}(Y_j, Y_k)$ for $j \neq k$.

Some general facts about matrix mechanics are the following:

$$\mathbf{A} + \mathbf{B} = \mathbf{B} + \mathbf{A} \qquad \text{(commutative law of addition)} \tag{5.3}$$

$$(\mathbf{A} + \mathbf{B}) + \mathbf{C} = \mathbf{A} + (\mathbf{B} + \mathbf{C}) \qquad \text{(associative law of addition)} \tag{5.4}$$

$$(\mathbf{AB})\mathbf{C} = \mathbf{A}(\mathbf{BC}) \qquad \text{(associative law of multiplication)} \tag{5.5}$$

$$\mathbf{C}(\mathbf{A} + \mathbf{B}) = \mathbf{CA} + \mathbf{CB} \qquad \text{(distributive law)} \tag{5.6}$$

$$\lambda(\mathbf{A} + \mathbf{B}) = \lambda\mathbf{A} + \lambda\mathbf{B} \qquad (\lambda = \text{any scalar}) \tag{5.7}$$

$$(\mathbf{A}')' = \mathbf{A} \tag{5.8}$$

$$(\mathbf{A} + \mathbf{B})' = \mathbf{A}' + \mathbf{B}' \tag{5.9}$$

$$(\mathbf{AB})' = \mathbf{B}'\mathbf{A}' \tag{5.10}$$

$$(\mathbf{ABC})' = \mathbf{C}'\mathbf{B}'\mathbf{A}' \tag{5.11}$$

$$(\mathbf{AB})^{-1} = \mathbf{B}^{-1}\mathbf{A}^{-1} \tag{5.12}$$

$$(\mathbf{ABC})^{-1} = \mathbf{C}^{-1}\mathbf{B}^{-1}\mathbf{A}^{-1} \tag{5.13}$$

$$(\mathbf{A}^{-1})^{-1} = \mathbf{A} \tag{5.14}$$

$$(\mathbf{A}')^{-1} = (\mathbf{A}^{-1})' \tag{5.15}$$

$$\mathbf{E}(\mathbf{A}) = \mathbf{A} \text{ if } \mathbf{A} \text{ is a constant matrix} \tag{5.16}$$

$$\mathbf{E}(\mathbf{AY}) = \mathbf{AE}(\mathbf{Y}) \quad \text{if } \mathbf{A} \text{ is a constant matrix and } \mathbf{Y} \text{ is a random vector} \tag{5.17}$$

$$\sigma^2(\mathbf{AY}) = \mathbf{A}[\sigma^2(\mathbf{Y})]\mathbf{A}' \quad \text{if } \mathbf{A} \text{ is a constant matrix and } \mathbf{Y} \text{ is a random vector} \tag{5.18}$$

$$\sigma^2(\mathbf{A} + \mathbf{Y}) = \sigma^2(\mathbf{Y}) \quad \text{if } \mathbf{A} \text{ is a constant vector and } \mathbf{Y} \text{ is a random vector} \tag{5.19}$$

POLYNOMIAL CURVE FITTING

We now proceed to the general problem of polynomial curve fitting, where we will use matrix methods to solve the least-squares equations and to derive error bounds. In polynomial least squares we seek to describe a function $Y(X)$ by an expression of the form $\sum_{k=0}^{M} c_k X^k$. We seek to minimize the sum of the squares of the deviations of $Y(X)$ from the expression $\sum_{k=0}^{M} c_k X^k$. In other words, we seek to minimize the expression $\sum_{j=1}^{N}[Y_j - \sum_{k=0}^{M} c_k X_j^k]^2$.

Differentiating this expression with respect to each c_k and setting the partial derivatives equal to 0 gives the following $M + 1$ equations:

$$\sum_{j=1}^{N} Y_j = c_0 N + c_1 \sum_{j=1}^{N} X_j + c_2 \sum_{j=1}^{N} X_j^2 + c_3 \sum_{j=1}^{N} X_j^3 + \cdots + c_M \sum_{j=1}^{N} X_j^M$$

$$\sum_{j=1}^{N} X_j Y_j = c_0 \sum_{j=1}^{N} X_j + c_1 \sum_{j=1}^{N} X_j^2 + c_2 \sum_{j=1}^{N} X_j^3 + c_3 \sum_{j=1}^{N} X_j^4 + \cdots + c_M \sum_{j=1}^{N} X_j^{M+1}$$

$$\vdots$$

$$\sum_{j=1}^{N} X_j^M Y_j = c_0 \sum_{j=1}^{N} X_j^M + c_1 \sum_{j=1}^{N} X_j^{M+1} + c_2 \sum_{j=1}^{N} X_j^{M+2}$$

$$+ c_3 \sum_{j=1}^{N} X_j^{M+3} + \cdots + c_M \sum_{j=1}^{N} X_j^{2M}$$

These $M + 1$ equations are generally referred to as the least-squares *normal equations*. If $M = 1$ we recover the familiar equations for model I linear regression. Now let us define the following matrices:

$$\mathbf{Y} = \begin{bmatrix} Y_1 \\ Y_2 \\ \vdots \\ Y_N \end{bmatrix} \quad \mathbf{X} = \begin{bmatrix} 1 & X_1 & X_1^2 & X_1^3 & \cdots & X_1^M \\ 1 & X_2 & X_2^2 & X_2^3 & \cdots & X_2^M \\ 1 & \cdot & \cdot & \cdot & \cdots & \cdot \\ 1 & \cdot & \cdot & \cdot & \cdots & \cdot \\ 1 & \cdot & \cdot & \cdot & \cdots & \cdot \\ 1 & X_N & X_N^2 & X_N^3 & \cdots & X_N^M \end{bmatrix} \quad \mathbf{C} = \begin{bmatrix} C_0 \\ C_1 \\ \vdots \\ C_M \end{bmatrix} \tag{5.20}$$

Note that $\mathbf{Y}$ and $\mathbf{C}$ are column vectors containing N and $M + 1$ rows, respectively, while $\mathbf{X}$ contains N rows and $M + 1$ columns. The left-hand side of the polynomial regression equations is easily shown to be $\mathbf{X'Y}$. The right-hand side of the regression equations can be shown to equal $\mathbf{X'XC}$. Thus the complete set of $M + 1$ equations can be written in the form

$$\mathbf{X'Y} = \mathbf{X'XC} \tag{5.21}$$

To solve Equation 5.21, we note that $\mathbf{X'X}$ is a square, symmetric matrix with $M + 1$ rows and columns. By premultiplying both sides of Equation 5.21 by $(\mathbf{X'X})^{-1}$, we obtain

$$(\mathbf{X'X})^{-1}\mathbf{X'Y} = (\mathbf{X'X})^{-1}\mathbf{X'XC} = \mathbf{IC} = \mathbf{C} \tag{5.22}$$

If we are dealing with controlled observations, then **X** is fixed. We assume the errors in **Y** are random. Hence our solution vector **C** has the form **C** = **AY**, where **A** is a fixed matrix equal to $(\mathbf{X}'\mathbf{X})^{-1}\mathbf{X}'$. Hence Equations 5.17 and 5.18 apply, and

$$\mathbf{E(C)} = \mathbf{E(AY)} = \mathbf{AE(Y)} \tag{5.23}$$

$$\boldsymbol{\sigma}^2(\mathbf{C}) = \boldsymbol{\sigma}^2(\mathbf{AY}) = \mathbf{A}\left[\boldsymbol{\sigma}^2(\mathbf{Y})\right]\mathbf{A}' = (\mathbf{X}'\mathbf{X})^{-1}\mathbf{X}'\left[\boldsymbol{\sigma}^2(\mathbf{Y})\right]\mathbf{X}(\mathbf{X}'\mathbf{X})^{-1} \tag{5.24}$$

Equation 5.23 tells us that **C** is an unbiased estimate of the true parameters as long as the errors in **Y** have an expectation value of 0 and the functional relationship between X and Y is a polynomial of order M. We had already discovered this fact in the case of model I linear regression in Chapter 2. We now see that it is also true in the more general case of model I polynomial regression. In Equation 5.24 we have taken advantage of the fact that $\mathbf{X}'\mathbf{X}$ is a symmetric matrix, so that $(\mathbf{X}'\mathbf{X})' = \mathbf{X}'\mathbf{X}$. If the errors in **Y** are random, then the off-diagonal elements of $\boldsymbol{\sigma}^2(\mathbf{Y})$ should all be 0. Furthermore, if the errors in **Y** all have the same variance, then the diagonal elements of $\boldsymbol{\sigma}^2(\mathbf{Y})$ should all be the same. This is the usual assumption in model I polynomial regression. Hence $\boldsymbol{\sigma}^2(\mathbf{Y}) = \sigma^2\mathbf{I}$. This fact greatly simplifies the right-hand side of Equation 5.24, which now becomes

$$\boldsymbol{\sigma}^2(\mathbf{C}) = \sigma^2(\mathbf{X}'\mathbf{X})^{-1}\mathbf{X}'\mathbf{X}(\mathbf{X}'\mathbf{X})^{-1} = \sigma^2(\mathbf{X}'\mathbf{X})^{-1} \tag{5.25}$$

We thus discover that the variances of the polynomial coefficients are equal to the corresponding diagonal elements of $(\mathbf{X}'\mathbf{X})^{-1}$ times σ^2. We had already discovered this fact for model I linear regression in Chapter 2, although we did not formulate the expression for the variances using matrices.

We will call the value of Y_j estimated from our least-squares polynomial $\hat{Y}_j$, that is, $\hat{Y}_j = \sum_{i=0}^{M} c_i X_j^i$. The difference between Y_j and the least-squares polynomial is $Y_j - \hat{Y}_j$. The column vector of these differences is $\mathbf{Y} - \mathbf{XC}$. From Equation 5.22 we have $\mathbf{C} = (\mathbf{X}'\mathbf{X})^{-1}\mathbf{X}'\mathbf{Y}$, so $\mathbf{Y} - \mathbf{XC} = \mathbf{Y} - \mathbf{X}(\mathbf{X}'\mathbf{X})^{-1}\mathbf{X}'\mathbf{Y} = (\mathbf{I} - \mathbf{X}(\mathbf{X}'\mathbf{X})^{-1}\mathbf{X}')\mathrm{Y}$. We therefore see that $\mathbf{Y} - \mathbf{XC}$ has the form $(\mathbf{I} - \mathbf{H})\mathbf{Y}$, where $\mathbf{I} - \mathbf{H}$ is a constant matrix and **Y** is a random vector. The matrix $\mathbf{H} = \mathbf{X}(\mathbf{X}'\mathbf{X})^{-1}\mathbf{X}'$ is referred to as the *hat matrix*, because $\mathbf{HY} = \hat{\mathbf{Y}}$. The **H** matrix is said to put the hat on the **Y** matrix. Using Equation 5.18, we see that

$$\boldsymbol{\sigma}^2[(\mathbf{I} - \mathbf{H})\mathbf{Y}] = (\mathbf{I} - \mathbf{H})\left[\boldsymbol{\sigma}^2(\mathbf{Y})\right](\mathbf{I} - \mathbf{H})' = \sigma^2(\mathbf{I} - \mathbf{H})(\mathbf{I} - \mathbf{H})' \tag{5.26}$$

where we have again assumed that the off-diagonal elements of $\boldsymbol{\sigma}^2(\mathbf{Y})$ are 0 and that the diagonal elements are all the same and equal to σ^2. We now observe that $\mathbf{H} = \mathbf{H}'$ and that $\mathbf{HH}' = \mathbf{X}(\mathbf{X}'\mathbf{X})^{-1}\mathbf{X}'\mathbf{X}(\mathbf{X}'\mathbf{X})^{-1}\mathbf{X}' = \mathbf{X}(\mathbf{X}'\mathbf{X})^{-1}\mathbf{X}^{\mathrm{t}} = \mathbf{H}$. Hence $(\mathbf{I} - \mathbf{H})(\mathbf{I} - \mathbf{H})' = \mathbf{I} - \mathbf{H}$ and

$$\boldsymbol{\sigma}^2(\mathbf{Y} - \mathbf{XC}) = \boldsymbol{\sigma}^2[(\mathbf{I} - \mathbf{H})\mathbf{Y}] = \sigma^2(\mathbf{I} - \mathbf{H}) \tag{5.27}$$

Now since $E[Z - E(Z)]^2 = E(Z^2) - [E(Z)]^2$, it follows that the diagonal elements of $\boldsymbol{\sigma}^2(\mathbf{Y} - \mathbf{XC})$ are the N expectation values of $(Y_j - \hat{Y}_j)^2 - [E(Y_j - \hat{Y}_j)]^2$, and the trace of $\boldsymbol{\sigma}^2(\mathbf{Y} - \mathbf{XC})$ is therefore the expectation value of $\Sigma_{j=1}^{N}\{(Y_j - \hat{Y}_j)^2 - [E(Y_j - \hat{Y}_j)]^2\}$. It can be shown (see the Exercises) that $\mathrm{tr}(\mathbf{H}) = M + 1$, and hence $\mathrm{tr}(\mathbf{I} - \mathbf{H}) = N - M - 1$. We therefore conclude that

$$E\left[\sum_{j=1}^{N}\left(Y_j - \hat{Y}_j\right)^2\right] - \left[E\left(Y_j - \hat{Y}_j\right)\right]^2 = (N - M - 1)\sigma^2 \qquad (5.28)$$

If Y is, in fact, an Mth-order polynomial function of X and the errors in Y are random, then Equation 5.22 gives us an unbiased estimate of the true $\mathbf{C}$ and $\mathbf{E(Y)} = \mathbf{XE(C)}$. In that case $E(Y_j - \hat{Y}_j) = 0$ and Equation 5.28 becomes

$$\sigma^2 = E\left[\frac{\sum_{j=1}^{N}\left(Y_j - \hat{Y}_j\right)^2}{N - M - 1}\right] \qquad (5.29)$$

We have seen this result before in the case of model I linear regression ($M = 1$), where we divided the sum of the squares of the deviations from the least-squares line by $N - 2$ to obtain an unbiased estimate of σ^2. The rule is to divide the sum of the squares of the deviations from the fitted line by the number of degrees of freedom to obtain an unbiased estimate of σ^2. The df is the number of observations in excess of the minimum required to determine the parameters of the fitted line. Equivalently, the df is N minus the number of coefficients which were determined by least squares. The rule applies whenever the dependent variable is a linear combination of the least-squares coefficients, because in such cases the solution vector will always have the form $\mathbf{C} = (\mathbf{X}'\mathbf{X})^{-1}\mathbf{X}'\mathbf{Y}$. The fact that $\mathrm{tr}(\mathbf{H}) = M + 1$ is not peculiar to the case of polynomial least squares. That $\mathrm{tr}(\mathbf{H}) = M + 1$ is true of any matrix $\mathbf{H}$ of the form $\mathbf{H} = \mathbf{X}(\mathbf{X}'\mathbf{X})^{-1}\mathbf{X}'$), where $\mathbf{X}$ is a matrix with N rows and $M + 1$ columns, and $N \geq M + 1$.

Equation 5.28 may be generalized to the case where we attempt to fit our data to a function $F(X) = \Sigma_{k=0}^{M} C_k G_k(X)$, where $G_k(X)$ is any function of X and C_k is a constant. In that case we have

$$E\left\{\sum_{j=1}^{N}\left[Y_j - F(X_j)\right]^2\right\} - \left[E\left(Y_j - F(X_j)\right)\right]^2 = (N - M - 1)\sigma^2 \quad (5.30)$$

We conclude that

$$\frac{E\left\{\sum\left[Y_j - F(X_j)\right]^2\right\}}{N - M - 1} = \sigma^2 + \frac{\sum\left\{E\left[Y_j - F(X_j)\right]\right\}^2}{N - M - 1} \tag{5.31}$$

In other words, the left-hand side of Equation 5.31 will give an unbiased estimate of σ^2 as long as the expectation value of the function we are trying to fit is equal to the expectation value of Y; that is, if we are trying to fit the right function, the left-hand side of Equation 5.31 will give us an unbiased estimate of σ^2. If the function we are trying to fit is the wrong function, the left-hand side of Equation 5.31 will overestimate σ^2 by an amount equal to $\Sigma[E(Y_j - F(X_j))]^2/(N - M - 1)$. The behavior of the left-hand side of Equation 5.31 underlies the use of analysis of variance in polynomial regression.

ANALYSIS OF VARIANCE IN POLYNOMIAL REGRESSION

We begin this discussion by considering the familiar sum of squares $\Sigma_{j=1}^{N}(Y_j - \bar{Y})^2$. We designate this sum of squares SSTO, for *total sum of squares*. In order to discuss this statistic in matrix notation, we need a matrix expression for $\bar{Y}$. We begin by defining a column matrix whose elements are all 1. We designate this column matrix $\mathbf{1}$. It will also be convenient at this time to define a second matrix equal to $\mathbf{11}'$. We designate this matrix $\mathbf{J}$. The matrix $\mathbf{J}$ is a square matrix whose elements are all 1. Now it is easy to show that $\bar{Y} = \mathbf{1}'\mathbf{Y}/N = \mathbf{Y}'\mathbf{1}/N$. The column matrix whose elements are $Y_j - \bar{Y}$ is therefore $\mathbf{Y} - \mathbf{11}'\mathbf{Y}/N = (\mathbf{I} - \mathbf{J}/\text{N})\mathbf{Y}$. We will call this column matrix $\boldsymbol{\Delta}$. We now recall that the column matrix containing the values of Y calculated from the least-squares polynomial is $\mathbf{XC} = \mathbf{X}(\mathbf{X}'\mathbf{X})^{-1}\mathbf{X}'\mathbf{Y} = \mathbf{HY}$. Obviously,

$$\boldsymbol{\Delta} = (\mathbf{I} - \mathbf{J}/N)\mathbf{Y} = (\mathbf{I} - \mathbf{J}/N)\mathbf{Y} - \mathbf{HY} + \mathbf{HY} = (\mathbf{I} - \mathbf{H} + \mathbf{H} - \mathbf{J}/N)\mathbf{Y} \tag{5.32}$$

The reason for subtracting and adding $\mathbf{HY}$ will soon become clear. Now it should be obvious that

$$\begin{aligned} \text{SSTO} &= \sum_{j=1}^{N}\left(Y_j - \bar{Y}\right)^2 \\ &= \boldsymbol{\Delta}'\boldsymbol{\Delta} = \mathbf{Y}'(\mathbf{I} - \mathbf{H} + \mathbf{H} - \mathbf{J}/N)(\mathbf{I} - \mathbf{H} + \mathbf{H} - \mathbf{J}/N)\mathbf{Y} \end{aligned} \tag{5.33}$$

Here we have used the fact that $\mathbf{I}$, $\mathbf{J}$, and $\mathbf{H}$ are all symmetric matrices, so that $\mathbf{I}' = \mathbf{I}$, $\mathbf{J}' = \mathbf{J}$, and $\mathbf{H}' = \mathbf{H}$. The right-hand side of Equation 5.33 can be expressed as the sum of two terms as follows:

$$\text{SSTO} = \mathbf{Y}'(\mathbf{I} - \mathbf{H})(\mathbf{I} - \mathbf{H})\mathbf{Y} + \mathbf{Y}'[(\mathbf{H} - \mathbf{J}/N)(\mathbf{I} - \mathbf{H}) + (\mathbf{I} - \mathbf{J}/N)(\mathbf{H} - \mathbf{J}/N)]\mathbf{Y} \quad (5.34)$$

Realizing that $\mathbf{XC} = \mathbf{HY}$ and $(\mathbf{XC})' = \mathbf{Y}'\mathbf{H}$, it should be clear that

$$\mathbf{Y}'(\mathbf{I} - \mathbf{H})(\mathbf{I} - \mathbf{H})\mathbf{Y} = (\mathbf{Y} - \mathbf{XC})'(\mathbf{Y} - \mathbf{XC}) = \sum_{j=1}^{N} \left(Y_j - \hat{Y}_j\right)^2 \quad (5.35)$$

We designate this term SSE, the error sum of squares. It is the sum of the squares of the deviations of the values of Y from the regression polynomial. The second term can be simplified by remembering that $\mathbf{HH} = \mathbf{H}$ and noting that $\mathbf{JJ} = N\mathbf{J}$. We discover that

$$\mathbf{Y}'[(\mathbf{H} - \mathbf{J}/N)(\mathbf{I} - \mathbf{H}) + (\mathbf{I} - \mathbf{J}/N)(\mathbf{H} - \mathbf{J}/N)]\mathbf{Y} = \mathbf{Y}'(\mathbf{H} - \mathbf{J}/N)\mathbf{Y} \quad (5.36)$$

In order to understand the meaning of the right-hand side of Equation 5.36, we note that $\mathbf{HX} = \mathbf{X}(\mathbf{X}'\mathbf{X})^{-1}\mathbf{X}'\mathbf{X} = \mathbf{X}$. Because the elements in the first column of $\mathbf{X}$ are all 1, the fact that $\mathbf{HX} = \mathbf{X}$ means that the sum of the elements in any row of $\mathbf{H}$ must equal 1. Similarly, the fact that $(\mathbf{HX})' = \mathbf{X}'\mathbf{H}' = \mathbf{X}'\mathbf{H} = \mathbf{X}^t$ implies that the sum of the elements in any column of $\mathbf{H}$ must equal 1. From these two facts we can easily show that $\mathbf{JH} = \mathbf{HJ} = \mathbf{J}$. Using these facts, we see that

$$\begin{aligned}
[(\mathbf{H} - \mathbf{J}/N)\mathbf{Y}]'(\mathbf{H} - \mathbf{J}/N)\mathbf{Y} &= \mathbf{Y}'(\mathbf{H} - \mathbf{J}/N)(\mathbf{H} - \mathbf{J}/N)\mathbf{Y} \\
&= \mathbf{Y}'(\mathbf{H} - \mathbf{HJ}/N - (\mathbf{J}/N)\mathbf{H} + \mathbf{J}/N)\mathbf{Y} \\
&= \mathbf{Y}'(\mathbf{H} - \mathbf{J}/N)\mathbf{Y} \qquad (5.37)
\end{aligned}$$

Thus we see that the right-hand side of Equation 5.36 equals $[(\mathbf{H} - \mathbf{J}/N)\mathbf{Y}]'(\mathbf{H} - \mathbf{J}/N)\mathbf{Y}$. However,

$$[(\mathbf{H} - \mathbf{J}/N)\mathbf{Y}]'(\mathbf{H} - \mathbf{J}/N)\mathbf{Y} = \sum_{j=1}^{N} \left(\hat{Y}_j - \bar{Y}\right)^2 \quad (5.38)$$

We call this sum of squares the regression sum of squares, or SSR. It is the sum of the squares of the differences between the mean value of Y and the values of Y estimated from the polynomial. We have now shown that

$$\text{SSTO} = \text{SSR} + \text{SSE} \quad (5.39)$$

We saw a similar result in Chapter 3, where we learned that when X and Y are linearly related, the variance of Y may be partitioned into two terms, the first of which, analogous to SSE, is the variance of Y about the regression line, and the second of which, analogous to SSR, is the portion of the variance accounted for by the regression line. We have now generalized the result in Chapter 3 to polynomial functions. It is important for us to keep in mind that SSTO, SSR, and SSE are statistics, that is, estimates of parameters, and it is obviously desirable to know just what they are estimating. In the case of SSE we have already seen from Equation 5.28 that

$$E(\text{SSE}) = \sum_{j=1}^{N} \left[E\left(Y_j - \hat{Y}_j\right)\right]^2 + (N - M - 1)\sigma^2 \tag{5.40}$$

To determine the expectation values of SSTO and SSR, we again examine the traces of the relevant variance–covariance matrices. For SSTO we want the trace of $\sigma^2(\mathbf{I} - \mathbf{J}/N)\mathbf{Y}$. Since $(\mathbf{I} - \mathbf{J}/N)\mathbf{Y}$ is a constant matrix times a random vector, we know from Equation 5.18 that

$$\begin{aligned}\boldsymbol{\sigma}^2(\mathbf{I} - \mathbf{J}/N)\mathbf{Y} &= (\mathbf{I} - \mathbf{J}/N)\boldsymbol{\sigma}^2(\mathbf{Y})(\mathbf{I} - \mathbf{J}/N)' \\ &= \sigma^2(\mathbf{I} - \mathbf{J}/N)(\mathbf{I} - \mathbf{J}/N)' = \sigma^2(\mathbf{I} - \mathbf{J}/N)\end{aligned} \tag{5.41}$$

The traces of both $\mathbf{I}$ and $\mathbf{J}$ are N, and hence $\text{tr}[\boldsymbol{\sigma}^2(\mathbf{I} - \mathbf{J}/N)\mathbf{Y}] = \sigma^2(N - 1)$. We conclude that

$$E(\text{SSTO}) = \sum_{j=1}^{N} \left[E\left(Y_j - \bar{Y}\right)\right]^2 + (N - 1)\sigma^2 \tag{5.42}$$

For SSR we want the trace of $\boldsymbol{\sigma}^2(\mathbf{H} - \mathbf{J}/N)\mathbf{Y}$. Proceeding as before, we have

$$\begin{aligned}\boldsymbol{\sigma}^2(\mathbf{H} - \mathbf{J}/N)\mathbf{Y} &= (\mathbf{H} - \mathbf{J}/N)\boldsymbol{\sigma}^2(\mathbf{Y})(\mathbf{H} - \mathbf{J}/N)' \\ &= \sigma^2(\mathbf{H} - \mathbf{J}/N)(\mathbf{H} - \mathbf{J}/N)' \\ &= \sigma^2[\mathbf{H} - \mathbf{HJ}/N - (\mathbf{J}/N)\mathbf{H} + \mathbf{J}/N] = \sigma^2(\mathbf{H} - \mathbf{J}/N)\end{aligned} \tag{5.43}$$

where we have again used the fact that $\mathbf{HJ} = \mathbf{JH} = \mathbf{J}$. Now $\text{tr}(\mathbf{H}) = M + 1$ and $\text{tr}(\mathbf{J}/N) = 1$, so $\text{tr}[\boldsymbol{\sigma}^2(\mathbf{H} - \mathbf{J}/N)\mathbf{Y}] = M\sigma^2$. We conclude that

$$E(\text{SSR}) = \sum_{j=1}^{N} \left[E\left(\hat{Y}_j - \bar{Y}\right)\right]^2 + M\sigma^2 \tag{5.44}$$

Our conclusions concerning SSTO, SSR, and SSE are summarized in Table 5.1.

TABLE 5.1 Degrees of Freedom and Expectation Values of SSTO, SSR, and SSE in Polynomial Regression

Sum of Squares	Degrees of Freedom	Expectation Value
SSTO	$N - 1$	$\sum_{j=1}^{N} \left[E\left(Y_j - \bar{Y}\right)\right]^2 + (N-1)\sigma^2$
SSR	M	$\sum_{j=1}^{N} \left[E\left(\hat{Y}_j - \bar{Y}\right)\right]^2 + M\sigma^2$
SSE	$N - M - 1$	$\sum_{j=1}^{N} \left[E\left(Y_j - \hat{Y}_j\right)\right]^2 + (N-M-1)\sigma^2$

An important point about Table 5.1 is that both the sums of squares and the degrees of freedom associated with the sums of squares are additive. This behavior is also true of differences of sums of squared errors and their associated df's. Note that the SSR has M degrees of freedom associated with it, one df for each of the coefficients $c_1, c_2, \ldots, c_M$. We can, in fact, partition the SSR among these coefficients by fitting polynomials containing all but one power of X. For example, suppose we had fit a third-order polynomial and wanted to know the contribution to the SSR from the quadratic term. We could obtain the contribution by fitting a polynomial of the form $Y = c_0 + c_1X + c_3X^3$ and calculating the SSR for that polynomial. The SSR for the complete cubic polynomial would be referred to as $\text{SSR}(X, X^2, X^3)$ to indicate the powers of X included in the polynomial. The SSR for the cubic polynomial minus the quadratic term would be $\text{SSR}(X, X^3)$. The difference between the two is designated $\text{SSR}(X^2|X, X^3)$, meaning the extra SSR when we add an X^2 term to a polynomial already containing a constant and an X and X^3 term. Mathematically, $\text{SSR}(X^2|X, X^3) = \text{SSR}(X, X^2, X^3) - \text{SSR}(X, X^3)$. It should be obvious that $\text{SSR}(X^2|X, X^3)$ is associated with 1 degree of freedom, since $\text{SSR}(X, X^2, X^3)$ has 3 df and $\text{SSR}(X, X^3)$ has 2.

One of the important questions that arises in polynomial regression is what terms to include in the polynomial. Trying to determine the answer can be confusing unless one has a clear idea of how to apply analysis of variance to polynomial curve fitting. As an example, we consider the data shown in Figure 5.1. The first step is always to look at the data. It seems clear from an examination of Figure 5.1 that we are dealing with a curvilinear function. Hence we will need a quadratic or higher-order polynomial to describe the data. Table 5.2 summarizes the results of fitting polynomials up to order 3 to the data.

If we simply assume that $Y = c_0$, we, of course, obtain $c_0 = \bar{Y}$. The associated SSE = SSTO = 2364. If we fit a function of the form $Y = c_0 +$

TABLE 5.2 Analysis of Fitting Polynomials Through Order 3 to the Data in Figure 5.1

Model		SSE
$Y = c_0$		2364 (= SSTO)
$Y = c_1X$		9442
$Y = c_0 + c_1X$		421
$c_0 = 157 \pm 10.25$	$c_0/s(c_0) = 15.3$	
$c_1 = -131 \pm 18.4$	$c_1/s(c_1) = -7.1$	
$Y = c_0 + c_2X^2$		301
$Y = c_1X + c_2X^2$		63
$Y = c_0 + c_1X + c_2X^2$		44
$c_0 = -47 \pm 22.2$	$c_0/s(c_0) = -2.1$	
$c_1 = 633 \pm 82.8$	$c_1/s(c_1) = 7.6$	
$c_2 = -694 \pm 74.8$	$c_2/s(c_2) = -9.3$	
$Y = c_0 + c_1X + c_3X^3$		49
$Y = c_0 + c_2X^2 + c_3X^3$		53
$Y = c_1X + c_2X^2 + c_3X^3$		47
$Y = c_0 + c_1X + c_2X^2 + c_3X^3$		39
$c_0 = -203 \pm 148$	$c_0/s(c_0) = -1.4$	
$c_1 = 1516 \pm 830$	$c_1/s(c_1) = 1.8$	
$c_2 = -2328 \pm 1531$	$c_2/s(c_2) = -1.5$	
$c_3 = 990 \pm 926$	$c_3/s(c_3) = 1.07$	

c_1X, we reduce the SSE to 421, and the SSR = 1943. Both c_0 and c_1 are judged to be highly significant by a t test. If we fit a quadratic function of the form $Y = c_0 + c_1X + c_2X^2$, we reduce the SSE to 44, and the SSR becomes 2320. Both c_1 and c_2 are highly significant by a t test, but c_0 is no longer significant at $\alpha = 0.05$. Note that the t test is two tailed, since rejecting the null hypothesis that $c_0 = 0$ implies that $|c_0| > 0$. If we go to a third-order polynomial of the form $Y = c_0 + c_1X + c_2X^2 + c_3X^3$, the SSE drops to 39, but none of the coefficients is significant at $\alpha = 0.05$ by a t test. How is it that both c_0 and c_1 are significant in the linear function, but only c_1 and c_2 are significant in the quadratic function, and none of the coefficients is significant in the cubic polynomial?

To answer this question, we must understand the relationship between the standard deviations of the coefficients and our analysis of variance. We will begin at the bottom and work our way up. In the linear case we know that SSR(X) = 1943. The SSE in this case is 421, and the associated df = 13 − 2 = 11. The mean SSE or MSE equals the SSE divided by the df. Hence MSE = 421/11 = 38.3. SSR(X) has only $M = 1$ degree of freedom associated with it. The ratio of the normalized sums of squares is therefore 1943/38.3 = 50.7. If the underlying functional relation were, in fact, Y = a constant, then both SSR(X) and MSE should be estimating the same

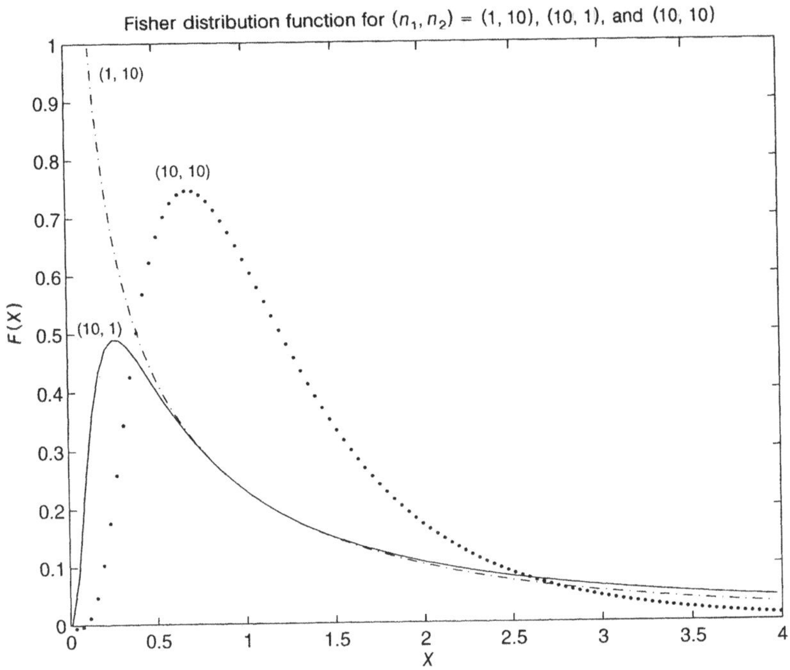

Figure 5.2 Fisher distribution function for (10, 10) ⋯ , (1, 10) —·—, and (10, 1) — degrees of freedom.

quantity, namely, σ^2. The null hypothesis is that $\text{SSR}(X) = \text{MSE}$. The alternative hypothesis is that $\text{SSR}(X) > \text{MSE}$, because if Y is not a constant $E(\text{SSR}) > E(\text{MSE})$. In this case, for example, if Y were, in fact, a linear function of X, then $E(\text{SSR}) = \sigma^2 + \beta^2\Sigma(X - \overline{X})^2$ (see Exercise 5.3) and $E(\text{MSE}) = \sigma^2$. If two distributions have the same variance and subsamples from each distribution are taken and the ratio of the variances estimated from the subsamples, the distribution of ratios will follow the so-called F distribution, named in honor of R. A. Fisher. The F distribution function depends on the df of the numerator (ν_1) and the df of the denominator (ν_2) and is asymmetric with a long tail to the right.[1] Figure 5.2 shows the F distribution function for several combinations of ν_1 and ν_2. When $\nu_1 < 3$ the probability density functions peak at $X = 0$; otherwise they peak at $0 < X < 1$. The expectation value of the ratio of sample variance

[1] The probability density function for the F distribution function is defined as follows:

$$p(x|\nu_1, \nu_2) = \Gamma[(\nu_1 + \nu_2)/2](\nu_1/\nu_2)^{\nu_1/2} x^{(\nu_1-2)/2} \Big/ \left\{\Gamma(\nu_1/2)\Gamma(\nu_2/2)[1 + (\nu_1/\nu_2)x]^{(\nu_1+\nu_2)/2}\right\}$$

estimates is undefined if $\nu_2 < 3$ and equals $\nu_2/(\nu_2 - 2)$ otherwise. Tabulated critical values of the F distribution are based on so-called upper-tail F tests, meaning that the alternative to the null hypothesis is that the expectation value of the numerator is greater than the expectation value of the denominator. This is the case here. Our numerator df is 1 and our denominator df is 11, and from a table of critical F values we find that $F_{1,11} = 19.7$ at $\alpha = 0.001$. Since $\mathrm{SSR}(X)/\mathrm{MSE} > 19.7$, we conclude with great confidence that $\mathrm{SSR}(X) > \mathrm{MSE}$. A very important point to note, however, is that $\sqrt{50.7} = 7.1$, which is identical to the magnitude of the ratio $c_1/s(c_1)$. The ratio $c_1/s(c_1)$ is, of course, the statistic we compare to the critical values of the t distribution function to determine whether c_1 is significant. The t test is two-tailed, because we want to know whether to accept the null hypothesis that $c_1 = 0$ or to conclude that $|c_1| > 0$. The F test and the t test are, in fact, equivalent in this case, because the critical values of the two distribution functions satisfy the relationship $[t(1 - \alpha/2; \mathrm{df})]^2 = F(1 - \alpha; 1, \mathrm{df})$.

Now we will move on to the quadratic function. The c_2 term is easy. We know that $\mathrm{SSR}(X^2|X) = \mathrm{SSR}(X, X^2) - \mathrm{SSR}(X)$. In this case $\mathrm{SSR}(X^2|X) = (2364 - 44) - (2364 - 421) = 2320 - 1943 = 377$. The MSE is now $44/(13 - 3) = 4.4$. Hence $\mathrm{SSR}(X^2|X)/\mathrm{MSE} = 377/4.4 = 85.7$, and $\sqrt{85.7} = 9.3$, which is identical to the magnitude of $c_2/s(c_2)$. For c_1 we know that $\mathrm{SSR}(X|X^2) = \mathrm{SSR}(X, X^2) - \mathrm{SSR}(X^2)$. For the function $Y = c_0 + c_2X^2$, $\mathrm{SSR} = 2364 - 301 = 2063$, so $\mathrm{SSR}(X^2) = 2063$, and $\mathrm{SSR}(X|X^2) = 2320 - 2063 = 257$. Hence $\mathrm{SSR}(X|X^2)/\mathrm{MSE} = 257/4.4 = 58.4$, and $\sqrt{58.4} = 7.6$, which is the same as $c_1/s(c_1)$.

In the case of the cubic polynomial, we know that $\mathrm{SSR}(X^3|X, X^2) = \mathrm{SSR}(X, X^2\ X^3) - \mathrm{SSR}(X, X^2) = (2364 - 39) - (2364 - 44) = 5$. The MSE in this case is $39/(13 - 4) = 4.33$, and $\mathrm{SSR}(X^3|X, X^2)/\mathrm{MSE} = 5/4.33 = 1.15$. We see that $\sqrt{1.15} = 1.07$, which is identical to $c_3/s(c_3)$. In the case of c_2 we know that $\mathrm{SSR}(X^2|X, X^3) = \mathrm{SSR}(X, X^2, X^3) - \mathrm{SSR}(X, X^3) = (2364 - 39) - (2364 - 49) = 10$. Hence $\mathrm{SSR}(X^2|X, X^3)/\mathrm{MSE} = 10/4.33 = 2.31$, and $\sqrt{2.31} = 1.5$, which is identical to the magnitude of $c_2/s(c_2)$. For c_1 we have $\mathrm{SSR}(X|X^2, X^3) = \mathrm{SSR}(X, X^2, X^3) - \mathrm{SSR}(X^2, X^3) = (2364 - 39) - (2364 - 53) = 14$. Hence $\mathrm{SSR}(X|X^2, X^3)/\mathrm{MSE} = 14/4.33 = 3.2$ and $\sqrt{3.2} = 1.8$, which is identical to $c_1/s(c_1)$.

You may have noticed by now that we have avoided discussing the statistical significance of c_0. There is a good reason for this. Until now we have taken advantage of the fact that SSTO = SSR + SSE (Eq. 5.39). Since SSTO is constant, any change in SSR must be accompanied by an equal and opposite change in SSE. The procedure for testing the importance of a given coefficient in the polynomial has been to calculate the difference of the SSRs with and without the given coefficient. However, when we toss c_0 out of the polynomial, the proof that SSTO = SSR + SSE falls apart. The reason can be traced to the derivation of Equation 5.37 from Equation 5.36. Here we took advantage of the fact that $\mathbf{JH} = \mathbf{HJ} = \mathbf{J}$. Recall that the proof of this equation rested in part on the fact that the first column of $\mathbf{X}$ consisted of 1's.

If c_0 is removed from the polynomial, the first column of **X** no longer contains 1's, and, in general, it is no longer true that **JH** = **HJ** = **J**.

Another way to appreciate the effect of removing c_0 from the polynomial is to note that

$$\text{SSTO} = \sum_{j=1}^{N} \left(Y_j - \bar{Y}\right)^2 = \sum_{j=1}^{N} \left(Y_j - \hat{Y}_j + \hat{Y}_j - \bar{Y}\right)^2 \tag{5.45}$$

When we expand the right-hand side of Equation 5.45, we obtain

$$\begin{aligned} \text{SSTO} &= \sum_{j=1}^{N} \left(Y_j - \hat{Y}_j\right)^2 + \sum_{j=1}^{N} \left(\hat{Y}_j - \bar{Y}\right)^2 + 2\sum_{j=1}^{N} \left(Y_j - \hat{Y}_j\right)\left(\hat{Y}_j - \bar{Y}\right) \\ &= \text{SSE} + \text{SSR} + 2\sum_{j=1}^{N} \left(Y_j - \hat{Y}_j\right)\left(\hat{Y}_j - \bar{Y}\right) \end{aligned} \tag{5.46}$$

The sum of cross terms $\Sigma_{j=1}^{N}(Y_j - \hat{Y}_j)(\hat{Y}_j - \bar{Y}) = 0$ if

$$\sum_{j=1}^{N} \left(Y_j - \hat{Y}_j\right)\hat{Y}_j = 0 \tag{5.47}$$

and

$$\sum_{j=1}^{N} \left(Y_j - \hat{Y}_j\right)\bar{Y} = 0 \tag{5.48}$$

That Equation 5.47 is true can easily be seen by going back to the normal equations. If we multiply the first normal equation by c_0, the second by c_1, the third by c_2, and so forth, we realize that the sum of all the expressions on the left-hand side equals $\Sigma_{j=1}^{N} Y_j \hat{Y}_j$ and the sum of all the expressions on the right-hand side equals $\Sigma_{j=1}^{N} \hat{Y}_j^2$. Thus $\Sigma_{j=1}^{N} Y_j \hat{Y}_j = \Sigma_{j=1}^{N} \hat{Y}_j^2$, and Equation 5.47 equals 0. With a little thought, we can see that this will be true even if c_0 is dropped from the polynomial. Equation 5.48, on the other hand, can only be true if $\Sigma_{j=1}^{N}(Y_j - \hat{Y}_j) = 0$. With c_0 in the polynomial, the first normal equation obviously guarantees that $\Sigma_{j=1}^{N} Y_j = \Sigma_{j=1}^{N} \hat{Y}_j$ and hence assures that Equation 5.48 equals 0. Without c_0 in the polynomial, this constraint disappears, and, in general, we cannot expect that SSTO = SSR + SSE. This fact is rather dramatically illustrated in the analysis of the data in Figure 5.1. If we attempt to fit a function of the form $Y = c_1 X$, the SSE is 9442, which is four times SSTO. Since SSTO, SSR, and SSE are all nonnegative numbers, it is impossible for SSE to exceed SSTO if SSTO = SSR + SSE.

Now what? Even without c_0, the expectation value of SSE is still as indicated in Table 5.1, with the understanding that $M + 1$ is the number of terms in the polynomial. Thus we can test the importance of c_0 from the difference in SSE with and without c_0 in the polynomial. The df associated with this difference in SSE values will be 1, since the polynomial with c_0 obviously contains one more term than the polynomial without c_0.

In the case of the linear function, the difference between the SSE of 9442 obtained with $Y = c_1X$ and the SSE of 421 obtained with $Y = c_0 + c_1X$ is 9021. The appropriate test statistic is this difference divided by the MSE, which for the polynomial $Y = c_0 + c_1X$ is 38.3. We therefore have $9021/38.3 = 235.5$. Now note that $\sqrt{235.5} = 15.3$, which is the same as the ratio $c_0/s(c_0)$.

To test c_0 in the quadratic polynomial, we need to know the SSE associated with the function $Y = c_1X + c_2X^2$. This turns out to be 63. The improvement in the SSE by adding the constant term is therefore $63 - 44 = 19$, and $19/\text{MSE} = 19/4.4 = 4.3$. We see that $\sqrt{4.3} = 2.1$, which is identical to the magnitude of $c_0/s(c_0)$.

For the cubic polynomial, the function $Y = c_1X + c_2X^2 + c_3X^3$ is associated with an SSE of 47, and $(47 - 39)/\text{MSE} = 1.85$. We see that $\sqrt{1.85} = 1.4$, which is identical to the magnitude of $c_0/s(c_0)$.

We can now more intelligently discuss the implications of the analysis in Table 5.2. The results from the four-term, third-order polynomial analysis tell us that we do not need all these terms to describe the data and furthermore that we could remove any one of the terms without having much effect on the SSE. The SSEs obtained with simple three-term polynomials range from 44 to 53; the SSE for the complete cubic polynomial is 39. The difference is considered not to be significant. The results do *not* tell us that we need none of these terms. They do tell us that we could remove any one of them; the others would, in effect, pick up the slack for the missing term.

When we go to simple three-term polynomials, the lowest SSE is obtained with the quadratic polynomial. In the case of the quadratic polynomial, we discover that the constant term is not quite significant at $\alpha = 0.05$, the critical t value being 2.2. It is clear that we need both the linear and the quadratic terms. Whether we include c_0 in the second-order polynomial is something of a judgment call in this case. The function $c_1X + c_2X^2$ actually does a good job of describing the data, with an SSE of 63. One might ask, why not just throw in everything but the kitchen sink? The reason is that beyond a certain point we begin to fit noise, and this is something we want to avoid, regardless of whether our purposes are prediction or identification of the characteristics of the true function.

REPLICATED DATA

When experimental observations have been replicated, we have additional information which can be incorporated into the analysis of variance. Consider

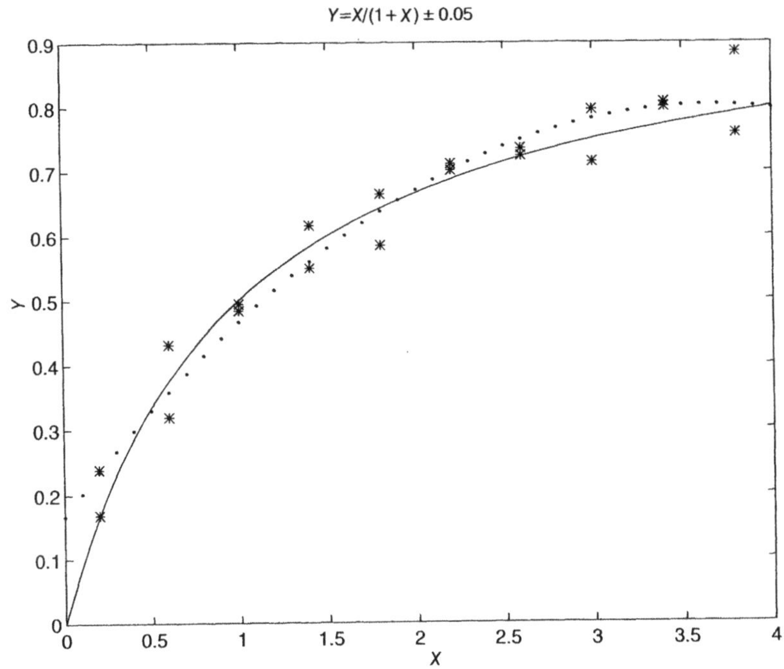

Figure 5.3 The function $Y = X/(1 + X)$ with paired noise-corrupted data ($*$) and a second-order polynomial fit ($\cdots$). The polynomial fit to the data is $Y(X) = 0.167 + 0.347X - 0.0473X^2$.

the data shown in Figure 5.3. The underlying function here is $Y = X/(1 + X)$. We have generated 10 pairs of noise-corrupted data, with the values of X ranging from 0.2 to 3.8 in increments of 0.4. The Y values were noise corrupted using normally distributed random numbers with a mean of 0 and a standard deviation of 0.05. The actual data are shown in Table 5.3.

Because we have replicated data, we can estimate σ^2 directly and independently of the polynomial curve-fitting process. For each pair of Y values, we calculate $(Y_1 - \bar{Y})^2 + (Y_2 - \bar{Y})^2$, where $\bar{Y} = (Y_1 + Y_2)/2$. Recall from Chapter 2 that this statistic has $2 - 1 = 1$ df associated with it. When we sum this statistic over the 10 pairs of Y values, we obtain a statistic which we call the SSPE, or pure error sum of squares:

$$\text{SSPE} = \sum_{j=1}^{R} \sum_{k=1}^{N/R} \left(Y_{jk} - \bar{Y}_j\right)^2 \tag{5.49}$$

where R is the number of replicated sets of observations, N is the total number of Y values, N/R is the number of replicates in each set, and $\bar{Y}_j$ is the mean of the Y values in the jth set. In this case $R = 10$ and $N/R = 2$.

TABLE 5.3 Data in Figure 5.3. $Y = X / (1 + X) \pm 0.05$

X	Y_1	Y_2	$(Y_1 + Y_2)/2$
0.2	0.2388	0.1686	0.2037
0.6	0.3188	0.4316	0.3752
1.0	0.4838	0.4943	0.4891
1.4	0.6164	0.5496	0.5830
1.8	0.6649	0.5850	0.6250
2.2	0.7113	0.7015	0.7064
2.6	0.7352	0.7239	0.7326
3.0	0.7955	0.7142	0.7549
3.4	0.7998	0.8064	0.8031
3.8	0.7585	0.8844	0.8215

The SSPE has $R(N/R - 1)$ df associated with it, which in this case is $10(2 - 1) = 10$.

Now we note that

$$\text{SSE} = \sum_{j=1}^{R} \sum_{k=1}^{N/R} \left(Y_{jk} - \bar{Y}_j + \bar{Y}_j - \hat{Y}_j\right)^2 \tag{5.50}$$

where $\hat{Y}_j$ is the value of Y estimated from our polynomial at the jth value of X. The SSE can be broken down into the following sums:

$$\text{SSE} = \sum_{j=1}^{R} \sum_{k=1}^{N/R} \left(Y_{jk} - \bar{Y}_j\right)^2 + \frac{N}{R} \sum_{j=1}^{R} \left(\bar{Y}_j - \hat{Y}_j\right)^2 + 2 \sum_{j=1}^{R} \sum_{k=1}^{N/R} \left(Y_{jk} - \bar{Y}_j\right)\left(\bar{Y}_j - \hat{Y}_j\right)$$

$$= \text{SSPE} + \text{SSLF} + 2 \sum_{j=1}^{R} \sum_{k=1}^{N/R} \left(Y_{jk} - \bar{Y}_j\right)\left(\bar{Y}_j - \hat{Y}_j\right) \tag{5.51}$$

where SSLF is the lack-of-fit sum of squares. The cross term equals 0 because $(N/R)\bar{Y}_j = \Sigma_{k=1}^{N/R} Y_{jk}$. Note that this conclusion is independent of $\hat{Y}_j$ and hence of what terms we include in the polynomial. We conclude that

$$\text{SSE} = \text{SSPE} + \text{SSLF} \tag{5.52}$$

Since $E(\text{SSE}) = E(\text{SSPE}) + E(\text{SSLF})$, it follows from Table 5.1 and the fact that $E(\text{SSPE}) = R(N/R - 1)\sigma^2 = (N - R)\sigma^2$ that

$$E(\text{SSLF}) = \sum_{j=1}^{N} \left[E\left(\bar{Y}_j - \hat{Y}_j\right)\right]^2 + (R - M - 1)\sigma^2 \tag{5.53}$$

We define the mean SSPE and the mean SSLF to be $\text{SSPE}/(N - R)$ and $\text{SSLF}/(R - M - 1)$, respectively. We call these MSPE and MSLF, respectively. If our polynomial is giving a good approximation of the underlying relationship between X and Y, then $E(\overline{Y}_j - \hat{Y}_j)$ should be close to 0 for all j, and MSLF should be close to MSPE. If the polynomial is giving a poor approximation of the underlying relationship, then MSLF will be greater than MSPE. The test statistic will be the ratio MSLF/MSPE, and we will compare this statistic to the critical F values with $\nu_1 = R - M - 1$ and $\nu_2 = N - R$ degrees of freedom.

Using Equation 5.49 and the data in Table 5.3, we find that SSPE = 0.0257. Hence MSPE = SSPE/10 = 0.00257. Since in this case $\sigma^2 = (0.05)^2 = 0.0025$, we indeed have a good estimate of σ^2. If we fit a linear function $Y = c_0 + c_1 X$ to the data, we obtain an SSE of 0.0942. Hence SSLF = 0.0942 − 0.0257 = 0.0685. The SSLF in this case has $R - M - 1 = 10 - 1 - 1 = 8$ df associated with it. Hence MSLF = 0.0685/8 = 0.00856. Our test statistic therefore becomes MSLF/MSPE = 0.00856/0.00257 = 3.33, and we compare this with the critical F values with 8 and 10 df. At $\alpha = 0.05$ the critical $F_{8,10}$ value is 3.07. We conclude that a significant portion of the SSLF is due to something other than noise. When we go to a quadratic polynomial $Y = c_0 + c_1 X + c_2 X^2$, we obtain an SSE of 0.0336. The SSLF is therefore 0.0336 − 0.0257 = 0.0079. The SSLF now has $10 - 2 - 1 = 7$ df, and the MSLF = 0.0079/7 = 0.00113. Our test statistic is therefore MSLF/MSPE = 0.00113/0.00257 = 0.44. At $\alpha = 0.05$ the critical $F_{7,10}$ = 3.14. Since our test statistic is far below this value, we accept the null hypothesis that SSLF = SSPE. The second-order polynomial fit is shown in Figure 5.3. The true function is not a second-order polynomial, and a systematic deviation between the polynomial and the true function is apparent. Nevertheless, these deviations are small compared to the scatter in the data. It is of interest that the MSLF/MSPE ratio in this case is less than 1, since the ratio of the expectation values is certainly greater than 1. However, the fact that the ratio of the expectation values is greater than 1 does not preclude the ratio being less than 1 in some cases.

Now suppose we had not calculated the SSPE and had instead relied on the methods used in the previous example to decide what polynomial to use. In this case the SSTO is 0.7504. Therefore SSR(X) = 0.7504 − 0.0942 = 0.6562, and the MSE for the linear function is 0.0942/(N − 2) = 0.0942/18 = 0.00523. The ratio SSR(X)/MSE = 0.7168/0.00523 = 137. At $\alpha = 0.05$ the critical $F_{1,18} = 4.41$. Hence the linear function is judged to be a big improvement over a constant function. For the quadratic test we have SSR($X^2|X$) = SSR(X, X^2) − SSR(X) = (0.7504 − 0.0336) − (0.7504 − 0.0942) = 0.7168 − 0.6562 = 0.0606. The MSE for the quadratic function is 0.0336/(N − 3) = 0.0336/17 = 0.00198. Therefore SSR($X^2|X$)/MSE = 0.0606/0.00198 = 30.7. At $\alpha = 0.05$ the critical $F_{1,17} = 4.45$, and we conclude that adding the quadratic term to the linear polynomial has significantly improved the fit. If we go to a cubic polynomial, we obtain SSE = 0.0275. Then SSR($X^3|X, X^2$) = SSR(X, X^2, X^3) − SSR(X, X^2) = (0.7504 − 0.0275) − (0.7504 − 0.0336) = 0.7229 − 0.7168 = 0.0061. The MSE for the cubic polynomial is $0.0275/(N - 4) = 0.0275/16 = 0.00172$. Therefore $SSR(X^3|X, X^2)/MSE = 0.0061/0.00172 = 3.55$. At $\alpha = 0.05$ the critical $F_{1,16} = 4.49$. Thus we have not significantly improved the fit by including a cubic term along with a linear and quadratic term. A quadratic polynomial seems to be the best choice for describing the data, the same conclusion we reached in our analysis using SSPE.

COMMENTS ABOUT POLYNOMIAL CURVE FITTING

In many cases, and indeed in the examples given here, high-order polynomials are not needed to describe experimental data. Often a quadratic or cubic polynomial will do the job. A common observation when higher-order polynomials are used is that the polynomial function displays "wiggles" which are probably not characteristic of the underlying functional relationship. Nor are wiggles likely to be a characteristic of a function we would like to use for predicting Y from X. Wiggles are the result of changes in the sign of the curvature of the function, in other words, changes in the sign of the second derivative. A linear function has a curvature of 0. A second-order polynomial has a constant curvature equal to $2c_2$. The curvature of a third-order polynomial equals $2c_2 + 6c_3X$ and hence changes sign when $X = -c_2/(3c_3)$. In general, the curvature of an nth-order polynomial can change sign $n - 2$ times as X ranges from $-\infty$ to $+\infty$. If it appears that the data we are trying to describe are going to require something more than a low-order polynomial, we may want to seriously consider fitting some other type of function. This may lead to a nonlinear least-squares problem (see page 103), but it may be the price we have to pay for avoiding the wiggles.

A second point about polynomial least squares is that serious roundoff errors can occur when the matrix $\mathbf{X}'\mathbf{X}$ is inverted. This can happen because (1) the determinant of $\mathbf{X}'\mathbf{X}$ is close to 0 and/or (2) the elements of $\mathbf{X}'\mathbf{X}$ differ greatly in magnitude. With respect to the second problem, we note that $(\mathbf{X}'\mathbf{X})_{ki} = \Sigma_{j=1}^{N} X_j^{(k+i-2)}$ if we include all coefficients $c_0, c_1, \ldots, c_M$ in the polynomial. In this case $(\mathbf{X}'\mathbf{X})_{1,1} = N$. If we are fitting a quadratic polynomial, then $M = 3$, and $(\mathbf{X}'\mathbf{X})_{3,3} = \Sigma_{j=1}^{N} X_j^4$. If the X values are on the order of 100, then $(\mathbf{X}'\mathbf{X})_{3,3}$ is likely to be on the order of 10^8 in magnitude. On the other hand, if the X values are on the order of 0.01, then $(\mathbf{X}'\mathbf{X})_{3,3}$ is likely to be on the order of 10^{-8} in magnitude. In order to keep the elements of $\mathbf{X}'\mathbf{X}$ of roughly comparable size, it is usually sufficient to simply divide all values of X by an appropriate constant. For example, if $|X|_{max}$ is the largest absolute value of any value of X, then simply dividing all values of X by $|X|_{max}$ or by a whole number comparable in magnitude to $|X|_{max}$ will usually solve the problem. One must, of course, keep in mind that the polynomial coefficients calculated with the scaled values of X must be back-transformed to obtain the coefficients appropriate to the unscaled values of X. For

TABLE 5.4 Ten Uniformly Distributed Random Numbers on the Interval [0, 10,000]

X_1	2190
X_2	470
X_3	6789
X_4	6793
X_5	9347
X_6	3835
X_7	5194
X_8	8310
X_9	346
X_{10}	535

example, if we decide to divide all values of X by 100 and $c_0, c_1, \ldots, c_M$ are the regression coefficients corresponding to the transformed values of X, then the coefficients corresponding to the original X values will be $c_k/100^k$.

The first problem arises when the powers of X are highly correlated. The nature of the problem is illustrated by the data in Tables 5.4 and 5.5. Table 5.4 lists 10 random numbers drawn from a uniformly distributed population on the interval [0, 10,000]. Table 5.5 lists product–moment correlation coefficients calculated between various powers of these 10 numbers. Note that the correlation coefficients are highest between similar powers of X, the magnitude of the correlation consistently dropping off as the difference between the powers of X increases. One can see, however, that there is perhaps a surprising degree of correlation between powers of X that differ by only 1, and that the magnitude of the correlation increases as the powers of X increase. With a little thought it should be obvious that when two variables

TABLE 5.5 Product–Moment Correlation Coefficients Between Powers of X for Data in Table 5.4

	X	X^2	X^3	X^4
X	1.0000	0.9671	0.9108	0.8559
X^2		1.0000	0.9846	0.9545
X^3			1.0000	0.9916
X^4				1.0000
	$X-\bar{X}$	$(X-\bar{X})^2$	$(X-\bar{X})^3$	$(X-\bar{X})^4$
$X-\bar{X}$	1.0000	0.0760	0.9336	0.2130
$(X-\bar{X})^2$		1.0000	0.2514	0.9533
$(X-\bar{X})^3$			1.0000	0.4285
$(X-\bar{X})^4$				1.0000

are perfectly correlated, from the standpoint of goodness of fit it makes no difference which independent variable we include in the regression. In such a case the solution to the least-squares equations is indeterminate, and the determinant of $\mathbf{X}'\mathbf{X} = 0$. Hence $\mathbf{X}'\mathbf{X}$ is singular and has no inverse. In the example given, the correlation coefficient between X^2 and X^3 is 0.9916, which suggests that including both X^2 and X^3 in a polynomial approximation of a function will probably be unnecessary, and that either one or the other will probably be sufficient and will give a comparable degree of goodness of fit. This is reminiscent of the conclusions we reached in analyzing the data in Figures 5.1 and 5.3.

The roundoff problem associated with this correlation can be dealt with in several ways. One may, for example, express the function in terms of orthogonal polynomials rather than powers of X. Orthogonal polynomials are polynomials whose coefficients have been chosen so that they are uncorrelated with each other over a particular range of X values. A simpler and usually sufficient solution to the problem is to simply express the polynomial in terms of powers of $X - \bar{X}$ rather than X. As is apparent from Table 5.5, this greatly reduces the correlation between odd and even powers of the independent variable, although it has little effect on the correlation between odd or even powers of X. In practice, the problem of roundoff errors is usually insignificant as long as one scales X so that the elements in $\mathbf{X}'\mathbf{X}$ are of comparable magnitude and one is not trying to fit a high-order polynomial. Most computer software packages for polynomial curve fitting use double-precision arithmetic and have been written to avoid roundoff problems. If you are writing your own program, however, you should be sensitive to the roundoff problem.

LINEAR MULTIPLE REGRESSION ANALYSIS

Our discussion of polynomial curve fitting leads logically into the subject of linear multiple regression analysis. Indeed, polynomial curve fitting is just a special case of linear multiple regression analysis. Our goal in both cases is often the same: to determine a simple analytical function which can be used to predict the value of a dependent variable. In polynomial curve fitting the independent variables are the various powers of X that appear in the polynomial function. These are clearly not independent variables as we defined independent variables in Chapter 2. The term *independent variables* in multiple regression analysis simply means the variables that are incorporated into the regression model. These variables may be and often are correlated. As we have seen, correlation between the independent variables can cause problems in polynomial curve fitting, and this is true for multiple regression analysis in general. However, the fact that two or more variables are correlated does not per se preclude their use as independent variables in multiple regression analysis. Linear multiple regression analysis is much more flexible than polynomial curve fitting, because it permits the independent variables to take virtually any analytical form; that is, the independent variables do not have to be powers of X. What is common to both techniques is that the dependent variable is a linear function of a set of coefficients which are determined by least squares. The matrix algebra equations for determining these coefficients and for testing their significance and the significance of the fitted function are the same in polynomial curve fitting and linear multiple regression analysis.

The basic linear multiple regression model takes the form

$$Y = C_0 + \sum_{k=1}^{M} C_k Z_k \tag{5.54}$$

where Z_k are the independent variables. In polynomial curve fitting $Z_k = X^k$, but in general linear multiple regression analysis, Z_k can be any analytical function. For example, we can have $Z_k = e^X$ or $Z_k = X/(1 + X)$. We may also postulate that Z_k are functions of more than one variable. For example, we could have $Z_k = X_1 X_2$ or $Z_k = X_1/X_2$. What analytical form the Z_k take and what variables are included in the model is up to the person trying to describe the dependent variable.

One of the most common linear multiple regression models is

$$Y = C_0 + C_1 X_1 + C_2 X_2 \tag{5.55}$$

For example, we might postulate that photosynthetic rate (Y) was a linear function of irradiance (X_1) and temperature (X_2) over some range of irradiance and temperature. We might carry out some systematic laboratory studies of photosynthetic rates at various irradiances and temperatures and then attempt to describe the data with Equation 5.55. If we got a good fit to the data, it might be appropriate to use the model to estimate photosynthetic rates in the field. An important restriction on the use of such models for predictive purposes is that the models should be used to interpolate rather than to extrapolate. It is improbable, for example, that photosynthetic rates will be a linear function of irradiance and temperature over a wide range of irradiance and temperature. Even if the Z_k are complex nonlinear functions, there is still no guarantee that the model will apply outside the region of observations, and defining the region of observations may be difficult, particularly if the model includes many independent variables. Figure 5.4 illustrates the nature of the problem. The ellipse outlines the set of X_1 and X_2 values that were used in a multiple regression model. The asterisk clearly lies within the range of X_1 and X_2 values used in developing the model, but it is well outside the ellipse. Thus one cannot be assured that the model is being applied within the region of observations by merely checking to see that each independent variable is within the range of values used for that variable in

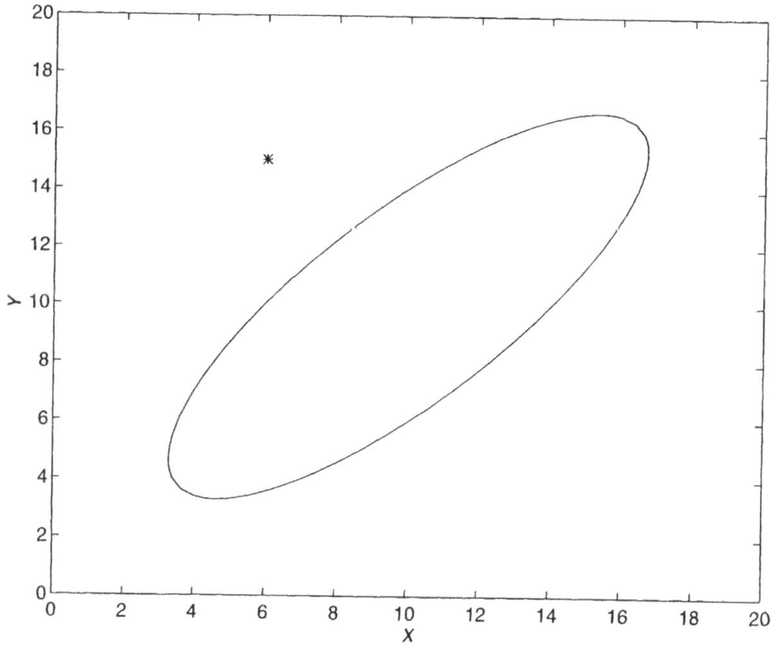

Figure 5.4 Region bounding the set of X_1 and X_2 values in a multiple regression model and a point (∗) lying within the range of both X_1 and X_2 but not within the ellipse.

developing the model. The problem of identifying the region of observations grows exponentially as one adds independent variables to the model.

Regardless of the analytical form of the independent variables, the strategy in linear multiple regression analysis is to choose the coefficients C_k so as to minimize

$$\sum_{k=1}^{N}\left(Y_k - C_0 - \sum_{m=1}^{M} Z_{km}C_m\right)^2$$

where Z_{km} is the kth value of the mth independent variable. Minimizing this expression leads directly to matrix equations similar to Equations 5.21 and 5.22:

$$\mathbf{Z}^t\mathbf{Y} = \mathbf{Z}^t\mathbf{Z}\mathbf{C} \tag{5.56}$$

$$(\mathbf{Z}^t\mathbf{Z})^{-1}\mathbf{Z}^t\mathbf{Y} = (\mathbf{Z}^t\mathbf{Z})^{-1}\mathbf{Z}^t\mathbf{Z}\mathbf{C} = \mathbf{I}\mathbf{C} = \mathbf{C} \tag{5.57}$$

where

$$\mathbf{Y} = \begin{bmatrix} Y_1 \\ Y_2 \\ \vdots \\ Y_N \end{bmatrix} \quad \mathbf{Z} = \begin{bmatrix} 1 & Z_{11} & Z_{12} & Z_{13} & \cdots & Z_{1M} \\ 1 & Z_{21} & Z_{22} & Z_{23} & \cdots & Z_{2M} \\ 1 & \cdot & \cdot & \cdot & \cdots & \cdot \\ 1 & \cdot & \cdot & \cdot & \cdots & \cdot \\ 1 & \cdot & \cdot & \cdot & \cdots & \cdot \\ 1 & Z_{N1} & Z_{N2} & Z_{N3} & \cdots & Z_{NM} \end{bmatrix} \quad \mathbf{C} = \begin{bmatrix} C_0 \\ C_1 \\ \vdots \\ C_M \end{bmatrix}$$

The treatment of errors, degrees of freedom, and analysis of variance in polynomial regression now carry over directly to linear multiple regression analysis, the only difference being the replacement of the **X** matrix with the **Z** matrix.

In multiple regression analysis it is customary to calculate a *coefficient of multiple determination*, R^2, which is defined by the equation

$$R^2 = \frac{\text{SSR}}{\text{SSTO}} \tag{5.58}$$

R^2 is analogous to the coefficient of simple determination, r^2, which we encountered in Chapter 3. As long as C_0 is included in the model, $0 \leq R^2 \leq 1$, and R^2 is a useful measure of the goodness of fit of the model to the data. However, if the model does not include C_0, it is quite possible for R^2 to exceed 1, even when the fit to the data is rather poor. The SSE, on the other hand, can always be used as a measure of goodness of fit, even when there is no C_0 in the model.

As an example of the application of multiple regression analysis, we consider phytoplankton pigment data collected from the mixed layer of the Hawaiian Ocean Time Series (HOTS) site (22°45′N, 158°W) from February 1989 to July 1993. The seven pigments used in the model in addition to chlorophyll *a* (chl *a*) are peridinin (per), 19′-butanoyloxyfucoxanthin (19′-but), fucoxanthin (fuco), 19′-hexanoyloxyfucoxanthin (19′-hex), prasinoxanthin (pras), zeaxanthin (zeax), and chlorophyll *b* (chl *b*). These seven pigments are associated with algal groups found in the North Pacific subtropical gyre as shown in Table 5.6. Chlorophyll *a* is a common measure of total algal biomass, and the total chl *a* concentration at the HOTS site can be partitioned into contributions from different algal groups based on the distribution of diagnostic pigments shown in Table 5.6 (Letelier et al., 1993). The goal of the multiple regression analysis is to develop an equation relating the concentration of chl *a* (the dependent variable) to the concentrations of diagnostic pigments (the independent variables), and in the process to determine which algal groups are most important to the structure of the phytoplankton community in this part of the ocean.

TABLE 5.6 Diagnostic Pigments Associated with Various Algal Groups in the North Pacific Subtropical Gyre

	Diagnostic Pigments						
Algal Group	per	19′-but	fuco	19′-hex	pras	zeax	chl *b*
Prochlorococcus spp.						X	X
Cyanobacteria						X	
Chrysophytes		X	X	X			
Prymnesiophytes		X	X	X			
Bacillariophytes			X				
Dinoflagellates	X						
Prasinophytes					X		X

The raw data are shown in Table 5.7. The first step in the analysis is to determine to what extent the independent variables are correlated. The simplest approach would be to calculate correlation coefficients between each pair of pigments. This approach will certainly identify pairs of pigments which are highly correlated, but the lack of such pairwise correlation does not guarantee the lack of a correlation problem between the independent variables. If one pigment, for example, can be expressed rather accurately as a linear combination of the other six pigments, then we may have a problem inverting the least-squares matrix. A more rigorous approach is therefore to calculate the coefficient of multiple correlation for each diagnostic pigment expressed as a linear combination of the other six pigments. The results of this exercise are shown in Table 5.8. The R^2 values range from 0.35 to 0.62. None of these values is high enough to cause a problem when the least-squares matrix is inverted. Problems with inversion of the least-squares matrix normally do not occur unless the maximum R^2 is greater than 0.9. The limitation of this approach is that it cannot distinguish between several simultaneous multicollinearities, but in most cases examination of R^2 values is sufficient to detect multicollinearity problems.

We now proceed with the multiple regression analysis. The model takes the form

$$Y_k = \sum_{m=1}^{M} C_m Z_{mk} \tag{5.59}$$

where Y_k is the kth chl a concentration and Z_{mk} is the kth value of the mth diagnostic pigment. The results of the linear regression analysis are shown in Table 5.9.

The model with all seven diagnostic pigments has an SSE of 1601 and an MSE of $1601/(30 - 7) = 69.6$. Since there are 30 chl a values and seven diagnostic pigments in the full model, the SSE has $30 - 7 = 23$ df. If we systematically eliminate one or another of the pigments from the model, the

TABLE 5.7 Mixed Layer Concentrations of chl *a* and Diagnostic Pigments (ng / l) Measured at the HOTS Site from 2/28/89–7/6/93

	Pigment							
Date	per	19′-but	fuco	19′-hex	pras	zeax	chl *b*	chl *a*
2/28/89	1.5	5.0	4.8	10.6	2.4	34.0	6.4	76.7
3/28/89	0.9	2.5	4.0	8.0	1.0	30.9	2.8	59.8
5/19/89	2.0	5.1	7.6	9.4	0.0	53.0	4.9	66.6
6/26/89	0.0	3.2	6.6	7.3	0.0	49.4	1.8	55.1
7/31/89	1.4	3.7	4.2	9.5	0.0	47.4	1.6	64.4
8/24/89	0.4	2.3	3.5	5.9	0.0	28.6	1.8	45.9
9/24/89	0.0	3.1	5.3	7.5	0.0	40.9	5.2	59.5
11/27/89	1.8	4.1	3.9	9.2	0.0	51.6	7.2	79.2
1/6/90	1.8	5.6	8.0	13.7	0.0	46.8	7.2	91.8
2/16/90	1.1	6.6	8.2	15.0	0.0	56.6	11.5	111.8
3/20/90	0.0	4.9	10.6	8.4	0.0	49.2	2.9	75.1
4/13/90	0.0	4.3	1.7	5.6	0.0	36.8	0.0	57.4
6/14/90	0.0	5.6	6.9	11.6	0.0	48.1	0.0	69.6
7/26/90	0.0	5.6	12.0	9.7	0.0	52.2	3.4	69.7
12/19/90	0.9	7.4	6.7	12.2	0.0	46.0	10.0	117.2
2/3/91	0.0	8.6	11.0	10.9	0.0	49.0	5.6	97.0
3/8/91	0.0	4.4	6.9	6.0	0.0	39.7	4.4	71.9
4/11/91	0.0	4.3	13.0	7.1	0.0	54.3	0.0	79.2
5/9/91	3.4	5.6	10.0	8.5	1.8	47.9	0.0	65.5
6/4/91	1.4	3.9	6.6	9.7	0.0	63.7	4.9	76.9
7/10/91	0.0	3.5	5.3	9.3	0.0	46.8	0.0	54.6
8/10/91	0.0	5.4	4.0	8.0	0.0	47.2	0.0	58.1
9/18/91	0.0	3.1	2.7	8.8	0.0	69.4	5.0	67.3
12/8/91	0.9	7.7	7.0	12.8	0.0	48.2	10.5	122.8
1/6/92	0.0	8.0	11.0	12.0	0.0	47.0	7.0	102.0
2/15/92	0.0	5.0	6.9	6.0	0.0	59.0	6.0	87.0
3/6/92	0.0	4.3	13.0	7.1	0.0	54.3	0.0	79.2
4/17/92	3.0	6.0	10.0	9.0	2.0	48.0	0.0	70.0
6/9/92	1.0	4.0	6.0	10.0	0.0	62.0	4.0	80.0
7/6/93	0.0	4.0	5.0	11.0	0.0	50.0	0.0	57.0

Data courtesy of R. Letelier.

SSEs range from 1617 to 3431. The F test for the significance of each pigment is, of course, made by dividing the difference of the SSEs by the MSE of 69.6. Thus for peridinin the test statistic is $(1625 - 1601)/69.6 = 0.34$. At $\alpha = 0.05$ the corresponding F value is $F_{1,23} = 4.28$. Thus peridinin, in the presence of the other six diagnostic pigments, is judged not to significantly improve the model. In fact, of the other pigments, only three, 19′-but, zeax, and chl b, are judged to significantly improve the model. Fucoxanthin, however, makes a significant contribution at $\alpha < 0.25$, and in order to be on the safe side, we include it in the next step of the analysis.

TABLE 5.8 Coefficients of Multiple Correlation for the Designated Diagnostic Pigment Expressed as a Linear Combination of the Other Six Diagnostic Pigments from the Data in Table 5.7

Diagnostic Pigment	R^2
Peridinin	0.502
19′-Butanoyloxyfucoxanthin	0.618
Fucoxanthin	0.446
19′-Hexanoyloxyfucoxanthin	0.557
Prasinoxanthin	0.550
Zeaxanthin	0.347
Chlorophyll *b*	0.441

For the next round, our model includes four diagnostic pigments, 19′-but, fuco, zeax, and chl *b*. The SSE for this model is 1642. Before proceeding any further, we should now ask whether adding per, 19′-hex, and pras together significantly improved this model. The appropriate test statistic is $(1642 - 1601)/69.6 = 0.59$. At $\alpha = 0.05$ the corresponding F value is $F_{3,\,23} = 3.03$. We conclude that per, 19′-hex, and pras collectively do not improve the

TABLE 5.9 Multiple Regression Analysis of Data in Table 5.7 with chl *a* as the Dependent Variable

Pigments Included in Model								
per	19′-but	fuco	19′-hex	pras	zeax	chl *b*	SSE	*F*
X	X	X	X	X	X	X	1601	
	X	X	X	X	X	X	1625	0.34
X		X	X	X	X	X	2647	15.0
X	X		X	X	X	X	1703	1.46
X	X	X		X	X	X	1617	0.23
X	X	X	X		X	X	1622	0.30
X	X	X	X	X		X	2871	18.2
X	X	X	X	X	X		3431	26.3
	X	X			X	X	1642	0.59
		X			X	X	3441	28.5
	X				X	X	1738	1.52
	X	X				X	3813	60.4
	X	X			X		3765	59.6
					X	X	5367	56.4
	X					X	4564	43.9
	X				X		3816	32.3

Final equation:

chl $a = (7.25 \pm 1.98)$(19′-but) $+ (0.60 \pm 0.19)$(zeax) $+ (2.78 \pm 1.00)$(chl b)

four-diagnostic-pigment model. Note that the difference of SSEs is associated in this case with 3 df, because we dropped three diagnostic pigments from the model.

If we now systematically eliminate one or another of the four remaining diagnostic pigments from the model, the SSEs range from 1738 to 3813. The MSE for the model with four diagnostic pigments is $1642/(30 - 4) = 63.2$. To test whether fucoxanthin is significant in this model, we calculate $(1738 - 1642)/63.2 = 1.52$. At $\alpha = 0.05$ the corresponding F value is $F_{1,26} = 4.23$. Thus fucoxanthin is not significantly improving the model, although the other three pigments are highly significant.

Finally, we go to a three-pigment model using 19′-but, zeax, and chl b. The SSE for this model is, of course, 1738 and the MSE is $1738/(30 - 3) = 64.4$. When we systematically eliminate one or another of these pigments from the model, the SSE increases dramatically. The F tests indicate that all three of these pigments are significant. The final multiple regression model is

$$\text{chl } a = 7.25(19'\text{-but}) + 0.60(\text{zeax}) + 2.78\,(\text{chl } b) \qquad (5.60)$$

The measured chl a concentrations and the values predicted from Equation 5.60 are shown in Figure 5.5. The model gives a good fit to the data, and the deviations appear to be random. The t statistic for each coefficient in the model is equal to the square root of the corresponding F statistic. Thus for 19′-but the t statistic is $\sqrt{56.4} = 7.5$. With 27 df the critical t statistic at $\alpha = 0.025$ (two-tailed test at $\alpha = 0.05$) is 2.05. Thus the 95% confidence limits for the 19′-but coefficient are $\pm(7.25/7.5)(2.05) = \pm 1.98$.

Now what does the model tell us? The presence of chl b in the equation tells us that we have a contribution from *Prochlorococcus* and/or prasinophytes. We can see from Table 5.7, however, that prasinoxanthin, which is found only in prasinophytes, was present in detectable concentrations on only 4 of the 30 sampling dates. It therefore seems reasonable to associate the chl b with *Prochlorococcus*. The presence of zeaxanthin in the model implies a contribution from *Prochlorococcus* and/or cyanobacteria. Two lines of reasoning suggest that the zeaxanthin is to be associated primarily with cyanobacteria. First, if most of the zeaxanthin were due to *Prochlorococcus*, we would expect the zeaxanthin and chl b concentrations to be highly correlated. As we have seen (Table 5.8), this is not the case. Second, the zeaxanthin/chl b ratio in *Prochlorococcus* is less than 1, perhaps no more than 0.3 (Letelier et al., 1993). The zeaxanthin concentrations measured in the mixed layer at the HOTS site are more than 10 times the chl b concentrations. Hence we can safely associate the zeaxanthin almost entirely with cyanobacteria. The 19′-but could be coming from chrysophytes and/or prymnesiophytes. However, the 19′-hex/19′-but ratio is roughly 50 in prymnesiophytes and less than 1 in chrysophytes (Letelier et al., 1993). The same ratio in the mixed layer at the HOTS site is only about 2. The implication is that most of the 19′-but is associated with chrysophytes. We conclude that

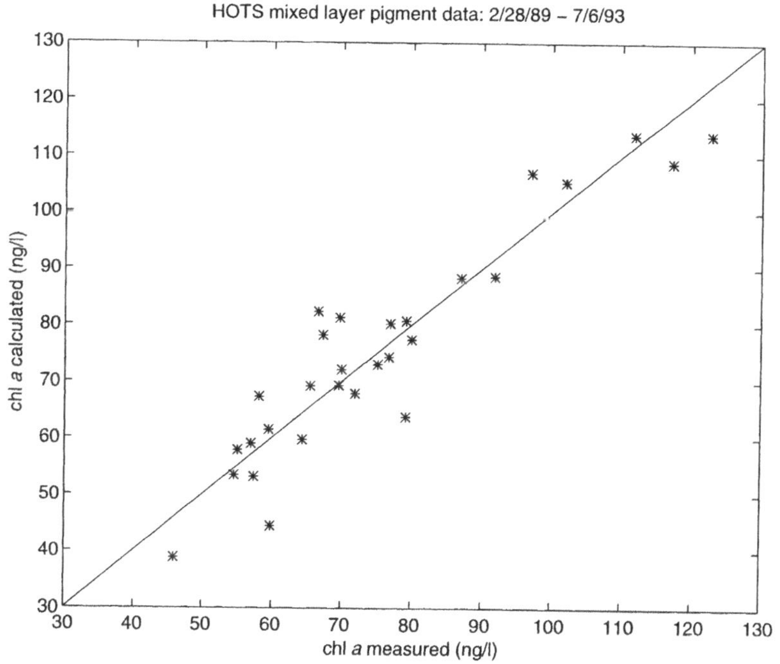

Figure 5.5 Measured HOTS mixed-layer chl *a* concentrations and values predicted from the multiple regression model chl a = 7.25(19′-but) + 0.60(zeax) + 2.78(chl *b*) based on the data in Table 5.7.

Prochlorococcus spp., cyanobacteria, and chrysophytes are the dominant algal groups in the mixed layer at the HOTS site.

NONLINEAR LEAST SQUARES

Up to this point we have dealt with models in which the dependent variable was a linear function of the model parameters. In some cases, however, we need a model in which the dependent variable is not a linear function of the model parameters. Such models fall into two basic categories, *intrinsically linear models* and *nonlinear models*. Intrinsically linear models appear to be nonlinear, but can be transformed into linear models. Consider, for example, the model

$$Y = (1 + \varepsilon) a \exp(bX) \tag{5.61}$$

where ε is a random error term. This model appears nonlinear, but if we take the logarithm of both sides we obtain

$$\ln(Y) = \ln(a) + bX + \ln(1 + \varepsilon) \tag{5.62}$$

We now see that $\ln(Y)$ is a linear function of $\ln(a)$ and b. If the error term $\ln(1 + \varepsilon)$ is independent of X and normally distributed with a mean of 0, all of the theory we have developed for linear least squares can be applied to the transformed model. We should realize, however, that the least-squares solution will select the function that gives the best fit to $\ln(Y)$ rather than to Y and that we will be getting unbiased estimates of $\ln(a)$ and b rather than a and b. If we can live with this or indeed want this to be the case, then transforming the intrinsically linear model to its linear form is appropriate.

Nonlinear models are models which cannot be transformed into a linear form. For example, the equation

$$Y = a[1 - \exp(-k(X - X_0))] \tag{5.63}$$

is sometimes used to model the size of an organism as a function of its age, the grazing rate of an organism as a function of the density of prey, or the photosynthetic rate of a plant as a function of light intensity. No transformation can linearize this model. How do we determine the best values of the parameters a, k, and X_0?

How we approach the problem depends on how many nonlinear parameters there are in the model. If there is only one nonlinear parameter, life is much simpler than if there are two or more nonlinear parameters. In many cases we can carry out much of the analysis as if there were only one nonlinear parameter. Equation 5.63, for example, would appear to contain two nonlinear parameters, k and X_0. However, we can rewrite Equation 5.63 in the form

$$Y = a + b\exp(-kX) \tag{5.64}$$

where $b = -a\exp(kX_0)$. Given k, we can solve for a and b using ordinary linear multiple regression analysis, where we treat $\exp(-kX)$ as the independent variable. The problem is finding k. We want to choose k so that $\Sigma_{n=1}^{N}[Y_n - a - b\exp(-kX_n)]^2$ is a minimum. Differentiating this expression with respect to k and setting the derivative equal to 0 gives

$$\sum_{n=1}^{N} Y_n X_n \exp(-kX_n) - a\sum_{n=1}^{N} X_n \exp(-kX_n) - b\sum_{n=1}^{N} X_n \exp(-2kX_n) = 0 \tag{5.65}$$

The procedure then is to guess a value of k and to solve for a and b using linear multiple regression analysis. The values of a and b are then plugged into Equation 5.65 and the value of the expression is calculated. If we have guessed the right value of k, the expression equals 0.

Trial-and-error guessing is not a very sophisticated procedure for determining parameter estimates. There are several more sophisticated methods, but all require an initial guess of k or at least a statement as to the possible range of values within which k may be found. One of the simplest methods is the so-called bisection rule, which requires that we specify an upper and lower bound on k. We call these upper and lower bounds k_{max} and k_{min}, respectively. The bounds must be finite to be useful. We then consider, for example, the function

$$F(k) = \sum_{n=1}^{N} Y_n X_n \exp(-kX_n) - a \sum_{n=1}^{N} X_n \exp(-kX_n) - b \sum_{n=1}^{N} X_n \exp(-2kX_n) \tag{5.66}$$

We are looking for the value of k that makes $F(k)$ equal 0. This value of k is called the root of $F(k)$. Assuming that there is only one root of $F(k)$ in the interval $[k_{min}, k_{max}]$, we can assume that $F(k)$ is positive on one side of the root and negative on the other side. A possible plot of $F(k)$ is shown in Figure 5.6. If we calculate the value of $F(k)$ at the left-hand side of the interval and at the right-hand side of the interval, the product of the two functions should be negative if there is one and only one root in the interval. Assuming this to be the case, we now divide the interval in half (bisect the interval) and calculate the value of $F(k)$ at the midpoint of the interval, k_{mid}.

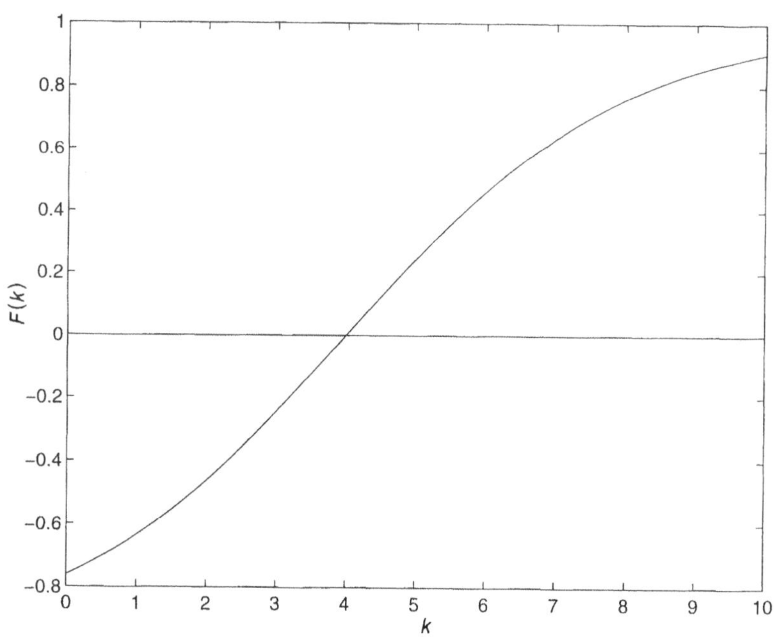

Figure 5.6 Behavior of a function $F(k)$ within a region where the function has only one root.

If the product of $F(k_{\text{min}})$ and $F(k_{\text{mid}})$ is negative, then the root lies between k_{min} and k_{mid}. In that case we make k_{mid} the new k_{max} and proceed to bisect again. On the other hand, if the product of $F(k_{\text{max}})$ and $F(k_{\text{mid}})$ is negative, then the root lies between k_{mid} and k_{max}. In that case we make k_{mid} the new k_{min} and bisect again. We proceed in this manner, bisecting the interval on each iteration, until k_{min} and k_{max} are close enough to satisfy us and/or $F(k)$ is sufficiently close to 0.

One possibility to consider during the iteration process is that one k_{mid} value may be very close to the root of $F(k)$. In that case it is possible that roundoff errors associated with the calculation of the various terms in $F(k)$ may cause large errors in the calculation of $F(k)$, so much so in fact that the sign of $F(k)$ is incorrect. To guard against this possibility, we need to have some criterion for stopping the iteration process when $F(k)$ becomes small enough. For the purposes of making this decision, it may be convenient to redefine $F(k)$. In the case of Equation 5.65, for example, we might define $F(k)$ as follows:

$$F(k) = a - \left[\sum_{n=1}^{N} Y_n X_n \exp(-kX_n) - b \sum_{n=1}^{N} X_n \exp(-2kX_n)\right] \Bigg/ \sum_{n=1}^{N} X_n \exp(-kX_n) \tag{5.67}$$

A reasonable criterion for stopping the iteration might then be that $|F(k)/a| < 0.001$.

Having determined the least-squares estimates of the parameters in the model, we would, of course, like some estimate of the uncertainty associated with those values. To derive the appropriate equations, consider the general case where we are trying to fit a function of the form $Y = f(\mathbf{Z}, \mathbf{C})$, where $\mathbf{Z}$ is a vector of independent variables and $\mathbf{C}$ is a vector of coefficients, at least some of which are nonlinear. Let $\mathbf{G}$ be the set of coefficients which minimize $\sum_{n=1}^{N}[Y_n - f(\mathbf{Z}_n, \mathbf{C})]^2$. As $\mathbf{C} \to \mathbf{G}$ it follows from the Taylor series expansion of the function $f(\mathbf{Z}, \mathbf{C})$ about $\mathbf{C} = \mathbf{G}$ that

$$f(\mathbf{Z}, \mathbf{C}) \to h(\mathbf{Z}, \mathbf{C}) = f(\mathbf{Z}, \mathbf{G}) + \sum_{m=1}^{M} \left[\frac{\partial f(\mathbf{Z}, \mathbf{C})}{\partial c_m}\right]_{\mathbf{C}=\mathbf{G}} (c_m - g_m) \tag{5.68}$$

where M is the number of coefficients in the model. When $\mathbf{C} = \mathbf{G}$ both $\sum_{n=1}^{N}[Y_n - f(\mathbf{Z}_n, \mathbf{C})]^2$ and $\sum_{n=1}^{N}[Y_n - h(\mathbf{Z}_n, \mathbf{C})]^2$ will be minimized. Since $h(\mathbf{Z}, \mathbf{C})$ is a linear function of the c_m coefficients, the values of the c_m which minimize $\sum_{n=1}^{N}[Y_n - h(\mathbf{Z}_n, \mathbf{C})]^2$ can therefore be determined by linear multiple regression methods. Let us define

$$D_{nm} = \left[\frac{\partial f(\mathbf{Z}_n, \mathbf{C})}{\partial c_m}\right]_{\mathbf{C}=\mathbf{G}}$$

Choosing c_m so as to minimize $\Sigma_{n=1}^{N}[Y_n - h(\mathbf{Z}_n, \mathbf{C})]^2$ leads directly to the following matrix equation:

$$\mathbf{D}'(\mathbf{Y} - \mathbf{F} + \mathbf{DG}) = \mathbf{D}'\mathbf{DC} \tag{5.69}$$

where $\mathbf{Y}$ and $\mathbf{F}$ are column vectors with elements Y_n and $f(\mathbf{Z}_n, \mathbf{G})$ and $\mathbf{D}$ is an $N \times M$ matrix with elements $\mathbf{D}_{nm}$. We conclude that $\mathbf{G}$ must satisfy the equation

$$\mathbf{G} = (\mathbf{D}'\mathbf{D})^{-1}\mathbf{D}'(\mathbf{Y} - \mathbf{F} + \mathbf{DG}) \tag{5.70}$$

This is not a useful equation for determining $\mathbf{G}$, since $\mathbf{G}$ appears on both sides of the equation and the elements of both $\mathbf{D}$ and $\mathbf{F}$ will, in general, depend on $\mathbf{G}$. The form of Equation 5.70, however, is very similar to the form of Equation 5.57. In fact, if $f(\mathbf{Z}, \mathbf{C})$ is a linear function of c_m, then Equation 5.68 is exact for all c_m and $\mathbf{F} = \mathbf{DG}$. Furthermore, in that case $\mathbf{D}$ is independent of $\mathbf{G}$, and we are back to linear multiple regression analysis. For example, in polynomial regression analysis $\boldsymbol{\sigma}^2(\mathbf{C}) = \sigma^2(\mathbf{X}'\mathbf{X})^{-1}$ if the errors in Y are random and have a common variance σ^2. The analogous equation commonly used in nonlinear least squares (Gallant, 1987; Neter et al., 1989) is

$$\mathbf{s}^2(\mathbf{G}) = \text{MSE}(\mathbf{D}'\mathbf{D})^{-1} \tag{5.71}$$

where

$$\text{MSE} = \frac{\text{SSE}}{N - M} = \frac{\sum_{n=1}^{N}[Y_n - f(\mathbf{Z}_n, \mathbf{G})]^2}{N - M} \tag{5.72}$$

The MSE is not an unbiased estimate of $\sigma^2(Y)$ in nonlinear least squares, but the bias is small when N is large. Clearly, Equation 5.71 is an approximation, but when the errors in Y are independent and normally distributed, then $(g_m - \gamma_m)/s(g_m)$ is approximately distributed as a standard normal variate (Appendix A), where γ_m is the true value of the parameter.

GAUSS–NEWTON ITERATION

We have now seen how to obtain approximate error bounds in nonlinear regression. We have also seen how to obtain parameter estimates in cases where there is effectively only one nonlinear parameter. Computer software packages designed to deal with nonlinear regression problems, however, normally do not distinguish between the linear and nonlinear coefficients in the model, but instead assume that all the parameters are nonlinear and use an iterative procedure for determining the parameter values that give the best fit to the data. This approach requires making an initial guess of all the parameter values. The power of this approach is that it can be programmed in a general way to deal with many different situations. The drawback is that the user must make an initial guess of all parameter values, and there is no guarantee that the iteration procedure will converge. In practice, however, it is often possible to make initial guesses that are close enough to the correct values that the iteration process has no difficulty converging. The iteration procedures commonly used are the method of *steepest descent*, the *Gauss–Newton* method, the *Newton* method, and the *Marquardt* method. We describe here the Gauss–Newton method as an example of this approach.

We assume as before that we are trying to fit a function $f(\mathbf{Z}, \mathbf{C})$, where $\mathbf{Z}$ is a vector of independent variables and $\mathbf{C}$ is a vector of coefficients. Our job is to choose the elements of $\mathbf{C}$ so that the SSE is a minimum, where the $\text{SSE} = \Sigma_{n=1}^{N}[Y_n - f(\mathbf{Z}_n, \mathbf{C})]^2$. Differentiating the SSE with respect to each of the M elements of $\mathbf{C}$ and setting the partial derivatives equal to 0 gives us the matrix equation

$$\mathbf{D}^t\mathbf{Y} = \mathbf{D}^t\mathbf{F} \tag{5.73}$$

where $\mathbf{F}$ is a column vector with elements $f(\mathbf{Z}_n, \mathbf{C})$, $\mathbf{D} = \partial\mathbf{F}/\partial\mathbf{C}$, and $\mathbf{D}_{nm} = \partial f(\mathbf{Z}_n, \mathbf{C})/\partial c_m$. Now let $\mathbf{G}_0$ be a vector holding our initial guess of $\mathbf{G}$. The Taylor series expansion of $\mathbf{F}$ about $\mathbf{C} = \mathbf{G}_0$ gives

$$\mathbf{F}(\mathbf{Z}, \mathbf{C}) = \mathbf{F}(\mathbf{Z}, \mathbf{G}_0) + \mathbf{D}_0(\mathbf{C} - \mathbf{G}_0) + \cdots \tag{5.74}$$

where the elements of $\mathbf{D}_0$ are $[\partial f(\mathbf{Z}_n, \mathbf{C})/\partial c_m]_{\mathbf{C}=\mathbf{G}_0}$. Substituting the right-hand side of Equation 5.74 for $\mathbf{F}$, we see that $\mathbf{D} = \partial\mathbf{F}/\partial\mathbf{C} = \mathbf{D}_0$ and Equation 5.73 becomes

$$\mathbf{D}_0^t(\mathbf{Y} - \mathbf{F}_0) = \mathbf{D}_0^t\mathbf{D}_0(\mathbf{C} - \mathbf{G}_0) \tag{5.75}$$

where $\mathbf{F}_0 = \mathbf{F}(\mathbf{Z},\mathbf{G}_0)$. Solving this equation for $\mathbf{C}$ gives

$$\mathbf{C} = \mathbf{G}_0 + \left(\mathbf{D}_0^t\mathbf{D}_0\right)^{-1}\mathbf{D}_0^t(\mathbf{Y} - \mathbf{F}_0) \tag{5.76}$$

The vector $\mathbf{C}$ calculated from Equation 5.76 now becomes our new $\mathbf{G}_0$, hopefully an improved estimate of $\mathbf{G}$. We now recalculate $\mathbf{F}_0$ and $\mathbf{D}_0$ and solve Equation 5.76 again. We continue this iterative process until we are satisfied that the process has converged. How rapidly the process converges depends on the nature of the functions and the accuracy of our initial guess of $\mathbf{G}$. If we make a very bad initial guess, it is possible that the process will not converge. The approximation in Equation 5.74 will be exact if we are dealing with a truly linear model, and it will be a good approximation if $\mathbf{G}_0$ is close to $\mathbf{G}$. Even if we make a bad initial guess, however, the Gauss–Newton method may still converge. In order to implement the Gauss–Newton method, we need to specify the partial derivatives of f with respect to each of the coefficients in the model, and, of course, we need to specify initial estimates for those coefficients.

As an example of the application of nonlinear least squares, consider the data in Table 5.10. The data are taken from a study of phosphate uptake by an experimental coral reef system. X is the concentration of phosphate and Y is the phosphate uptake rate. The student carrying out this study hypothesized that the relationship between X and Y could be described by the hyperbolic equation $Y = AX/(B + X)$. His problem was to choose the values of A and B that give the best fit to the data.

This is an interesting problem, since there are several different ways to analyze the data. The problem is intrinsically linear. For example,

$$\frac{1}{Y} = \frac{1}{A} + \frac{B}{A}\frac{1}{X} \tag{5.77}$$

or

$$\frac{X}{Y} = \frac{B}{A} + \frac{1}{A}X \tag{5.78}$$

Thus we can treat the problem with linear regression analysis if we are willing to consider $1/Y$ or X/Y the dependent variable. If we treat Y as the dependent variable, we must use nonlinear least squares and try to find the values of A and B that minimize

$$\sum_{n=1}^{N} \left(Y_n - \frac{AX_n}{B + X_n} \right)^2 \tag{5.79}$$

TABLE 5.10 Phosphate Uptake Rate (Y) Versus Phosphate Concentration (X) in an Artificial Coral Reef System

X^a	Y^b
0.2	0.2
1.6	1.0
2.2	1.3
5.0	1.8
8.8	3.2
23.0	6.4

[a] Units are μM.

[b] Uptake rate was normalized to respiration rate and has units of moles phosphate per 100 moles oxygen.

Differentiating Equation 5.79 with respect to A and B and setting the derivatives equal to 0 leads to the following equations:

$$\sum_{n=1}^{N} \frac{Y_n X_n}{B + X_n} = A \sum_{n=1}^{N} \left(\frac{X_n}{B + X_n} \right)^2 \tag{5.80}$$

$$\sum_{n=1}^{N} \frac{Y_n X_n}{(B + X_n)^2} = A \sum_{n=1}^{N} \frac{X_n^2}{(B + X_n)^3} \tag{5.81}$$

Equations 5.80 and 5.81 can easily be solved using the bisection rule. The function $F(B)$ whose root we seek can be

$$F(B) = \frac{\sum \dfrac{Y_n X_n}{(B + X_n)^2}}{A \sum \dfrac{X_k^2}{(B + X_k)^3}} - 1 \tag{5.82}$$

where A is determined from Equation 5.80. We can also, of course, use Gauss–Newton or some other iteration scheme.

The results of solving the problem by various schemes are shown in Table 5.11. Treating Y^{-1} as the dependent variable, we obtain what appears to be an outstanding fit to the data. The correlation coefficient between Y^{-1} and X^{-1} is 0.998, which means that the SSR is 99.6% of the SSTO. The intercept and slope are 0.282 ± 0.059 and 0.948 ± 0.028, respectively, where the error bounds are ± 1 standard deviation. Using Equation 5.77 to transform these gives $A = 3.54$ and $B = 3.36$. When we plug in these values of A and B to

TABLE 5.11 Results of Various Least-Squares Fits to Data in Table 5.10

	Model		
	$Y = \dfrac{AX}{B + X}$	$\dfrac{1}{Y} = \dfrac{1}{A} + \dfrac{B}{A}\dfrac{1}{X}$	$\dfrac{X}{Y} = \dfrac{B}{A} + \dfrac{1}{A}X$
Observed Y		Predicted Y	
0.2	0.09	0.20	0.34
1.0	0.72	1.14	1.67
1.3	0.97	1.40	1.97
1.8	2.05	2.12	2.71
3.2	3.27	2.56	3.10
6.4	6.37	3.09	3.51
		SSE	
	0.26	11.5	10.1

determine our estimates of Y, we find, perhaps to our surprise, that the fit to the data is not all that good. In fact, we have underestimated the largest value of Y by more than a factor of 2. What is the explanation for this? The explanation is that by fitting Y^{-1} we have converted the smallest values of Y into the largest values of Y, and vice versa. The least-squares fitting procedure gives equal weight to all squared deviations, and an error by a factor of 2 in a small number may look no more important than a much smaller percentage error in a large number. Thus while it seems appealing to transform the data to obtain a linear model, there is a price associated with the transformation. If the values of Y cover a wide range, as is the case here, we may dramatically shift the emphasis of the errors.

If we use X/Y as the dependent variable, we give more emphasis to the larger values of Y, since X and Y are positively correlated in this case. The correlation coefficient between X and X/Y is 0.993, which again suggests a good fit to the data. The value of r in this case, however, is statistically meaningless, since X appears in both variables. Transforming from the slope and intercept of this fit, we obtain $A = 3.83$ and $B = 2.08$. Calculating the values of Y estimated from these coefficients, we obtain a somewhat improved fit, but we still underestimate the largest value of Y by 45%. Note that we also overestimate the two smallest values of Y by about 70%. Given these results, the initial guess of A and B for the Gauss–Newton iteration was $A = 3$ and $B = 3$. The iteration rapidly converged and in five cycles settled down on the values of $A = 15.4 \pm 3.2$ and $B = 32.6 \pm 10.1$. In using the bisection rule, the initial guess was that B lay between 0 and 50. The bisection process requires 10 cycles to reduce the uncertainty in the estimate by a factor of about 1000 (actually $2^{10} = 1024$). Both methods converged on the same answers, as they should. Two aspects of the nonlinear fit are noteworthy. First, in the case of the Gauss–Newton method, the iteration procedure converged even though the initial guesses were not very close to the final values. Second, the final values were, in fact, quite different from the values obtained from the linearizations of the model, and the fit to the Y data was much better. The SSE for the nonlinear fit is about 40 times smaller than the SSEs for the transformed linear fits. This is due mainly to the much better fit to the largest Y value. The original data and nonlinear fit are shown in Figure 5.7. Although the fit looks quite good, the nonlinear model, in fact, underestimates the smallest Y value by about a factor of 2, comparable to the percentage error of the linearized models in fitting the largest value of Y. If one believes that the variances of the errors in all the Y data are similar and equal to the MSE of $0.26/(6 - 2) = 0.065$, then the standard deviation of the errors is about $\sqrt{0.065} = 0.25$, and the lowest Y value is not significantly different from 0. Hence the nonlinear fit to the lowest Y value is acceptable. The experimentalist must decide, in this case and in general, whether the assumption of a common variance in the errors is justified. An important point to learn from this example is that the weight one gives to the deviations of the model from the dependent variable can have a dramatic

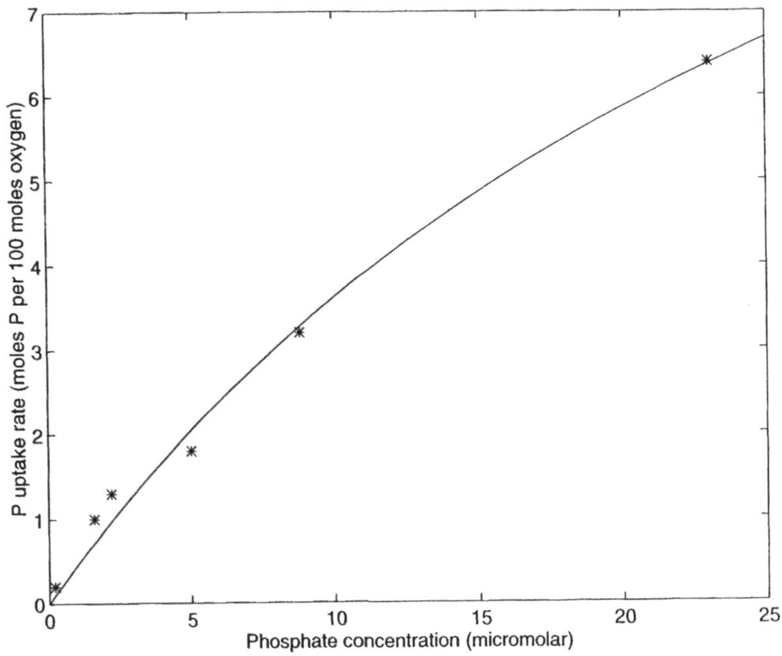

Figure 5.7 Phosphate uptake rate normalized to respiration rate on a coral reef versus phosphate concentration. Data courtesy M. Atkinson.

effect on parameter estimates and the fit of the model to the data. Transforming the data effectively alters the weights and hence can have a large effect on the goodness of fit to the original data.

THE CREATION OF SPURIOUS RELATIONSHIPS THROUGH TRANSFORMATIONS

Transforming variables to linearize results or for other purposes is a common practice in data analysis. There are several potential problems associated with this practice. First, as we have seen, transformations may dramatically change the significance of errors. Second, certain kinds of transformations have the potential for creating apparently significant relationships when, in fact, none exists. We have already noted, for example, that the correlation between transformed variables such as X/Y and X is statistically meaningless, because both variables are a function of X. The following example will hopefully clarify the nature of the problem.

Geochemist Giacomo collects data on the concentrations of beryllium and aluminum in the Atlantic Ocean. The data are shown in Figure 5.8A. There

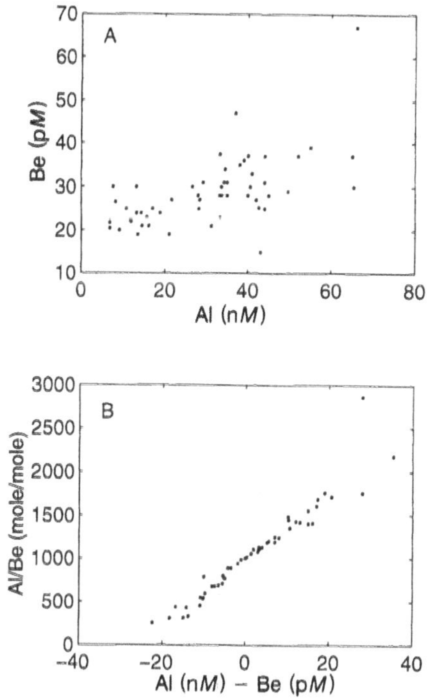

Figure 5.8 Aluminum and beryllium concentrations measured at various stations in the Atlantic Ocean. Data courtesy Dr. Chris Measures.

is no apparent relationship between the concentrations of the two elements. For reasons best known to himself, Giacomo decides to plot Al/Be versus Al – Be, with the Al concentrations in nanomolar units and the Be concentrations in picomolar units. The results are shown in Figure 5.8*B*. There is a remarkably linear relationship between the two variables, Al/Be and Al (nM) – Be (pM). Giacomo thinks he has found something very interesting, and in a sense he has, but what he has discovered has nothing to do with marine geochemistry. Inspection of Figure 5.8*A* shows that in this data set the Be concentrations are far less variable than the Al concentrations. Almost all the Be values lie between 20 and 40 pM. The Al concentrations are more-or-less uniformly distributed between 6 and 60 nM. The function Al – Be can be written as Be(Al/Be – 1). Thus what Giacomo has plotted in Figure 5.8*B* amounts to a graph of X versus Be(X – 1), where X = Al/Be. Since Al in this case is far more variable than Be, the variance in Al is driving the correlation in Figure 5.8*B*, with Be playing the role of a small-amplitude noise signal on the function $X - 1$. Relationships of the sort seen in Figure 5.8*B* can be produced using random numbers, as we will see in Exercise 5.10. Spurious relationships of this sort can be avoided by ensuring that the variables plotted on the ordinate and abscissa represent truly independent measurements. This is not to say that one should absolutely avoid plotting functions such as X/Y versus X, but such plots should not be used to decide whether significant relationships exist in a data set. That determination must be made by other means.

WEIGHTED LEAST SQUARES

One possible resolution of the problem of a nonrandom distribution of errors in the dependent variable is to perform a weighted least squares. In other words, we seek to minimize $\Sigma_{n=1}^{N} w_n [Y_n - f(\mathbf{Z}_n, \mathbf{C})]^2$, where w_n are weighting factors chosen by the scientist. This approach is equally valid for both linear and nonlinear least squares. The w_n are chosen to be inversely proportional to the variance of the errors associated with each Y_n. In other words, if σ_n^2 is the variance of the error associated with Y_n, then

$$w_n = \frac{\sigma^2}{\sigma_n^2} \tag{5.83}$$

where σ^2 is a proportionality constant. The application of this method obviously assumes that one has some basis for estimating the variance of each error term. Once the weights have been chosen, it is convenient to define a matrix **W** whose diagonal elements are the weights and whose off-diagonal elements are 0. In other words, $W_{nn} = w_n$ and $W_{nm} = 0$ for $m \neq n$. The weighted least-squares analogs of Equations 5.22, 5.57, and 5.76 are then

$$\mathbf{C} = (\mathbf{X}'\mathbf{W}\mathbf{X})^{-1}\mathbf{X}'\mathbf{W}\mathbf{Y} \tag{5.84}$$

$$\mathbf{C} = (\mathbf{Z}'\mathbf{W}\mathbf{Z})^{-1}\mathbf{Z}'\mathbf{W}\mathbf{Y} \tag{5.85}$$

$$\mathbf{C} = \mathbf{G}_0 + (\mathbf{D}_0'\mathbf{W}\mathbf{D}_0)^{-1}\mathbf{D}_0'\mathbf{W}(\mathbf{Y} - \mathbf{F}_0) \tag{5.86}$$

The weighted least-squares analog of the estimated variance–covariance matrix is

$$s^2(\mathbf{C}) = \mathrm{MSE}_w(\mathbf{X}'\mathbf{W}\mathbf{X})^{-1} \tag{5.87}$$

where **X** is replaced by **Z** or **D** in the case of linear multiple regression or nonlinear least squares, respectively, and MSE_w is the weighted analog of the MSE and is given by

$$\mathrm{MSE}_w = \frac{\sum_{n=1}^{N} w_n [Y_n - f(\mathbf{Z}_n, \mathbf{C})]^2}{N - M} \tag{5.88}$$

MSE_w is an estimate of the proportionality constant σ^2.

EXERCISES

5.1 Let $\mathbf{H} = \mathbf{X}(\mathbf{X}^t\mathbf{X})^{-1}\mathbf{X}^t$, where $\mathbf{X}$ is an $N \times (M + 1)$ matrix whose first column contains all 1's. Show that $\mathbf{HX} = \mathbf{X}$ and that $\mathbf{X}^t\mathbf{H} = \mathbf{X}^t$. Show that the fact that $\mathbf{HX} = \mathbf{X}$ and that the first column of $\mathbf{X}$ contains all 1's imply that the sum of the elements in any row of $\mathbf{H}$ must be 1, that is, $\sum_{n=1}^{N} H_{kn} = 1$ for all k. Similarly, show that the fact that $\mathbf{X}^t\mathbf{H} = \mathbf{X}^t$ and that the first column of $\mathbf{X}$ contains all 1's imply that the sum of the elements in any column of $\mathbf{H}$ must equal 1, that is, $\sum_{n=1}^{N} H_{nk} = 1$ for all k. Show that the fact that the sum of the elements in each row and each column of $\mathbf{H}$ equals 1 implies that $\mathbf{JH} = \mathbf{HJ} = \mathbf{J}$, where $\mathbf{J}$ is a matrix whose elements are all 1's.

5.2 Show that if $\mathbf{X}$ is any $N \times (M + 1)$ matrix with $N \geq M + 1$ and $\mathbf{H} = \mathbf{X}(\mathbf{X}^t\mathbf{X})^{-1}\mathbf{X}^t$, then $\text{tr}(\mathbf{H}) = M + 1$. *Hint*: Let c_{mk} be the element in row m and column k of $\mathbf{X}^t\mathbf{X}$ and let a_{jm} be the element in row j and column m of $(\mathbf{X}^t\mathbf{X})^{-1}$. Then

$$c_{mk} = \sum_{n=1}^{N} X_{mn}^t X_{nk} = \sum_{n=1}^{N} X_{nm} X_{nk}$$

and

$$\sum_{m=1}^{M+1} a_{jm} c_{mk} = \sum_{m=1}^{M+1} a_{jm} \sum_{n=1}^{N} X_{nm} X_{nk} = \delta_{j,k}$$

where $\delta_{j,k} = 1$ if $j = k$ and $\delta_{j,k} = 0$ if $j \neq k$. Now what is the diagonal element in row and column p of $\mathbf{H}$ in terms of the elements in $\mathbf{X}$ and $\mathbf{A} = (\mathbf{X}^t\mathbf{X})^{-1}$? Show that the sum of these diagonal elements equals $M + 1$.

5.3 Use Equation 5.44 to show that in model I linear regression when the observations are controlled

$$E(\text{SSR}) = \sigma^2 + \beta^2 \sum_{n=1}^{N} \left(X_n - \bar{X}\right)^2$$

5.4 In our discussion of linear and nonlinear least squares, we have made frequent use of matrix inverses without explaining how to calculate the inverse of a matrix. Now we will consider that problem. We begin with a square $N \times N$ matrix $\mathbf{A}$. We know that if $\mathbf{A}^{-1}$ exists, then $\mathbf{A}^{-1}\mathbf{A} = \mathbf{AA}^{-1} = \mathbf{I}$. Without proof, we will now use the fact that $\mathbf{A}^{-1}$ can be expressed as the product of N^2 matrices, each of which performs an elementary row operation. An elementary row operation is defined as (1) multiplying a row of a matrix by a constant or (2) adding a constant multiple of one row of a matrix to another row. We will consider the case $N = 2$. The matrix $\begin{bmatrix} K & 0 \\ 0 & 1 \end{bmatrix}$, when multiplying on the left, multiplies every element in the first row of any 2×2 matrix by K and leaves the second row unchanged. Similarly, the matrix $\begin{bmatrix} 1 & 0 \\ K & 1 \end{bmatrix}$, when multiplying on the left, adds K times the first row of any 2×2 matrix to the second row, leaving the first row unchanged.

Now consider the matrix

$$\mathbf{A} = \begin{bmatrix} 1 & 2 \\ 3 & 4 \end{bmatrix}$$

which we discussed briefly before. To find $\mathbf{A}^{-1}$, we want to identify a series of elementary row operations which will convert $\mathbf{A}$ to $\mathbf{I}$. We will work on $\mathbf{A}$ one column at a time, converting each column into the corresponding column of $\mathbf{I}$. We will perform the same elementary row operations on $\mathbf{I}$, because we know that $\mathbf{A}^{-1}\mathbf{I} = \mathbf{A}^{-1}$. Thus the same set of elementary row operations that convert $\mathbf{A}$ to $\mathbf{I}$ should convert $\mathbf{I}$ to $\mathbf{A}^{-1}$. Perhaps the simplest way to set up the problem is to write $\mathbf{A}$ and $\mathbf{I}$ side by side, that is, $\begin{bmatrix} 1 & 2 & 1 & 0 \\ 3 & 4 & 0 & 1 \end{bmatrix}$.

We begin with the first column, and we always work on the diagonal element in the column first. In this case the diagonal element in column 1 of $\mathbf{A}$ is already equal to the corresponding element of $\mathbf{I}$, so no elementary row operation is required. Equivalently, we could say that we multiply the first row by 1. We now go to the other element in column 1. We want to convert the 3 to 0. We can do this by adding -3 times the first row to the second row. In other words, we multiply on the left by $\begin{bmatrix} 1 & 0 \\ -3 & 1 \end{bmatrix}$, and we do this to $\mathbf{I}$ as well. The result is $\begin{bmatrix} 1 & 2 & 1 & 0 \\ 0 & -2 & -3 & 1 \end{bmatrix}$. We now move to the second column and go to the diagonal element. We need to convert the -2 to a 1. We can do this by multiplying the second row by -0.5. In other words, we multiply on the left by $\begin{bmatrix} 1 & 0 \\ 0 & -0.5 \end{bmatrix}$. The result is $\begin{bmatrix} 1 & 2 & 1 & 0 \\ 0 & 1 & 1.5 & -0.5 \end{bmatrix}$. Finally, we need to convert the 2 in the first row and second column to a 0. This can be done by adding -2 times the second row to the first row, that is, by multiplying on the left by $\begin{bmatrix} 1 & -2 \\ 0 & 1 \end{bmatrix}$. The result is $\begin{bmatrix} 1 & 0 & -2 & 1 \\ 0 & 1 & 1.5 & -0.5 \end{bmatrix}$. We conclude that the inverse of $\begin{bmatrix} 1 & 2 \\ 3 & 4 \end{bmatrix}$ is $\begin{bmatrix} -2 & 1 \\ 1.5 & -0.5 \end{bmatrix}$. Note that

$$\begin{bmatrix} -2 & 1 \\ 1.5 & -0.5 \end{bmatrix} = \begin{bmatrix} 1 & -2 \\ 0 & 1 \end{bmatrix} \begin{bmatrix} 1 & 0 \\ 0 & -0.5 \end{bmatrix} \begin{bmatrix} 1 & 0 \\ -3 & 1 \end{bmatrix} \begin{bmatrix} 1 & 0 \\ 0 & 1 \end{bmatrix} \tag{5.89}$$

The four matrices on the right-hand side of this equation are the four elementary row operation matrices that we needed to convert $\mathbf{A}$ to $\mathbf{I}$.

Now that you know how to calculate the inverse of a matrix, calculate $(\mathbf{X}^t\mathbf{X})^{-1}$ for the simple linear regression model. In this case

$$\mathbf{X}^t\mathbf{X} = \begin{bmatrix} N & \sum_{n=1}^{N} X_n \\ \sum_{n=1}^{N} X_n & \sum_{n=1}^{N} X_n^2 \end{bmatrix}$$

Using this result, show that Equations 2.64 and 2.65 are consistent with Equation 5.25.

5.5 Let

$$\mathbf{X}_0 = \begin{bmatrix} 1 \\ X_0 \\ X_0^2 \\ \vdots \\ X_0^M \end{bmatrix}$$

Then $Y_0 = \mathbf{X}_0^t\mathbf{C} = \mathbf{X}_0^t(\mathbf{X}^t\mathbf{X})^{-1}\mathbf{X}^t\mathbf{Y}$ is the estimated mean value of Y at $X = X_0$ from a polynomial regression model. What is $\sigma^2(Y_0) = V(Y_0)$? Show that your answer is consistent with Equation 2.66 when $M = 1$.

5.6 What is the analog in matrix notation of Equation 2.67 in the case where Y_0 is estimated from a polynomial?

5.7 The pigment data collected from the deep chlorophyll maximum layer at the HOTS site from 12/3/88 to 7/6/93 are listed in Table 5.12. Calculate R^2 values for the seven diagnostic pigments (do not include chl *a*). The R^2 for one pigment is greater than 0.9. With which two other pigments is it principally correlated? Perform a linear multiple regression analysis treating chl *a* as the dependent variable, assuming that the chl *a* concentration can be expressed as a sum of coefficients times concentrations of diagnostic pigments, that is, chl $a = \sum_{m=1}^{7} c_m X_m$, where X_m is the concentration of the mth diagnostic pigment. You will find that a model containing three diagnostic pigments is needed to describe the chl *a* data, and that removal of any one of those pigments from the model significantly increases the SSE. What are the three pigments?

5.8 Use the Gauss–Newton method or the bisection method to fit the data in Table 5.10 to a function of the form $Y = a(1 - e^{-kX})$. Estimate the standard deviation of a and k using Equation 5.71.

TABLE 5.12 Concentrations of chl *a* and Diagnostic Pigments (ng/l) Measured in the Chlorophyll Maximum Layer at the HOTS Site from 12/3/88–7/6/93

	Pigment							
Date	per	19′-but	fuco	19′-hex	pras	zeax	chl *b*	chl *a*
12/3/88	0.0	21.7	5.0	32.7	1.3	13.9	46.3	134.9
1/9/89	0.0	5.2	2.2	8.0	0.0	8.2	20.2	61.8
2/28/89	2.1	21.1	3.6	37.0	1.6	26.0	45.4	173.4
3/28/89	1.1	25.5	5.0	40.9	2.0	43.7	78.0	248.0
5/19/89	3.4	66.3	11.7	89.2	1.3	60.6	176.8	349.1
6/26/89	2.8	24.4	4.7	34.4	1.0	24.4	86.2	193.4
7/31/89	1.8	30.9	5.1	36.6	0.9	38.0	112.3	218.7
8/24/89	2.0	32.7	6.0	56.0	0.0	38.3	101.8	271.4
10/18/89	0.8	41.3	4.1	38.2	0.0	34.2	126.0	230.3
11/27/89	1.0	29.2	4.1	35.7	0.0	31.2	121.1	213.7
1/6/90	2.5	51.8	7.7	56.8	0.0	34.8	162.7	285.0
2/16/90	3.4	59.9	14.0	91.6	5.6	41.0	141.7	330.9
3/20/90	0.0	50.2	0.0	42.8	3.6	46.8	170.2	358.6
4/13/90	2.2	14.9	6.2	39.3	0.0	40.7	53.7	236.1
6/14/90	3.0	56.5	15.4	57.4	0.0	66.8	187.7	350.2
7/26/90	2.5	26.3	8.2	40.5	0.0	66.1	62.0	257.9
12/19/90	2.6	47.9	8.9	49.3	1.0	25.6	127.0	233.8
2/3/91	1.4	47.6	9.8	60.6	2.2	41.6	119.1	279.9
3/8/91	4.1	44.8	13.2	64.9	2.3	47.0	88.7	299.5
4/11/91	1.5	41.6	8.0	55.9	1.9	57.7	96.6	293.9
5/9/91	1.3	34.8	3.8	27.5	1.3	12.4	66.9	127.3
6/4/91	2.4	65.6	10.9	67.5	1.6	43.3	183.2	333.2
7/10/91	0.0	37.2	8.2	45.5	0.0	41.1	161.5	293.7
8/10/91	4.9	24.5	6.9	38.6	0.0	58.1	71.2	279.1
9/18/91	0.0	41.2	7.2	55.1	0.0	41.2	182.9	292.6
12/8/91	2.8	48.7	9.1	50.4	0.6	26.0	130.2	239.1
1/6/92	1.7	56.3	12.1	58.3	1.8	37.1	132.5	312.4
2/15/92	4.2	45.5	12.8	65.2	2.2	46.6	89.1	299.8
3/6/92	2.2	42.3	8.3	56.3	2.4	58.2	96.7	302.1
4/17/92	0.4	50.8	11.1	62.7	2.8	41.7	160.3	281.8
6/9/92	2.9	70.5	9.8	73.8	3.1	42.1	201.0	342.0
7/6/93	1.1	38.2	8.2	41.9	0.9	32.0	134.4	238.4

Data courtesy R. Letelier.

5.9 Consider the model I case in which there is no error associated with the **X** matrix, but the observations are not controlled; that is, the **X** matrix varies from one repetition of an experiment to the next. Assume that the errors in **Y** are independent of **X** and have an expectation value of 0. Show that under these conditions the **C** matrix calculated from Equation 5.22 is an unbiased estimate of the true **C**. To prove this fact, you will need to know that for two independent matrices **X** and **Y** whose product is defined, **E(XY)** = **E(X)E(Y)**.

5.10 Generate 100 uniformly distributed random numbers on the interval 6 to 60. Generate another set of 100 uniformly distributed random numbers on the interval 20 to 40. Call the first set of numbers Al and the second set Be. Plot Al/Be versus Al − Be. Compare the result to Figure 5.8*B*.

REFERENCES

Gallant, R. A. 1987. *Nonlinear Statistical Models*. Wiley, New York.

Jerlov, N. G. 1968. *Optical Oceanography*. Elsevier, Amsterdam.

Letelier, R. M., R. R. Bidigare, D. V. Hebel, M. Ondrusek, C. D. Winn, and D. M. Karl. 1993. Temporal variability of phytoplankton community structure based on pigment analysis. *Limnol. Oceanogr.* **38** 1420–1437.

Neter, J., W. Wasserman, and M. H. Kutner. 1989. *Applied Linear Regression Models*, 2nd ed. Irwin, Boston.

6

NUMERICAL INTEGRATION

We begin our discussion of numerical integration by considering the following example. A student wanted to determine the amount of carbon being fixed by phytoplankton in an estuary at a location where the water column was 30 m deep. The student measured the photosynthetic rates of the phytoplankton at a series of depths and obtained the results shown in Table 6.1 and Figure 6.1.

Having obtained these results, the student asks the question, "How much carbon is being fixed per square meter of surface area?" In the upper 5 m of the water column, we can see that the amount of carbon fixed per square meter of surface area is (2 mg C m^{-3} h^{-1})(5 m) = 10 mg C m^{-2} h^{-1}. Between 5 and 10 m the amount of carbon fixed per square meter of surface area is (5 mg C m^{-3} h^{-1})(5 m) = 25 mg C m^{-2} h^{-1}. Proceeding in this manner from the surface to 30 m, we see that the total amount of carbon fixed in the water column is (2)(5) + (5)(5) + (10)(5) + (7)(5) + (4)(5) + (2)(5) = 150 mg C m^{-2} h^{-1}. Looking at Figure 6.1, we notice that what we have calculated is just the area between the abscissa and the graph of photosynthetic rate. Between 0 and 5 m, for example, the length of the abscissa is 5 m and the height of the function is 2 mg C m^{-3} h^{-1}. The area of the rectangle between 0 and 5 m is therefore (5 m)(2 mg C m^{-3} h^{-1}) = 10 mg C m^{-2} h^{-1}. Between 5 and 10 m the length of the abscissa is 10 − 5 = 5 m and the height of the function is 5 mg C m^{-3} h^{-1}. The area of the rectangle from 5 to 10 m is therefore (5 mg C m^{-3} h^{-1})(5 m) = 25 mg C m^{-2} h^{-1}.

We call the area bounded by the abscissa and a function $F(X)$ between $X = A$ and $X = B$ the integral of $F(X)$ from A to B. If $F(X)$ is negative

TABLE 6.1 Photosynthetic Rates of Phytoplankton in an Estuary as a Function of Depth

Depth Interval (m)	Photosynthetic Rate (mg C m^{-3} h^{-1})
0–5	2
5–10	5
10 15	10
15–20	7
20–25	4
25–30	2

anywhere in the interval $[A, B]$, the area below the abscissa is considered to be negative area. The integral of the function $F(X)$ from A to B may also be thought of as the average value of the function from A to B times $(B - A)$. For the data in Table 6.1, $\overline{F(X)} = (2 + 5 + 10 + 7 + 4 + 2)/6 = 5$ mg C m^{-3} h^{-1}, and $B - A = 30$ m. Hence the integral equals (5 mg C m^{-3} h^{-1})(30 m) = 150 mg C m^{-2} h^{-1}.

Integrating the function in Table 6.1 is easy, because the function is constant over each 5 m interval. In general, we cannot expect the function to

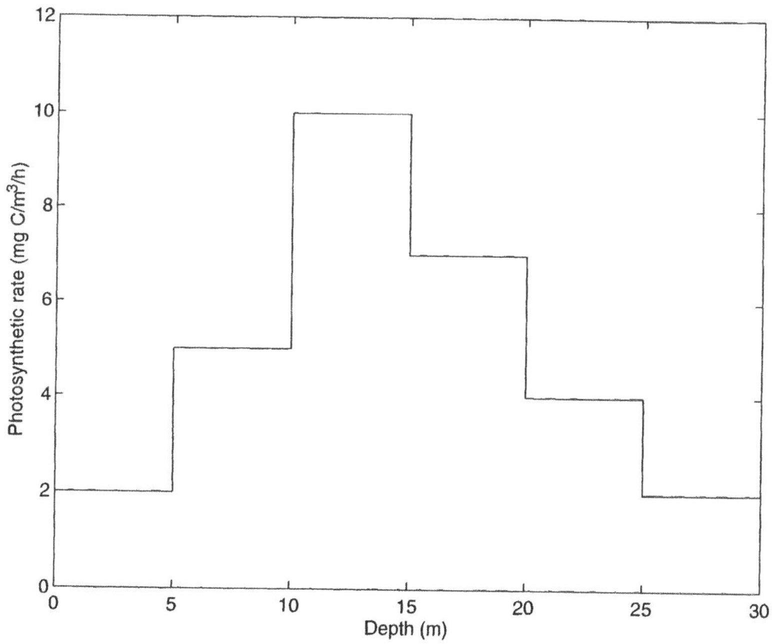

Figure 6.1 Representation of the data in Table 6.1. The area under the photosynthetic rate function is the integral of the function from 0 to 30 m depth.

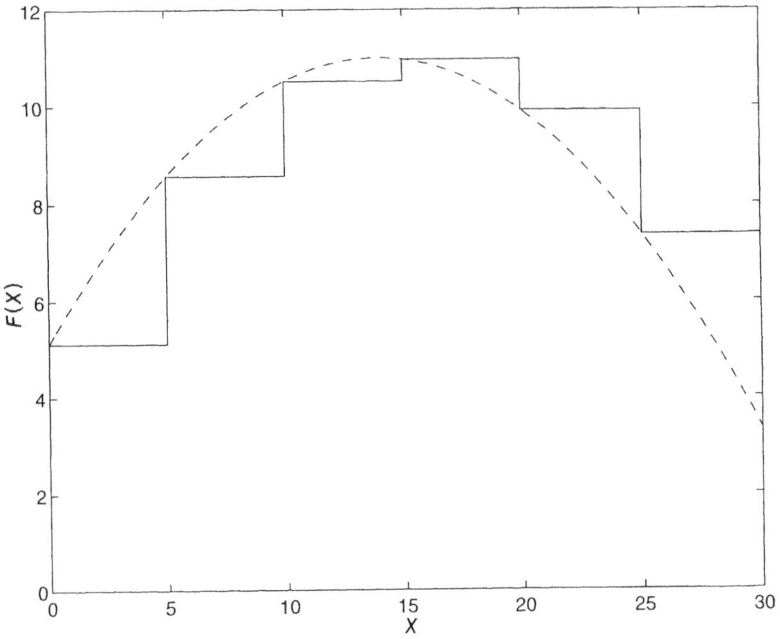

Figure 6.2 A hypothetical function $F(X)$ on the X interval $[0, 30]$. The dashed curve is the true function, and the area under the curve is the integral of $F(X)$ from 0 to 30. The integral may be approximated by summing the areas of a series of rectangles whose widths are 5 and whose heights are the value of $F(X)$ at the left-hand side of each interval. The integral of the rectangles is the area under the solid line in the figure.

behave in such a simple way. Consider, for example, the function shown by the dashed line in Figure 6.2. It is not clear how to calculate the area under this curve, but we could certainly approximate the area by dividing up the range of integration $[0, 30]$ into small intervals and approximating the function over each interval by its value at the left-hand side of the interval. We call the width of each interval ΔX. In Figure 6.2 we have chosen ΔX to be 5. Our approximate integration scheme therefore amounts to calculating the sum of the areas of $(B - A)/\Delta X$ rectangles, where in this case $(B - A)/\Delta X = 30/5 = 6$. The height of the nth rectangle is $F[A + (n - 1)\Delta X]$. In the example given, the area of the rectangles would underestimate the area under the function in the interval $[0, 15]$ and overestimate the area under the function in the interval $[15, 30]$. Now it is reasonable to assume that the accuracy of the approximation will improve as we make ΔX smaller and smaller and that the approximation should become exact in the limit as $\Delta X \to 0$. We designate the integral of the function $F(X)$ from A to B as

$\int_A^B F(X)\,dX$ and claim that

$$\begin{aligned}\int_A^B F(X)\,dX &= \lim_{\Delta X \to 0} \sum_{n=1}^{(B-A)/\Delta X} F[A + (n-1)\Delta X]\,\Delta X \\ &= \lim_{\Delta X \to 0} [F(A) + F(A + \Delta X) + \cdots \\ &\qquad + F(B - 2\Delta X) + F(B - \Delta X)]\Delta X \qquad (6.1)\end{aligned}$$

If the limit on the right-hand side of Equation 6.1 is not finite and unique, then the function $F(X)$ is not integrable from A to B. For example, the function $F(X) = 1/X^2$ is not integrable over any region that includes $X = 0$. Henceforth we will confine our attention to functions that are integrable.

Now suppose that there exists a function $G(X)$ such that

$$F[A + (n-1)\Delta X] = \frac{G(A + n\Delta X) - G[A + (n-1)\Delta X]}{\Delta X} \qquad (6.2)$$

for $n = 1$ to $(B - A)/\Delta X$. In other words,

$$F(A) = \frac{G(A + \Delta X) - G(A)}{\Delta X},$$

$$F(A + \Delta X) = \frac{G(A + 2\Delta X) - G(A + \Delta X)}{\Delta X}, \ldots,$$

$$F(B - \Delta X) = \frac{G(B) - G(B - \Delta X)}{\Delta X}$$

Substituting the right-hand side of Equation 6.2 for $F[A + (n - 1)\Delta X]$, it is straightforward to show that

$$\sum_{n=1}^{(B-A)/\Delta X} F[A + (n-1)\Delta X]\,\Delta X = G(B) - G(A)$$

This, of course, is assuming that we can identify a function $G(X)$ that satisfies Equation 6.2. In fact, there are an infinite number of such functions, and all of them are determined by the equation

$$G(A + m\Delta X) = G(A) + \Delta X \sum_{n=1}^{m} F[A + (n-1)\Delta X]$$

where $G(A)$ can have any value. In other words, the function $G(X)$ is determined to within an arbitrary constant by requiring that it satisfy Equation 6.2. The value of the constant is immaterial for the purposes of evaluating $\sum_{n=1}^{(B-A)/\Delta X} F[A+(n-1)\Delta X]\,\Delta X$, since the value of the constant cancels out when we calculate $G(B) - G(A)$. Letting $X = A + (n-1)\Delta X$, we see that Equation 6.2 is equivalent to

$$F(X) = \frac{G(X+\Delta X) - G(X)}{\Delta X}$$

Now we have seen that

$$\int_A^B F(X)\,dX = \lim_{\Delta X \to 0} \sum_{n=1}^{(B-A)/\Delta X} F[A+(n-1)\Delta X]\,\Delta X$$

However, in the limit as $\Delta X \to 0$ it should now be obvious that $F(X) = dG/dX$. We conclude that

$$\int_A^B F(X)\,dX = G(B) - G(A) \tag{6.3}$$

where $dG/dX = F(X)$. Here $G(X)$ is called the *antiderivative* of $F(X)$. The problem of integrating an analytical function therefore amounts to being able to determine the antiderivative of the function and evaluate it at the end points of the integral. In some cases, however, it is impossible to express the antiderivative of the integrand as an analytical function, or it may be impossible to obtain a complete representation of the function over the range of integration. In such cases we may use numerical methods to approximate the value of the integral, and by sampling the integrand at enough points between A and B, we should be able to determine the value of the integral as accurately as we desire or as measurement will permit.

Let us consider how we might go about deciding where to sample the integrand to obtain as much accuracy as possible in our answer. In deriving Equation 6.3 we assumed that the integrand was sampled at the left-hand side of each interval. Suppose we were dealing with a finite interval $[a, b]$ and approximated the integral as $F(a)(b-a)$. Clearly, this approximation would be exact if the function $F(X)$ were constant on the interval $[a, b]$. For example, if $F(X) = K$, where K is a constant, then $G(X) = KX + C$, where C is any constant. We then have $G(b) - G(a) = Kb + C - Ka - C = K(b-a)$. However, since $F(a) = K$, $F(a)(b-a) = K(b-a)$. Note that the constant C in $G(X)$ cancels out when we calculate $G(b) - G(a)$. We will henceforth assume that $C = 0$, since its value has no effect on the value of the integral.

Now assume that $F(X) = X$. In this case $G(X) = X^2/2$ and $G(b) - G(a) = (b^2 - a^2)/2$. However, $F(a) = a$, and $F(a)(b-a) = a(b-a)$. These two expressions are equal only if $b = a$, which is obviously not a very interesting case. In fact, approximating the integral by multiplying $(b - a)$ by the value of the integrand at $X = a$ will automatically give us the exact value of the integral only if $F(X)$ is constant on the interval. Hence this is not a very powerful numerical integration technique.

Now suppose that instead of evaluating the integrand at one end of the interval, we evaluate it at the midpoint of the interval. In other words, we approximate $\int_a^b F(X)\,dX$ by $(b - a)F((b + a)/2)$. If $F(X) = K$, then $F((b + a)/2) = K$, and $(b - a)F((b + a)/2) = (b - a)K$, which is the exact value of the integral. Now suppose $F(X) = X$. In this case

$$F\left(\frac{b+a}{2}\right) = \frac{b+a}{2}$$

and

$$(b-a)F\left(\frac{b+a}{2}\right) = (b-a)\left(\frac{b+a}{2}\right) = \frac{b^2-a^2}{2}$$

which is the exact value of the integral. Now suppose $F(X) = X^2$. In this case $G(X) = X^3/3$, and the exact value of the integral is $(b^3 - a^3)/3$. At the midpoint

$$F\left(\frac{b+a}{2}\right) = \left(\frac{b+a}{2}\right)^2$$

and

$$(b-a)F\left(\frac{b+a}{2}\right) = \frac{b^3 - ab^2 + a^2b - a^3}{4}$$

This is not the exact value of the integral, so we have not identified a method which will exactly integrate a quadratic function. We have, however, identified a method which will exactly integrate any linear function of X between two points. This method is referred to as the *midpoint rule*.

Another simple approach which might occur to us is to evaluate the function at the two end points of the interval and then multiply the width of the interval by the average value of the function at the two end points. In other words, we approximate $\int_a^b F(X)\,dX$ by $(b - a)(F(b) + F(a))/2$. If $F(X) = K$, then $F(b) = F(a) = K$, and

$$(b-a)\left(\frac{F(b)+F(a)}{2}\right) = (b-a)\left(\frac{K+K}{2}\right) = (b-a)K$$

which is the exact value of the integral. If $F(X) = X$, then $F(a) = a$ and

$F(b) = b$. Hence

$$(b-a)\left(\frac{F(b)+F(a)}{2}\right) = (b-a)\left(\frac{b+a}{2}\right) = \frac{b^2-a^2}{2}$$

which is the exact value of the integral. Now suppose $F(X) = X^2$. In that case $F(a) = a^2$ and $F(b) = b^2$. Hence

$$(b-a)\left(\frac{F(b)+F(a)}{2}\right) = (b-a)\left(\frac{b^2+a^2}{2}\right) = \frac{b^3-ab^2+a^2b-a^3}{2}$$

This is not the same as $(b^3 - a^3)/3$, so we still do not have a method that will exactly integrate a quadratic function. However, we have identified another method which will exactly integrate a linear function. This method is called the *trapezoid rule*, because the form of the equation is similar to the equation used to calculate the area of a trapezoid.

Both the midpoint rule and the trapezoid rule give comparable accuracy in numerical integration, in the sense that they will both exactly integrate a linear function and, in general, will fail to exactly integrate more complex functions. The midpoint rule would seem to be more efficient, because it requires sampling at only one point in each interval $[a, b]$, whereas the trapezoid rule requires sampling at two points. In most cases, however, the difference in the efficiency of the two methods is not as great as a factor of 2. The reason is that in most cases we subdivide the range of integration into several intervals and apply the midpoint or trapezoid rule to each interval. For example, if we are integrating photosynthetic rates from the surface to 150 m, we might subdivide the range of integration into six equal intervals of 25 m. To apply the midpoint rule, we would sample at six depths, 12.5 m, 37.5 m, . . . , 137.5 m. To apply the trapezoid rule, we would sample at seven depths, 0 m, 25 m, 50 m, . . . , 150 m. To achieve comparable accuracy with the trapezoid rule and midpoint rule, we need to sample the integrand at only one more point within the range of integration for the trapezoid rule. The reason is that interior points are the right-hand side of one interval and the left-hand side of the next interval for the trapezoid rule. Thus interior points serve double duty for the trapezoid rule. One important point about both methods is that they do not require dividing the range of integration into intervals of equal size. The intervals can be any size, as long as we sample the integrand at appropriate points. Going back to the photosynthetic rate integration problem, for example, we might choose to subdivide the upper 100 m into five equal intervals of 20 m and let the bottom 50 m be the last interval. To apply the midpoint rule in this case, we would need to sample at 10 m, 30 m, 50 m, 70 m, 90 m, and 125 m. To apply the trapezoid rule, we would need to sample at 0 m, 20 m, 40 m, 60 m, 80 m, 100 m, and 150 m. One feature about the trapezoid rule is that as long as one samples the integrand at the end points of the integral, one can always apply the trapezoid rule, regardless of where the interior points may lie. For example, if we measured photosynthetic rates at the surface and at 25 m, 40 m, 60 m, 90 m, and 150 m, we could integrate using the trapezoid rule with interval widths of 25 m, 15 m, 20 m, 30 m, and 60 m. The midpoint rule is much less forgiving with respect to lack of planning. For example, if we measured photosynthetic rates at 20 m, 35 m, 60 m, 90 m, and 130 m, we could not integrate the results from the surface to 150 m using the midpoint rule.[1]

TAYLOR SERIES

Up to this point in our discussion, we have judged numerical integration schemes on the basis of their ability to exactly integrate low-order polynomials. We have identified two methods which will exactly integrate linear functions, and we will soon be introduced to methods which will integrate higher-order polynomials. Why are we interested in methods which will exactly integrate polynomials?

The rationale for being able to integrate simple polynomials stems from the assumption that the integrand can, with a fair degree of accuracy, be described by a low-order polynomial within each interval of integration. The basis for this expectation stems from the following line of reasoning.

We learned in Chapter 1 that for two functions F and G, $(FG)' = F'G + FG'$. It follows that

$$\int_A^B \frac{d[F(X)G(X)]}{dX}\,dX = F(B)G(B) - F(A)G(A)$$

$$= \int_A^B G(X)\frac{dF(X)}{dX}\,dX + \int_A^B F(X)\frac{dG(X)}{dX}\,dX \tag{6.4}$$

Rearranging Equation 6.4, we obtain

$$\int_A^B F(X)\frac{dG(X)}{dX}\,dX = F(B)G(B) - F(A)G(A) - \int_A^B G(X)\frac{dF(X)}{dX}\,dX \tag{6.5}$$

[1] However, if we change one of the depths, we can perform the integration using the midpoint rule. Which depth needs to be changed and to what value?

Now suppose $G(X) = X - B$ and $F(X) = W'(X)$. Then $G(B) = 0$,

$$F(X)\frac{dG(X)}{dX} = \frac{dW(X)}{dX}$$

and

$$\frac{dF(X)}{dX} = \frac{d^2W(X)}{dX^2}$$

Equation 6.5 then becomes

$$W(B) = W(A) + (B - A)W'(A) - d\int_A^B (X - B)\frac{d^2W(X)}{dX^2}\,dX \quad (6.6)$$

We can apply Equation 6.5 to the integral on the right-hand side of Equation 6.6, this time setting $G(X) = (X - B)^2/2$ and $F(X) = d^2W(X)/dX^2$. We then have $G(B) = 0$,

$$F(X)\frac{dG(X)}{dX} = (X - B)\frac{d^2W(X)}{dX^2}$$

and

$$\frac{dF(X)}{dX} = \frac{d^3W(X)}{dX^3}$$

Equation 6.6 then becomes

$$W(B) = W(A) + (B - A)W'(A) + \frac{(B - A)^2}{2}W''(A)$$

$$+\frac{1}{2}\int_A^B (X - B)^2\frac{d^3W(X)}{dX^3}\,dX \quad (6.7)$$

We can apply Equation 6.5 to the integral on the right-hand side of Equation 6.7, this time setting $G(X) = (X - B)^3/3$ and $F(X) = d^3W(X)/dX^3$. We then have $G(B) = 0$,

$$F(X)\frac{dG(X)}{dX} = (X - B)^2\frac{d^3W(X)}{dX^3}$$

and

$$\frac{dF(X)}{dX} = \frac{d^4W(X)}{dX^4}$$

Equation 6.7 then becomes

$$W(B) = W(A) + (B - A)W'(A) + \frac{(B-A)^2}{2}W''(A) + \frac{(B-A)^3}{6}W'''(A) - \frac{1}{6}\int_A^B (X-B)^3 \frac{d^4W(X)}{dX^4}\,dX \quad (6.8)$$

The pattern by now should be clear, and indeed we can show by induction that

$$W(B) = W(A) + \sum_{n=1}^{\infty} \frac{W^{(n)}(A)(B-A)^n}{n!} \quad (6.9)$$

where $W^{(n)}(A)$ is the nth derivative of $W(X)$ evaluated at $X = A$. Equation 6.9 is the Taylor series expansion of the function $W(X)$ about the point $X = A$.

The Taylor series representation of a function is of practical use in evaluating the function only if the series converges rapidly, that is, only if the first few terms are significant. This may, in fact, be the case if the interval of integration is not too large. Consider, for example, the function $F(X) = X/(1 + X)$ which we approximated with a polynomial in Figure 5.3. If we differentiate this function n times, we find that $F^{(n)}(X) = n!(-1)^{n+1}/(1 + X)^{n+1}$. If we were interested in integrating this function over the interval $[0, 4]$, as in Figure 5.3, it would seem reasonable to consider an expansion of the function about $X = 2$. We therefore want to determine whether a representation of the form

$$F(X) = F(2) + \sum_{n=1}^{N} \frac{F^{(n)}(2)(X-2)^n}{n!}$$

might give a good approximation of the function on the interval $[0, 4]$. We choose to expand about $X = 2$ so that we can be assured that $|X - 2|$ will never be larger than 2 for any X in the interval. The Taylor series expansion of $F(X) = X/(1 + X)$ then takes the form

$$F(X) = \frac{2}{3} + \sum_{n=1}^{\infty} \frac{(X-2)^n}{(-3)^{n+1}}$$

$$= \frac{2}{3} + \frac{(X-2)}{9} - \frac{(X-2)^2}{27} + \frac{(X-2)^3}{81} - \frac{(X-2)^4}{243} + \cdots \quad (6.10)$$

It is easy to show that the absolute value of the ratio of successive terms in this series is at most $\frac{2}{3}$ on the interval [0, 4]. Hence the series converges rapidly. If we integrate the function $F(X) = X/(1 + X)$ from 0 to 4, we find that the exact value of the integral is $4 - \ln(5) = 2.39$. If we approximate $F(X)$ by taking terms through only second order in its Taylor series expansion about $X = 2$, we find that the integral of the approximate function equals

$$\int_0^4 \left(\frac{2}{3} + \frac{(X-2)}{9} - \frac{(X-2)^2}{27} \right) dX = \frac{200}{81} = 2.47$$

The discrepancy of $2.47 - 2.39 = 0.08$ is only about 3% of the exact value of the integral. The rationale then is that if the integral of a function $F(X)$ can be well approximated by the integral of the first few terms in its Taylor series expansion about an appropriate value of X, then a numerical integration scheme that can exactly integrate the first few terms in the Taylor series expansion of the function should give a good approximation to the integral of $F(X)$. More generally, if the function $F(X)$ can be well approximated by a low-order polynomial on the interval $[a, b]$, then a numerical integration scheme that can exactly integrate all the terms in the low-order polynomial should give a good approximation to the integral of $F(X)$ from a to b.

SIMPSON'S RULE

So far we have identified two numerical integration schemes which will exactly integrate a linear function of X. One of the simplest and most commonly used higher-order numerical integration schemes is Simpson's rule. With Simpson's rule we use the value of the integrand at the two end points and at the midpoint of the interval of integration. Simpson's rule then says that

$$\int_a^b F(X)\, dX \cong \left[F(a) + 4F\left(\frac{a+b}{2} \right) + F(b) \right] \left(\frac{b-a}{6} \right) \qquad (6.11)$$

Let us see how well Simpson's rule does. If $F(X)$ is a constant, it should be obvious by inspection that Simpson's rule will give us the exact integral, which is $b - a$. If $F(X) = X$, Simpson's rule gives us

$$\left[a + 4\left(\frac{a+b}{2} \right) + b \right] \left(\frac{b-a}{6} \right) = \frac{b^2 - a^2}{2}$$

which is again the exact integral. If $F(X) = X^2$, Simpson's rule gives us

$$\left[a^2 + 4\left(\frac{a+b}{2}\right)^2 + b^2\right]\left(\frac{b-a}{6}\right) = (a^2 + ab + b^2)\left(\frac{b-a}{3}\right) = \frac{b^3 - a^3}{3}$$

the exact value of the integral. If $F(X) = X^3$, Simpson's rule gives us

$$\left[a^3 + 4\left(\frac{a+b}{2}\right)^3 + b^3\right]\left(\frac{b-a}{6}\right)$$

$$= (a^3 + a^2b + ab^2 + b^2)\left(\frac{b-a}{4}\right) = \frac{b^4 - a^4}{4}$$

again the exact value of the integral. If $F(X) = X^4$, Simpson's rule gives us

$$\left[a^4 + 4\left(\frac{a+b}{2}\right)^4 + b^4\right]\left(\frac{b-a}{6}\right)$$

$$= (5a^4 + 4a^3b + 6a^2b^2 + 4ab^3 + 5b^2)\left(\frac{b-a}{24}\right)$$

The exact value of the integral in this case is $(b^5 - a^5)/5$. These two expressions are not equal, and hence Simpson's rule cannot be expected to exactly integrate a fourth-order polynomial. It does, however, exactly integrate cubic polynomials, and this is a significant improvement over the midpoint and trapezoid rules. There is, however, a price associated with using Simpson's rule. We must sample the integrand at three points in each interval. If we divide our range of integration into n intervals, we will need to sample the integrand at a total of $2n + 1$ points to be able to use Simpson's rule on each interval. Assume, for example, that we want to measure photosynthetic rates at five depths in the water column and estimate the integral of the photosynthetic rate from the surface to 150 m. If we use Simpson's rule, we would divide the upper 150 m of the water column into two intervals. The intervals would not have to be equal in length. The first interval could be from the surface to 50 m, for example, and the second interval from 50 to 150 m. In this case we would take our measurements at the surface and at 25 m, 50 m, 100 m, and 150 m. If we used the trapezoid rule, we could sample at the same depths, or we could sample at the surface, 150 m, and any three intermediate depths. With the trapezoid rule we would consider the range of integration to be divided into four intervals and apply the trapezoid rule to each interval. With the midpoint rule we would, of course, divide the range of integration into five intervals, not necessarily of the same length, and sample at the midpoint of each interval.

The midpoint and trapezoid rules give us more flexibility in choosing our intervals than does Simpson's rule, because for a given number of measurements we will be working with roughly twice as many intervals. In addition, it does not automatically follow that being able to exactly integrate a cubic polynomial over the interval [a, b] will give us a more accurate estimate of the integral of a function than chopping the interval into two smaller pieces and using the midpoint rule on each of the smaller intervals. If the function in question can be well approximated by a piecewise linear representation, the latter approach may, in fact, give the more accurate answer. Which numerical integration scheme one chooses to use will depend in part on one's intuitive feelings about the nature of the function being integrated. In reaching a decision, however, it is important to have a clear picture of the strengths and weaknesses of the various methods available.

THE METHOD OF MOMENTS

The numerical integration schemes we have considered so far follow the general equation

$$\int_a^b \rho(X)F(X)\,dX \cong \sum_{n=1}^{N} W_n F(X_n) \tag{6.12}$$

where $\rho(X)$ is a weighting function, the W_n are weights, and the X_n are abscissas. For example, with the midpoint rule $\rho(X) = 1$, $N = 1$, $W_1 = b - a$, and $X_1 = (b + a)/2$. We have argued that it is desirable for the numerical integration scheme to be able to exactly integrate low-order polynomials. It follows from our discussion of Taylor series that for a fixed N, the higher the order of the polynomial that the numerical integration scheme is able to integrate exactly, the more accurate it is likely to be in estimating the value of integrals. Until now we have simply tested the ability of numerical integration methods to integrate powers of X, but we can go one step further. We can require that the numerical integration method be able to exactly integrate $\rho(X)$ times all powers of X from 0 to $2N - 1$. We can constrain the numerical integration method in this way because, having selected N, we are free to choose N weights and N abscissas. We can require that the N weights and N abscissas satisfy a total of $2N$ constraint equations, and the constraint equations are going to be that the numerical integration method exactly integrate $\rho(X)$ times powers of X from 0 to $2N - 1$. The integral $\int_a^b \rho(X)X^m\,dX$ is called the mth moment of the weighting function $\rho(X)$ on the interval $[a, b]$, hence the name method of moments.

We begin by considering the simple case where $\rho(X) = 1$. If we choose $N = 1$, then we must pick one weight and one abscissa. Hence we can satisfy two constraint Equations and can require that the method be able to exactly integrate $X^0 = 1$ and X. It is sufficient for our purposes to choose $a = -1$ and $b = 1$. Results obtained for the interval $[-1, 1]$ can easily be generalized to any finite interval $[a, b]$, as we will see. On the interval $[-1, 1]$ we have

$$\int_{-1}^{1} dX = 2 \tag{6.13}$$

$$\int_{-1}^{1} X\,dX = 0 \tag{6.14}$$

Hence 2 is the zeroth moment and 0 is the first moment of the weighting function $\rho(X) = 1$ on the interval $[-1, 1]$. Now for the zeroth moment $F(X) = 1$ and hence $F(X_1) = 1$. For the first moment, $F(X) = X$ and hence $F(X_1) = X_1$. For $N = 1$ we therefore require that

$$2 = W_1 \tag{6.15}$$

$$0 = W_1 X_1 \tag{6.16}$$

The solutions to Equations 6.15 and 6.16 are obviously $W_1 = 2$ and $X_1 = 0$. Hence the method of moments tells us that, for $N = 1$,

$$\int_{-1}^{1} F(X)\,dX \cong 2F(0) \tag{6.17}$$

It should be obvious that Equation 6.17 is, in fact, the midpoint rule applied to the interval $[-1, 1]$. Thus the midpoint rule is the simplest ($N = 1$) form of the method of moments.

Now consider the case $N = 2$. With $N = 2$ we should be able to exactly integrate a cubic polynomial. We already know the zeroth and first moments. The second and third moments are as follows:

$$\int_{-1}^{1} X^2\,dX = \frac{2}{3} \tag{6.18}$$

$$\int_{-1}^{1} X^3\,dX = 0 \tag{6.19}$$

For $N = 2$ we now require that

$$2 = W_1 + W_2 \tag{6.20}$$

$$0 = W_1 X_1 + W_2 X_2 \tag{6.21}$$

$$\tfrac{2}{3} = W_1 X_1^2 + W_2 X_2^2 \tag{6.22}$$

$$0 = W_1 X_1^3 + W_2 X_2^3 \tag{6.23}$$

The solutions to these equations may not be immediately obvious, but we can use symmetry arguments to help find them. Since we are dealing with an interval that is symmetric about the origin and since our weighting function is 1, we can expect the abscissas to occur in + and − pairs, with one abscissa at the origin if N is odd. We can also expect the weights to be equal for each abscissa pair. In this case we anticipate that $W_1 = W_2$ and that $X_2 = -X_1$. Assuming this to be the case, Equations 6.21 and 6.22 are automatically satisfied, and Equation 6.20 implies that $W_1 = W_2 = 1$. Equation 6.22 then tells us that $X_2 = -X_1 = \sqrt{\frac{1}{3}}$.

We could proceed in this manner to higher values of N, but the problem of finding the weights and abscissas would become much harder. Fortunately, the solutions for larger values of N have been worked out and have been tabulated with 30 significant figures up to $N = 512$. It turns out that the abscissas are the roots of the Legendre polynomials $P_n(X)$, which are defined by

$$P_n(X) = \frac{1}{2^n n!}\left(\frac{d}{dX}\right)^n (X^2 - 1)^n \tag{6.24}$$

and this particular numerical integration scheme is known as Gauss–Legendre quadrature. For $n = 1$ we have $P_1(X) = X$, the root of which is obviously $X = 0$. For $n = 2$, $P_2(X) = \frac{3}{2}X^2 - \frac{1}{2}$, the roots of which are $\pm\sqrt{\frac{1}{3}}$.

We noted before that results obtained on the interval $[-1, 1]$ could be generalized to any finite interval $[a, b]$. Let us consider how this is done. We know that the sum of the weights must be $b - a$. In Gauss–Legendre quadrature the sum of the weights is always 2. To generalize the Gauss–Legendre weights to the interval $[a, b]$, we simply multiply the weights by $(b - a)/2$. The abscissas on the interval $[a, b]$ must be centered on $(b + a)/2$ rather than 0, and their range must be $b - a$ rather than 2. The transformation for the abscissas is therefore

$$X_{[a,b]} = \frac{b + a}{2} + \left(\frac{b - a}{2}\right) X_{[-1,1]}$$

where $X_{[-1,1]}$ is a Gauss–Legendre abscissa on the interval $[-1, 1]$ and $X_{[a,b]}$ is the transformed abscissa on the interval $[a, b]$.

To see how we might apply the method of moments, consider the earlier problem of estimating areal photosynthetic rates in a system where the euphotic zone is 150 m deep. Assume that we are working on a low budget and can afford to sample at only three depths. To try to get the most accurate answer from our data, we decide to integrate the results using Gauss–Legendre quadrature. At what depths should we sample, and how do we integrate the data? The third Legendre polynomial is $P_3(X) = \frac{5}{2}X^3 - \frac{3}{2}X$. The roots of this polynomial are $X = 0$ and $X = \pm\sqrt{\frac{3}{5}}$. The associated weights are $\frac{8}{9}$ and $\frac{5}{9}$. Transforming the abscissas to the interval [0, 150], we find that $X = 0$ transforms to X = 75, X = $-\sqrt{\frac{3}{5}}$ transforms to $75 - 75\sqrt{\frac{3}{5}} = 17$, and $X = \sqrt{\frac{3}{5}}$ transforms to $75 + 75\sqrt{\frac{3}{5}} = 133$. The weight for the sample at 75 m is $75(\frac{8}{9}) = 66\frac{2}{3}$, and the weights for the other two samples are $75(\frac{5}{9}) = 41\frac{2}{3}$. Thus if $F(X)$ is the photosynthetic rate (or indeed any function of depth) at depth X, then

$$\int_0^{150} F(X)\, dX \cong 41\tfrac{2}{3}F(17) + 66\tfrac{2}{3}F(75) + 41\tfrac{2}{3}F(133) \qquad (6.25)$$

Equation 6.25 will exactly integrate any polynomial of order 5 or less and should do a good job of approximating the integral of any function which can be adequately described by a fifth-order polynomial on the interval [0, 150].

In some cases it may be advisable to include a weighting function $\rho(X)$ into the numerical integration scheme. One possible application would be a case in which we wish to evaluate an indefinite integral of the form $\int_0^\infty F(X)\, dX$. Obviously, we cannot approximate this integral using the midpoint rule, the trapezoid rule, or Gauss–Legendre quadrature directly. There are basically two approaches to numerically integrating such an integral. One is to transform the integration variable so that the range of integration becomes finite. For example, if the range of integration is $[0, \infty]$, we might let $Y = (X - 1)/(X + 1)$ or $Y = e^{-X}$. In the former case the range of integration becomes $[-1, 1]$ and in the latter case [0, 1]. The back transformations are $X = (1 + Y)/(1 - Y)$ and $dX = 2dY/(1 - Y)^2$ in the former case and $X = -\ln(Y)$ and $dX = -dY/Y$ in the latter case. Whether such transformations are useful or not depends on the nature of the function being integrated. In the case of the transformation $Y = (X - 1)/(X + 1)$, for example, the X interval [0, 1] is mapped into the Y interval $[-1, 0]$, and the X interval $[1, \infty]$ is mapped into the Y interval [0, 1]. This may not be a very useful mapping if the X interval $[1, \infty]$ contributes a large portion of the integral. Similarly, the transformation $Y = e^{-X}$ maps the X interval [0, ln(2)] into the Y interval $[\frac{1}{2}, 1]$ and the X interval $[\ln(2), \infty]$ into the Y interval $[0, \frac{1}{2}]$. This may not be a very useful mapping if the X interval $[\ln(2), \infty]$ contributes a large portion of the integral. If one contemplates making such transformations to convert the range $[0, \infty]$, for example, to a finite range, it is important to have some idea of what portion of the range $[0, \infty]$ is most important and to be sure that the transformation gives adequate emphasis to that portion of the range. For example, if one feels that the X intervals $[0, A]$ and $[A, \infty]$ are of roughly equal importance, then the transformations $Y = (X - A)/(X + A)$ or $Y = 2^{-X/A}$ may do the job.

The second approach to the problem of integrating an indefinite integral is to assume that the integrand $F(X)$ can be expressed as $F(X) = G(X)\exp(-kX)$ or $G(X)\exp[-(kX)^2]$. In the former case, for example, we have

$$\int_0^\infty F(X)\, dX = \int_0^\infty G(X)\exp(-kX)\, dX = \frac{1}{k}\int_0^\infty G\left(\frac{Y}{k}\right)\exp(-Y)\, dY \quad (6.26)$$

In evaluating the integral on the right-hand side of Equation 6.26, we consider $\exp(-Y)$ to be the weighting function $\rho(Y)$. If $G(Y/k))$ can be well approximated by a low-order polynomial, then we can use the method of moments to calculate the integral. The moments of the weighting function $\exp(-Y)$ on the interval $[0, \infty]$ are $n!$, that is

$$\int_0^{\infty} Y^n \exp(-Y)\, dY = n! \tag{6.27}$$

and the abscissas for numerical integration are the roots of the Laguerre polynomials, which are defined by

$$L_n(X) = \exp(X)\left(\frac{d}{dX}\right)^n [X^n \exp(-X)] \tag{6.28}$$

For example, $L_1(X) = 1 - X$ and $L_2(X) = X^2 - 4X + 2$. Estimating integrals on the interval $[0, \infty]$ using the weighting function $\exp(-Y)$ is referred to as Gauss–Laguerre quadrature. The estimation proceeds as follows:

$$\int_0^{\infty} F(X)\, dX = \int_0^{\infty} G(X)\exp(-kX)\, dX = \frac{1}{k}\int_0^{\infty} G\left(\frac{Y}{k}\right)\exp(-Y)\, dY$$

$$\cong \frac{1}{k}\sum_{n=1}^{N} W_n G\left(\frac{X_N}{k}\right) = \frac{1}{k}\sum_{n=1}^{N} W_n F\left(\frac{X_n}{k}\right)\exp(X_n) \tag{6.29}$$

where the X_n are the roots of the Nth Laguerre polynomial and the W_n are the corresponding weights. If one chooses to use the weighting function $\exp[-(kX)^2]$, the calculations proceed in a similar manner, the final equation being

$$\int_0^{\infty} F(X)\, dX \cong \frac{1}{k}\sum_{n=1}^{N} W_n F\left(\frac{X_n}{k}\right)\exp(X_n^2) \tag{6.30}$$

where in this case the abscissas are the roots of the Hermite polynomials. This form of numerical integration is known as Gauss–Hermite quadrature. As in the case of Gauss–Legendre quadrature, the abscissas and weights for Gauss–Laguerre and Gauss–Hermite quadrature have been tabulated to 30 significant figures.

As in many fields of endeavor, there is room for individualism in numerical integration. Suppose we are not happy with anything anyone else has done and want to develop our own numerical integration scheme. We will consider a case where this might be a reasonable thing to do. We have discussed the problem of integrating photosynthetic rates over the depth of the euphotic zone, which near Hawaii might extend to a depth of 150 m. We know that photosynthetic rates are roughly a hyperbolic function of light intensity and that light intensity decays exponentially in the water column. As a first approximation it might therefore be reasonable to assume that photosynthetic rates are constant in the upper, light-saturated portion of the water column and that they decline monotonically with increasing depth in the light-limited region of the water column, reaching 0 at the base of the euphotic zone. Assume that the division between the light-saturated and the light-limited regions of the water column occurs at a depth equal to 20% of the depth of the euphotic zone. The simplest possible representation of the behavior we have described would be a piecewise linear function having a constant value in the upper 20% of the euphotic zone and declining linearly with depth to a value of 0 at the bottom of the euphotic zone. By using this behavior as our weighting function, we might be able to develop a better numerical integration scheme than one which ignores what we know about the photosynthetic characteristics of marine phytoplankton.

We will be dealing with a finite range of integration, and we have seen that in such cases it is sufficient to calculate weights and abscissas for the range of integration $[-1, 1]$, since these can easily be transferred to any other finite range of integration $[A, B]$. In this case we will map the range $[-1, 1]$ into the range $[0, 150]$.

The upper 20% of the euphotic zone is roughly the upper 30 m of the water column near Hawaii, and this maps into the interval $[-1, -0.6]$. The constant value we assume for our weighting function on this interval is arbitrary, and to make matters simple, we will assume that $\rho(X) = 1$ on the interval $[-1, -\frac{3}{5}]$. If $\rho(X)$ is going to decline linearly to 0 at $X = 1$, then $\rho(X) = \frac{5}{8}(1 - X)$ on the interval $[-\frac{3}{5}, 1]$. The moments of $\rho(X)$ on the interval $[-1, 1]$ are then given by the equation

$$\int_{-1}^{1} \rho(X) X^n \, dX = \frac{1}{(n+1)(n+2)} \left[\frac{5}{8} + (n+2)(-1)^n + \frac{3}{8}\left(-\frac{3}{5}\right)^{n+1} \right] \tag{6.31}$$

We will designate the nth moment of the weighting function as M_n. The problem now is to determine the weights and abscissas for the numerical integration scheme. It turns out that the N abscissas for any numerical integration scheme using the method of moments are the roots of the polynomial $P(X)$ obtained by expanding the following determinant:

$$\begin{vmatrix} 1 & X & X^2 & \cdot & \cdot & X^N \\ M_0 & M_1 & M_2 & \cdot & \cdot & M_N \\ M_1 & M_2 & M_3 & \cdot & \cdot & M_{N+1} \\ \vdots & \vdots & \vdots & \vdots & \vdots & \vdots \\ M_{N-1} & M_N & M_{N+1} & \cdot & \cdot & M_{2N-1} \end{vmatrix} = P(X) \tag{6.32}$$

For example, in the case $N = 1$, the single abscissa is the solution of the equation $M_1 - M_0 X = 0$, where $M_1 - M_0 X$ is the determinant

$$\begin{vmatrix} 1 & X \\ M_0 & M_1 \end{vmatrix}$$

In exercise 6.7 you are asked to derive the corresponding equation for the case $N = 2$. The problem of finding the abscissas therefore becomes the problem of determining the roots of a polynomial, and this can be done with software available for virtually any computer. Note that the coefficient of the constant term in the polynomial is simply the determinant

$$\begin{vmatrix} 1 & 0 & 0 & \cdot & \cdot & 0 \\ M_0 & M_1 & M_2 & \cdot & \cdot & M_N \\ M_1 & M_2 & M_3 & \cdot & \cdot & M_{N+1} \\ \vdots & \vdots & \vdots & \vdots & \vdots & \vdots \\ M_{N-1} & M_N & M_{N+1} & \cdot & \cdot & M_{2N-1} \end{vmatrix}$$

The coefficient of X in the polynomial is the determinant

$$\begin{vmatrix} 0 & 1 & 0 & \cdot & \cdot & 0 \\ M_0 & M_1 & M_2 & \cdot & \cdot & M_N \\ M_1 & M_2 & M_3 & \cdot & \cdot & M_{N+1} \\ \vdots & \vdots & \vdots & \vdots & \vdots & \vdots \\ M_{N-1} & M_N & M_{N+1} & \cdot & \cdot & M_{2N-1} \end{vmatrix}$$

The coefficient of X^2 in the polynomial is the determinant

$$\begin{vmatrix} 0 & 0 & 1 & \cdot & \cdot & 0 \\ M_0 & M_1 & M_2 & \cdot & \cdot & M_N \\ M_1 & M_2 & M_3 & \cdot & \cdot & M_{N+1} \\ \vdots & \vdots & \vdots & \vdots & \vdots & \vdots \\ M_{N-1} & M_N & M_{N+1} & \cdot & \cdot & M_{2N-1} \end{vmatrix}$$

and so forth. Assuming that one has access to a software package that calculates determinants and roots of polynomials, it is straightforward to calculate the abscissas for the method of moments given the moments of the appropriate weighting function. The N weights are then calculated by solving the matrix equation

$$\mathbf{XW} = \mathbf{M} \tag{6.33}$$

where

$$\mathbf{X} = \begin{pmatrix} 1 & 1 & 1 & 1 & 1 \\ X_1 & X_2 & \cdot & \cdot & X_N \\ X_1^2 & X_2^2 & \cdot & \cdot & X_N^2 \\ \vdots & \vdots & \vdots & \vdots & \vdots \\ X_1^{N-1} & X_2^{N-1} & \cdot & \cdot & X_N^{N-1} \end{pmatrix} \quad \mathbf{W} = \begin{pmatrix} W_1 \\ W_2 \\ \vdots \\ W_N \end{pmatrix} \quad \mathbf{M} = \begin{pmatrix} M_0 \\ M_1 \\ \vdots \\ M_{N-1} \end{pmatrix}$$

Two interesting facts about the abscissas calculated by determining the roots of the polynomial $P(X)$ are the following:

1. The abscissas will all lie within the range of integration, although only the moments of the distribution function are explicitly provided as input.
2. The abscissas for the case $N = \text{M} + 1$ interleave the abscissas for the case $N = M$.

The weights and abscissas for the problem under consideration, calculated from Equations 6.32 and 6.33, are shown in Table 6.2. If $F(X)$ is the photosynthetic rate at depth X, then our numerical integration scheme is to approximate $\int_0^{150} F(X)\,dX$ as follows:

$$\int_0^{150} F(X)\,dX \cong 75 \sum_{n=1}^{N} W_n \frac{F[75(1 + X_n)]}{\rho(X_n)} \tag{6.34}$$

Note that it is the function F/ρ which is assumed to be well approximated by a low-order polynomial, and it is this function which appears in the summation on the right-hand side of Equation 6.34. Note also that the weights are $75\ W_n$ and that the abscissas have been transformed as previously discussed. Now assume as an example that we decide to measure the photosynthetic rates at five depths. Using Equation 6.34 and the data in Table 6.2, we would estimate the integral of the photosynthetic rates as follows:

$$\int_0^{150} F(X)\,dX \cong 15.8F(6) + 30.4F(30) + 38.9F(66) + 36.2F(105) + 23.9F(135) \tag{6.35}$$

TABLE 6.2 Weights (W) and Abscissas (X) on the Interval $[-1, 1]$ for the weighting Function $\rho(X) = 1$ for $-1 \leq X \leq -\frac{3}{5}$ and $\rho(X) = \frac{5}{8}(1 - X)$ for $-\frac{3}{5} \leq X \leq 1$

W	X
$N = 1$	
1.2	−0.3111
$N = 2$	
0.7532	−0.6714
0.4468	0.2961
$N = 3$	
0.4615	−0.8115
0.5659	−0.1741
0.1726	0.5777
$N = 4$	
0.3012	−0.8795
0.5002	−0.4382
0.3214	0.1712
0.0772	0.7218
$N = 5$	
0.2101	−0.9167
0.4051	−0.5949
0.3625	−0.1192
0.1832	0.3934
0.0391	0.8037
$N = 6$	
0.1547	−0.9390
0.3205	−0.6954
0.3490	−0.3212
0.2449	0.1204
0.1091	0.5400
0.0217	0.8544
$N = 7$	
0.1185	−0.9534
0.2541	−0.7640
0.3152	−0.4634
0.2661	−0.0910
0.1650	0.2968
0.0681	0.6406
0.0130	0.8878
$N = 8$	
0.0936	−0.9633
0.2045	−0.8124
0.2761	−0.5658
0.2633	−0.2525
0.1969	0.0929
0.1130	0.4279
0.0444	0.7120
0.0082	0.9110

INTEGRATION OVER TIME

Up to this point we have considered numerical integration over space. We integrate over space when we want to know either the average or the total quantity of something between two points, within a given area or within a given volume. When we integrate over time, we are interested in studying or predicting the temporal behavior of some function, for example, the amount of carbon dioxide in the atmosphere, the size of a fish population, or the $^{15}N/^{14}N$ ratio in ammonium during an incubation in which both uptake and regeneration of ammonium occur. In some cases the integration may be carried out analytically, that is, without resort to numerical methods. In general, it is best to use analytical solutions to study and predict the temporal behavior of systems, because the nature of the relationship between the independent and dependent variables is clearly summarized in the formulation of the analytical solution. In other cases, however, the differential equations describing the rate of change of the function(s) of interest do not have simple analytical solutions, either because of the inherent nature of the equations or because the independent variables which appear in the equations do not have a simple analytical representation over the relevant time frame. In such cases we must resort to numerical methods to integrate the functions from one point in time to the next.

Let us consider a simple case in which we are interested in studying the temporal behavior of a function F. The rate of change of F is governed by an equation of the form

$$\frac{dF}{dt} = f(X) \tag{6.36}$$

where X is a set of variables, often including F itself, which influence the rate of change of F. Assume that we know the value of $F(t)$ at the present time, which we will call t_0. We would like to predict the value of F at some time t in the future. From our discussion of integration we know that

$$F(t) = F(t_0) + \int_{t_0}^{t} \frac{dF}{dt}\, dt = \int_{t_0}^{t} f\, dt \tag{6.37}$$

If f depends on F, as is often the case, the value of the integrand will be known only at time $t = t_0$. In such cases it would seem that we can evaluate the integral in Equation 6.37 only by assuming f to be constant, in which case the value of the integral is $(t - t_0)f$. This assumption may not lead to serious errors if the difference $t - t_0$ is small. Given this limitation, projecting F over long time periods requires dividing the range of integration into many small intervals and projecting F one step at a time from one interval to the next. The more sophisticated forms of numerical integration overcome this limitation, but not, as we will see, without paying a price.

To appreciate the implications of assuming f to be constant, expand F in its Taylor series about $t = t_0$. We obtain

$$F(t_0 + \Delta t) = F(t_0) + \left[\frac{dF}{dt}\right]_{t=0} \Delta t + \left[\frac{d^2F}{dt^2}\right]_{t=t_0} \frac{(\Delta t)^2}{2} + \cdots$$

$$= F(t_0) + f(t_0)\,\Delta t + \left[\frac{d^2F}{dt^2}\right]_{t=t_0} \frac{(\Delta t)^2}{2} + \cdots \tag{6.38}$$

If we project using only $f(t_0)$, we are ignoring all terms in the Taylor series expansion except $F(t_0) + f(t_0)\,\Delta t$. This may be valid if the neglected terms are small compared to $f(t_0)\,\Delta t$. In other words, we assume that most of the change in F between time t_0 and $t_0 + \Delta t$ is accounted for by $f(t_0)\,\Delta t$. As a rule of thumb, it is common to require that the absolute value of $[d^2F/dt^2]_{t=t_0}(\Delta t)^2/2$ be much smaller than the absolute value of $[dF/dt]_{t=t_0}\,\Delta t$, or, in other words, that

$$\Delta t \ll \left| \frac{2\left[\dfrac{dF}{dt}\right]_{t=t_0}}{\left[\dfrac{d^2F}{dt^2}\right]_{t=t_0}} \right| \tag{6.39}$$

If the numerical integration scheme is to accurately project the function between times t_0 and t, then Equation 6.39 must be satisfied for all values of t within the range of integration. Integrating a function with the equation

$$F(t_0 + \Delta t) = F(t_0) + f(t_0)\,\Delta t \tag{6.40}$$

is referred to as the *Euler method* of integration. A person using the Euler method should try to ensure that $\Delta t \ll |2F'/F''|$ for all values of t within the range of integration.

Even when this condition is satisfied, however, there is still a possibility that the Euler method, and indeed any numerical integration method, may greatly overestimate or underestimate the true value of the function after a sufficient period of time. Consider, for example, the simple differential equation

$$\frac{dF}{dt} = kF \tag{6.41}$$

with $F(0) = F_0$. In this case we can see that $F'' = kF' = k^2F$. According to the line of reasoning we have followed, we should choose $\Delta t \ll 2F'/F'' = 2/k$. Assume that we have done this and begin the integration at time $t = 0$.

Then at time Δt the predicted value of the function will be

$$F(\Delta t) = F_0 + kF_0\,\Delta t = (1 + k\Delta t)F_0 \tag{6.42}$$

Projecting to time $2\Delta t$, we have

$$F(2\Delta t) = F(\Delta t) + kF(\Delta t)\,\Delta t = (1 + k\Delta t)F(\Delta t) = (1 + k\Delta t)^2 F_0 \tag{6.43}$$

Proceeding with this reasoning, we see that at time $n\Delta t$ the projected value of the function will be $(1 + k\Delta t)^n F_0$. The true value of the function at time $n\Delta t$ is $F_0 \exp(kn\Delta t)$. The ratio of the value projected by the Euler method to the true value is therefore

$$\frac{(1 + k\Delta t)^n}{\exp(kn\Delta t)} = \left[\frac{1 + k\Delta t}{\exp(k\Delta t)}\right]^n$$

However, the Taylor series expansion of $\exp(k\Delta t) = 1 + k\Delta t + (k\Delta t)^2/2 + \cdots$. It is therefore clear that $(1 + k\Delta t)/\exp(k\Delta t) < 1$. Hence as n becomes large the ratio $(1 + k\Delta t)^n/\exp(kn\Delta t)$ approaches 0, and this will be true no matter how small we make Δt.

The example cited here is one of a function which is accelerating at all times, that is, a function whose curvature F'' is always positive. A similar line of reasoning can be used to show that the Euler method of integration will systematically overestimate a function which is decelerating at all times. These two types of functions may not occur often in oceanographic applications, but they certainly do not exhaust the list of equations or systems of equations which may be handled poorly by low-order numerical integration schemes. As a general rule it is a good idea when projecting a function or functions with numerical integration techniques to repeat the exercise with a higher-order numerical integration method and the same Δt, or the same numerical integration method and a smaller Δt. In this way one can hopefully spot artifacts which arise due to the approximations inherent in finite-difference numerical integration methods. We will have a chance to explore this issue a little further in the Exercises.

We mentioned earlier that there were more sophisticated methods of numerical integration which overcome some of the shortcomings of the Euler method. These methods either explicitly or implicitly include more terms in the Taylor series expansion of the function before projecting to time $t + \Delta t$. We will consider the explicit approach first.

Competition between two species is sometimes modeled by the equations

$$\frac{dX}{dt} = X' = (a - bX - cY)X \tag{6.44}$$

$$\frac{dY}{dt} = Y' = (e - fX - gY)Y \tag{6.45}$$

Here a and e are growth rate constants which, in the absence of other terms, would lead to the exponential growth of both X and Y. The constants b and g reflect the tendency of X and Y, respectively, to compete with themselves, and the constants c and f reflect the impact of interspecies competition on X and Y, respectively. Given values of X and Y at time 0, we could use the right-hand sides of Equations 6.44 and 6.45 to project X and Y using the Euler method. However, by differentiating both sides of each equation, we obtain

$$X'' = (a - bX - cY)X' - (bX' + cY')X \tag{6.46}$$

$$Y'' = (e - fX - gY)Y' - (fX' + gY')Y \tag{6.47}$$

where the values of X' and Y' are obtained from Equations 6.44 and 6.45, respectively. If we go one step further, we obtain

$$X''' = (a - bX - cY)X'' - 2(bX' + cY')X' - (bX'' + cY'')X \tag{6.48}$$

$$Y''' = (e - fX - gY)Y'' - 2(fX' + gY')Y' - (fX'' + gY'')Y \tag{6.49}$$

where the values of X', X'', Y', and Y'' are obtained from Equations 6.44 to 6.47. At this point we have all the information we need to project both functions using terms through $(\Delta t)^3$ in the Taylor series expansion:

$$X(t + \Delta t) = X(t) + X'\Delta t + X''\frac{(\Delta t)^2}{2} + X'''\frac{(\Delta t)^3}{6} \tag{6.50}$$

$$Y(t + \Delta t) = Y(t) + Y'\Delta t + Y''\frac{(\Delta t)^2}{2} + Y'''\frac{(\Delta t)^3}{6} \tag{6.51}$$

We now have an approach which will give us the exact answer for any function which can be expressed as a cubic function of time. The price we have paid to achieve this accuracy is the need to calculate three times as many functions (i.e., X', X'', and X''' versus only X') at each point in time. Could we have achieved comparable accuracy by using Euler's method with Δt reduced by a factor of 3? We will have a chance to explore this question in the Exercises.

RUNGE–KUTTA METHODS

The implicit methods used to include more terms in the Taylor series before making a projection are called Runge–Kutta methods. The classical Runge–Kutta method accounts for terms through $(\Delta t)^4$ in the Taylor series expansion. For the purposes of our discussion, we will examine the rationale behind second-order Runge–Kutta methods and then simply state the equations used in the classical fourth-order Runge–Kutta method.

We will assume that we are dealing with a differential equation of the form

$$\frac{dY}{dt} = Y' = f(Y) \tag{6.52}$$

By applying the chain rule for differentiation of a function of a function, we conclude that

$$Y'' = \frac{df}{dY}\frac{dY}{dt} = f\frac{df}{dY} \tag{6.53}$$

$$Y''' = \frac{df}{dY}\frac{dY}{dt}\frac{df}{dY} + f\frac{dY}{dt}\frac{d^2f}{dY^2} = \left(\frac{df}{dY}\right)^2 f + f^2\frac{d^2f}{dY^2} \tag{6.54}$$

The Taylor series expansion of the function Y is therefore

$$Y(t+\Delta t) = Y(t) + f\Delta t + \left(f\frac{df}{dY}\right)\frac{(\Delta t)^2}{2} + \left(\left(\frac{df}{dY}\right)^2 f + f^2\frac{d^2f}{dY^2}\right)\frac{(\Delta t)^3}{6} + \cdots \tag{6.55}$$

Now suppose we project the function $Y(t)$ forward in time by an amount $\alpha\Delta t$ using the Euler method. We will denote this estimate of the function at time $t + \alpha\Delta t$ by $\hat{Y}(t + \alpha\Delta t)$. Then $\hat{Y}(t + \alpha\Delta t) = Y(t) + f\alpha\Delta t$ and

$$f\left[\hat{Y}(t+\alpha\Delta t)\right] = f(Y(t) + f\alpha\Delta t) = f(Y) + \frac{df}{dY}f\alpha\Delta t + \frac{d^2f}{dY^2}\frac{(f\alpha\Delta t)^2}{2} + \cdots \tag{6.56}$$

Now let us make an estimate of Y at time $t + \Delta t$. We will call this estimate $\hat{Y}(t + \Delta t)$ and calculate it from the following equation:

$$\hat{Y}(t+\Delta t) = Y(t) + \Delta t[Af + Bf(Y(t) + f\alpha\Delta t)] \tag{6.57}$$

Substituting the right-hand side of Equation 6.56 for $f(Y(t) + f\alpha\,\Delta t)$, we obtain

$$\hat{Y}(t+\Delta t) = Y(t) + \Delta t\left[Af + Bf + B\frac{df}{dY}f\alpha\Delta t + B\frac{d^2f}{dY^2}\frac{(f\alpha\Delta t)^2}{2} + \cdots\right] \tag{6.58}$$

Comparing terms in Equations 6.55 and 6.58, we see that $\hat{Y}(t + \Delta t)$ will be correct through second order in Δt if the following conditions are satisfied:

$$A + B = 1 \tag{6.59}$$

$$\alpha B = \frac{1}{2} \tag{6.60}$$

It is impossible for Equation 6.58 to be correct through third order in Δt, because Equation 6.58 is missing the term containing $(df/dY)^2 f$. Equations 6.59 and 6.60 contain three parameters and define only two constraints. Therefore there are an infinite number of ways to satisfy these two conditions, and indeed there are an infinite number of second-order Runge–Kutta schemes. The two most commonly used methods are to set $A = B = \frac{1}{2}$ and $\alpha = 1$ or to set $A = 0$, $B = 1$, and $\alpha = \frac{1}{2}$. The former method is called the *improved Euler method* and the latter the *modified Euler method*. In the case of the improved Euler method, we have

$$\hat{Y}(t + \Delta t) = Y(t) + \frac{\Delta t}{2}[f(t) + f(Y(t) + f(t)\,\Delta t)] \tag{6.61}$$

an equation which should remind us of the trapezoid rule. In the case of the modified Euler method, we have

$$\hat{Y}(t + \Delta t) = Y(t) + \Delta t f\left(Y(t) + f(t)\frac{\Delta t}{2}\right) \tag{6.62}$$

an equation which should remind us of the midpoint rule. The classical Runge–Kutta method is analogous to Simpson's rule and is accurate through fourth order in Δt. The projection is made using the following equation:

$$\hat{Y}(t + \Delta t) = Y(t) + \frac{\Delta t}{6}[f_1 + 2f_2 + 2f_3 + f_4] \tag{6.63}$$

where

$$f_1 = f(Y(t))$$

$$f_2 = f\left(Y(t) + \frac{\Delta t}{2} f_1\right)$$

$$f_3 = f\left(Y(t) + \frac{\Delta t}{2} f_2\right)$$

$$f_4 = f(Y(t) + \Delta t f_3)$$

f_1 is obviously the slope of the function at time t. f_2 and f_3 are both estimates of the slope of the function at time $t + \Delta t/2$, in the former case by projecting Y from time t to $t + \Delta t/2$ using the original slope f_1 and in the latter case using f_2. f_4 is an estimate of the slope of the function at time $t + \Delta t$ obtained by projecting Y from time t to $t + \Delta t$ using the slope f_3.

EXERCISES

6.1 We have shown (Eq. 6.8) that

$$W(B) = W(A) + (B - A)W'(A) + \frac{(B-A)^2}{2}W''(A) + \frac{(B-A)^3}{6}W'''(A) + \frac{1}{6}\int_A^B \frac{(B-X)^3 d^4W(X)}{dX^4}\,dX$$

Let us see if we can derive the rest of the Taylor series by induction. Show that if

$$W(B) = W(A) + \sum_{n=1}^{N} \frac{W^{(n)}(A)(B-A)^n}{n!} + \frac{1}{N!}\int_A^B (B-X)^N \left(\frac{d}{dX}\right)^{N+1} W(X)\,dX$$

then

$$W(B) = W(A) + \sum_{n=1}^{N+1} \frac{W^{(n)}(A)(B-A)^n}{n!} + \frac{1}{(N+1)!}\int_A^B (B-X)^{N+1} \left(\frac{d}{dX}\right)^{N+2} W(X)\,dX$$

6.2 Show that

$$\int_0^4 \frac{X}{1+X}\,dX = 4 - \ln(5)$$

What would be the estimated value of this integral if we used Equation 6.10 to approximate the integrand and included terms through $(X-2)^4$ in the Taylor series? Why do odd powers of $X - 2$ in the expansion contribute nothing to the value of the integral?

6.3 In Equation 6.42 we considered the problem of numerically integrating the equation $dF/dt = kF$. We also saw in Equations 6.46 to 6.51 how we could include more terms in the Taylor series expansion for a simple competing species model. We then asked whether including these extra terms was worth the effort, or whether it might be sufficient to simply use the Euler method with a smaller Δt. Let us explore this question with the equation $dF/dt = F' = F$.

Calculate F'' and F''' for this model. Let $F(0)$ be 1. Let $\Delta t = X$. What is the value of $F(X)$ projected using the first three terms in the Taylor series expansion of $F(X)$ about $F(0)$? If we repeat this process n times, what will be the predicted value of $F(nX)$? What will be the ratio of the predicted value to the true value of $F(nX)$. [The true value of $F(nX)$ is e^{nX}.) Suppose $X = 0.3$ and $n = 100$. Then $nX = 30$. What is the ratio of the predicted to the true value of $F(30)$?

Now suppose we project F using the Euler method, but with a time step three times smaller. In this case $X = 0.1$, and $nX = 30$ implies $n = 300$. Does the choice $X = 0.1$ satisfy the condition that $\Delta t \ll |2F'/F''|$? What is the ratio of the approximate to the true value of $F(30)$ if we use the Euler method to project F 300 times with $X = 0.1$ each time? What value of X gives about the same accuracy as the third-order Taylor series projection method for estimating $F(30)$? In answering this last question, remember to keep the product of n and X equal to 30.

6.4 We noted that the modified and improved Euler methods bear a remarkable similarity to the midpoint and trapezoid rules, respectively, and that the classical fourth-order Runge–Kutta method is very similar to Simpson's rule. We have seen that the midpoint and trapezoid rules will exactly integrate a linear function of X and that Simpson's rule will exactly integrate a cubic polynomial. The modified and improved Euler methods are, however, second-order Runge–Kutta methods, and the classical Runge–Kutta method is a fourth-order method. How would you account for the fact that Runge–Kutta methods for making temporal projections of a function seem to be one order better than their counterparts for carrying out spatial integrations?

6.5 The *logistic equation* is sometimes used to model the dynamics of animal populations. The equation takes the form

$$\frac{dN}{dt} = N' = rN\left(1 - \frac{N}{K}\right) \tag{6.64}$$

Here K is the nonzero equilibrium value of N (i.e., $dN/dt = 0$ when $N = K$) and is referred to as the carrying capacity of the system. The parameter r is the exponential growth rate of N and has units of inverse time. $dN/dt \cong rN$ when $N \ll K$. Suppose we wanted to integrate this equation using the simple Euler method. In this case we would want to ensure that $\Delta t \ll |2N'/N''|$. Assume that N_0 is the initial value of N. What will be the minimum value of $|2N'/N''|$ during the course of the integration if the integration continues until $N \approx K$? (*Note:* N_0 can be larger or smaller than K. Does your answer to this question depend on whether N_0 is larger or smaller than K?)

We can nondimensionalize this problem by letting $\tau = rt$ and $X = N/K$. Making these substitutions, the differential equation becomes

$$\frac{dX}{d\tau} = X(1 - X) \tag{6.65}$$

Let $X_0 = 0.1$ be the initial value of X. Try integrating Equation 6.65 from $\tau = 0$ to $\tau = 10$ using the simple Euler method with a $\Delta\tau$ of 0.1, 0.5, and 1. Which of these values of $\Delta\tau$ would you say satisfy the criterion that $\Delta\tau \ll |2X'/X''|$. Graph your results and compare them with the exact solution, which is

$$X = \frac{X_0 e^{\tau}}{1 + X_0(e^{\tau} - 1)} \tag{6.66}$$

How would you account for the difference in the comparisons of approximate and exact solutions in this case versus Exercise 6.3?

6.6 Predator–prey models are sometimes used to study/understand the population dynamics of animal populations. A simple predator–prey model is the following:

$$\frac{dX}{dt} = X' = (a - cY)X \tag{6.67}$$

$$\frac{dY}{dt} = Y' = (-e + fX)Y \tag{6.68}$$

In these Equations a is the exponential growth rate of the prey, X, cY is the grazing rate of the predator, Y, on the prey, e is the death rate of the predator, and fX is the rate of increase of predator biomass. Assume that we wish to model the dynamics of X and Y and are told that $a = 1$, $c = 0.1$, $e = 0.5$, and $f = 0.02$. Assume that at time 0, $X = 20$, and $Y = 4$. What are the nonzero equilibrium values of X and Y? What are the values of X' and Y' at time 0? Derive the equations for X'' and Y'' and evaluate them at time 0. What are the values of $|2X'/X''|$ and $|2Y'/Y''|$ at time 0? Integrate Equations 6.67 and 6.68 from time 0 to time 80 using the simple Euler method and $\Delta t = 0.04$. Now integrate the same equations using Equation 6.63, again with $\Delta t = 0.04$. Graph the results in both cases. How would you account for the difference in the results you have obtained with the two methods of numerical integration?

6.7 Given the following criteria:

$$W_1 + W_2 = M_0$$
$$W_1X_1 + W_2X_2 = M_1$$
$$W_1X_1^2 + W_2X_2^2 = M_2$$
$$W_1X_1^3 + W_2X_2^3 = M_3$$

show that the abscissas X_1 and X_2 are the roots of the quadratic equation

$$(M_0M_2 - M_1^2)X^2 + (M_1M_2 - M_0M_3)X + (M_1M_3 - M_2^2) = 0.$$

Hint: Multiply the first equation by X_1 and subtract it from the second equation. Multiply the second equation by X_1 and subtract it from the third equation. Multiply the third equation by X_1 and subtract it from the fourth equation. By dividing the first two equations you have obtained in this way, you should be able to show that

$$X_2 = \frac{M_1X_1 - M_2}{M_0X_1 - M_1}$$

By similarly dividing the second and third equations you obtained in this way, you should be able to show that

$$X_2 = \frac{M_2X_1 - M_3}{M_1X_1 - M_2}$$

Equating these two expressions for X_2, you should be able to prove that X_1 solves the given quadratic equation.

6.8 Assume that the photosynthetic rate varies as a function of irradiance according to the equation

$$P = \frac{11.5I}{10.8 + I}$$

where P is the photosynthetic rate in mg C m^{-3} d^{-1} and I is the irradiance in mol-quanta m^{-2} d^{-1}. Assume that light intensity varies as a function of depth in the water column according to the equation

$$I = 50e^{-kz}$$

where $k = 0.04$ m^{-1} and z is the depth in meters. Estimate the value of the integral $\int_0^{200\ \mathrm{m}} P(z)\, dz$ using the following numerical integration methods:

a. trapezoid rule with five equally spaced intervals

b. midpoint rule with five equally spaced intervals

c. Legendre–Gauss quadrature with five sampling depths

Find the exact value of the integral to three significant figures by integrating using the midpoint rule with successively smaller values of Δz until the answer converges to three significant figures.

7

BOX MODELS

Box models are representations of a system in which quantities or concentrations of materials are depicted with boxes or similar closed figures and the flow or flux of materials between boxes is represented with arrows. Figure 7.1, for example, is a simple representation of the flow of an essential nutrient through a food chain. The boxes in this case represent trophic levels, and the arrows depict the direction of transfer of the nutrient associated with various processes. The solid lines depict grazing, the dashed lines depict respiration, and the dash–dot lines depict excretion.

Diagrams such as Figure 7.1 help to make it clear what components and processes a modeler has incorporated into a model. This may seem like a very simple procedure, but making a decision about what to include and what to exclude from a model can be a very important and nontrivial exercise. All models are simplifications of the real world. The first task of the modeler is to decide what questions he or she will address with the model. The modeler must then design a model which is simple enough to be solved with a reasonable amount of time and effort and not so simple that some important characteristics of the system are omitted or improperly represented. A model which is hopelessly complicated is of no more use than a model which is overly simplified.

Once the modeler has decided what components and processes to include in the model, he or she must formulate mathematical equations to describe the transfers depicted by the arrows in the model. This is a step where scientific understanding and lack of understanding about the system in question become an issue. The modeler may discover, for example, that there is little or no information about some of the critical processes which he or

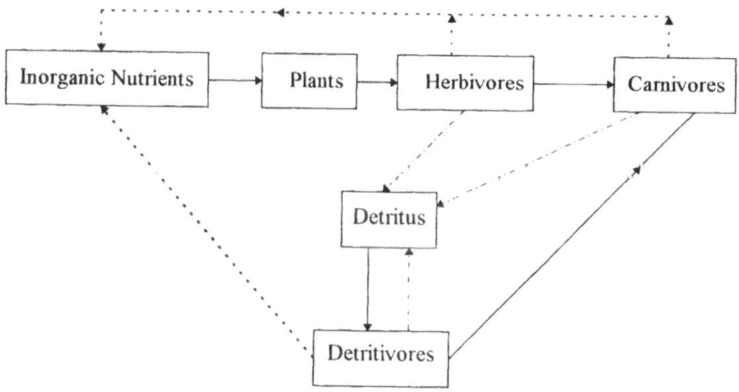

Figure 7.1 Box model of the flow of an essential nutrient through a food chain.

she hopes to describe in the model. In fact, one of the important functions of model building may sometimes be to help focus attention on areas where more understanding is needed.

Assuming the modeler has enough information to formulate mathematical equations describing the processes in the model, what does he or she do with the model? Models are usually formulated for one of two reasons. The first is to gain some insight, and, in particular, to acquire some predictive ability, about the dynamics of a system. Box models are commonly used, for example, to predict the impact of fishing pressure on the stocks of commercially important fish. The implications of such models have been used to guide management decisions with respect to fisheries (Pauly et al., 1984). On a smaller scale box models have been used to describe the temporal pattern of concentration and isotope ratio of ammonium as a result of uptake and regeneration by microorganisms (Laws, 1984). Such models are used to analyze and interpret experimental data. The second major purpose of box models has been to predict the flows and distribution of materials or concentrations in a system at steady state. Such models have been used, for example, to examine the relative importance of grazing and nutrient limitation in controlling the standing crop and growth rates of phytoplankton in the ocean (Frost and Franzen, 1992). As we will see, they may also be used to examine the resiliency of a system to perturbations.

GEOCARB II

The example of a box model for studying the dynamics of a system is taken from the geochemical literature. The model, developed by Robert Berner and called GEOCARB, has been used to examine the importance of various natural processes in controlling the concentration of CO_2 in the atmosphere.

This is a timely subject, since anthropogenic emissions of CO_2, primarily the result of fossil fuel burning, are presently increasing the CO_2 concentration of the atmosphere at an unprecedented rate. There is concern that this increase in atmospheric CO_2 could lead to global warming of several degrees Celsius due to a greenhouse effect. It is therefore of interest to know what natural processes are at work to regulate atmospheric CO_2 and how these processes have influenced the concentration of CO_2 in the atmosphere and the temperature of the earth over geologic time.

The original version of the GEOCARB model was described by Berner in 1991 (Berner, 1991). The model we will examine is an updated version described in 1994 (Berner, 1994). Figure 7.2 is a box model depicting the components and fluxes in the system. There are three reservoirs of carbon in the model: (1) carbon in the form of carbonate minerals (calcite, aragonite, and dolomite) precipitated in the oceans; (2) carbon in the form of CO_2 in the atmosphere and (primarily) HCO_3^- in the oceans, and (3) carbon in the form of organic matter in sediments. The fundamental processes involved in the model are (1) the uptake of CO_2 from the atmosphere and its transformation to dissolved HCO_3^- during the weathering of silicate rocks, with the HCO_3^- soon precipitated in the ocean as carbonate minerals, (2) upon burial at great depths, the thermal breakdown of carbonate minerals via metamorphism and magmatism, with the transfer of the resulting CO_2 back to the ocean–atmosphere system, and (3) the burial of organic matter in sediments, its weathering on the continents, and its thermal breakdown via diagenesis, metamorphism, and magmatism followed by oxidation of reduced gases to CO_2 (Berner, 1991). These processes can be summarized by the following reactions:

$$CO_2 + CaSiO_3 \underset{\substack{\text{metamorphism}\\ \text{magmatism}}}{\overset{\text{weathering}}{\rightleftharpoons}} CaCO_3 + SiO_2 \tag{7.1}$$

$$CO_2 + MgSiO_3 \underset{\substack{\text{metamorphism}\\ \text{magmatism}}}{\overset{\text{weathering}}{\rightleftharpoons}} MgCO_3 + SiO_2 \tag{7.2}$$

$$CO_2 + H_2O \underset{\substack{\text{weathering}\\ \text{metamorphism}\\ \text{plus air oxidation}}}{\overset{\text{organic burial}}{\rightleftharpoons}} CH_2O + O_2 \tag{7.3}$$

Equations 7.1 and 7.2 constitute the major controls on atmospheric CO_2 on a geologic time scale and are referred to as the Urey reactions (Urey, 1952). When the Urey reactions proceed from left to right, they describe the weathering of silicate rocks, which removes carbon from the ocean–atmosphere system by fixing it in carbonates. The return of this carbon to the ocean–atmosphere via metamorphism and volcanism (the Urey reactions

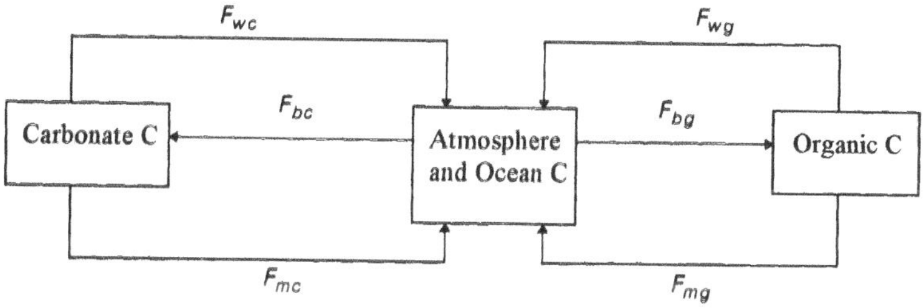

Figure 7.2 Box model representing the reservoirs and processes in the GEOCARB model.

proceeding from right to left) may take tens of millions to hundreds of millions of years. This delay can result in large changes in atmospheric CO_2 (Berner, 1991).

An important point about the carbon cycle depicted in Figure 7.2 is that the amount of carbon stored in the ocean–atmosphere system is small compared to the amount of carbon in the carbonate and organic carbon reservoirs. Current estimates indicate, for example, that the amounts of carbon in the ocean and atmosphere are less than 1% and no more than about 0.01%, respectively, of the carbon stored as either carbonates or organic carbon (Longhurst, 1991). It is reasonable to assume that the carbonate and organic carbon reservoirs have been large compared to the carbon stored in the ocean–atmosphere system over Phanerozoic time, that is, since the beginning of the Cambrian, about 570 million years ago. In formulating the carbon budget within the context of the model (Figure 7.2), one therefore assumes that changes in the amount of carbon in the ocean–atmosphere box are small compared to the changes in carbonate and organic carbon. In other words, one assumes that any change in the amount of organic carbon is balanced by an equal and opposite change in the amount of carbonate carbon. This does not mean that changes in the amount of carbon in the ocean–atmosphere system are unimportant. Changes in the concentration of atmospheric CO_2 are believed to have an important influence on the climate of the earth and on the rates of the Urey reactions, as we will see. Important as these changes may be in determining chemical reaction rates, they are still small compared to the changes in organic and carbonate carbon which occur on a time scale of one million years.

In formulating the mathematical equations describing the model, Berner (1994) considered changes in both the amount of ^{12}C and ^{13}C in the organic and carbonate carbon reservoirs. Since information on the isotopic composition of carbonates undergoing burial and the fractionation factor between organic matter and carbonates as a function of time are available to some extent on a geologic time scale, inclusion of both ^{12}C and ^{13}C in the mass balance equations proved a useful way to constrain the model. The basic equations of the GEOCARB II model are as follows:

$$F_{wc} + F_{mc} + F_{wg} + F_{mg} = F_{bc} + F_{bg} \tag{7.4}$$

$$\delta_c(F_{wc} + F_{mc}) + \delta_g(F_{wg} + F_{mg}) = \delta_{bc}F_{bc} + (\delta_{bc} - \alpha_c)F_{bg} \tag{7.5}$$

$$F_{wc} = f_{BB}(CO_2)f_{LA}(t)f_D(t)f_E(t)k_{wc}C \tag{7.6}$$

$$F_{wg} = f_D(t)f_R(t)k_{wg}G \tag{7.7}$$

$$F_{mc} = f_G(t)f_C(t)F_{mc}(0) \tag{7.8}$$

$$F_{mg} = f_G(t)F_{mg}(0) \tag{7.9}$$

$$\frac{dC}{dt} = F_{bc} - (F_{wc} + F_{mc}) \tag{7.10}$$

$$\frac{dG}{dt} = F_{bg} - (F_{wg} + F_{mg}) \tag{7.11}$$

$$\frac{d(\delta_C C)}{dt} = \delta_{bc}F_{bc} - \delta_c(F_{wc} + F_{mc}) \tag{7.12}$$

$$\frac{d(\delta_g G)}{dt} = (\delta_{bc} - \alpha_c)F_{bg} - \delta_g(F_{wg} + F_{mg}) \tag{7.13}$$

$$F_{wsi} = F_{bc} - F_{wc} = f_B(CO_2)f_R(t)f_E(t)f_D(t)^{0.65}F_{wsi}(0) \tag{7.14}$$

where:

F_{wc}; F_{wg} = rate of release of carbon to the ocean–atmosphere–biosphere system via the weathering of carbonates (c) and organic matter (g)

F_{mc}; F_{mg} = rate of release of carbon to the ocean–atmosphere–biosphere system via the metamorphic/volcanic/diagenetic breakdown of carbonates (c) and organic matter (g)

F_{bc}; F_{bg} = burial rate of carbon as carbonates (c) and organic matter (g) in sediments

F_{wsi} = rate of uptake of atmospheric CO_2 via the weathering of Ca and Mg silicates followed by precipitation of the Ca and Mg as carbonates (Urey reaction)

$f_{BB}(CO_2)$ = dimensionless feedback function for carbonates expressing the dependence of weathering on atmospheric CO_2

$f_B(CO_2)$ = dimensionless feedback function for silicates expressing the dependence of weathering on atmospheric CO_2

$f_{LA}(t)$ = carbonate land area(t)/carbonate land area(0) equal to $f_A(t)f_L(t)$, where $f_A(t)$ = total land area(t)/total land area(0) and $f_L(t)$ = [carb/total](t)/[carb/total](0)

$f_D(t)$ = river runoff(t)/river runoff(0) due to changes in paleogeography. Runoff is per unit area [dependence of runoff on atmospheric CO_2 is included in $f_B(CO_2)$]

$f_E(t)$ dimensionless parameter expressing the dependence of weathering rate on soil biological activity due to land plants [$f_E(t) = 1$ at present]

$f_G(t)$ = degassing rate(t)/degassing rate(0) due to tectonics

$f_C(t)$ = dimensionless parameter expressing the dependence of degassing rate on the relative proportions of carbonates deposited in shallow water and in the deep sea [$fC(t) = 1$ at present]

$f_R(t)$ = uplift factor; mean continental relief(t)/mean continental relief(0)

δ = δ^{13}C value (‰); δ_c average value for all carbonates, δ_g average value for all organic matter, and δ_{bc} value for carbonates being buried at time t ($\approx$ shallow oceanic value)

α_c = isotope fractionation factor between organic matter and carbonates (‰)

k_{wc}; k_{wg} = rate constants for weathering of carbonates and organics

C; G = masses of carbon as carbonates and as organic matter

Equations 7.10 and 7.11 simply state that the rates of change of C and G equal the net flux of carbon into the C and G reservoirs, respectively. The equations follow directly from the box model, Figure 7.2. Equation 7.4 in effect states that the sum of the carbon in C and G is constant. This is the standard GEOCARB assumption, but as Berner (1994) has noted, the behavior of the model changes very little if this assumption is relaxed, that is, if the total amount of carbon in the crustal system is allowed to vary.

Equations 7.4, 7.10 and 7.11 apply to ^{12}C, which accounts for about 99% of naturally occurring carbon. Equations 7.5, 7.12 and 7.13 are the analogs of Equations 7.4, 7.10 and 7.11 for ^{13}C. If the ratio $^{13}C/^{12}C$ were the same in all forms of carbon, there would be no point in including additional equations describing the fluxes of ^{13}C. The fact is, however, that the $^{13}C/^{12}C$ ratio varies from one form of carbon to another, since biological and geochemical processes tend to alter or fractionate this ratio; and we do have some information on how the ratio has changed in different forms of carbon over geologic time. To derive Equation 7.5, for example, we note that the balance of ^{13}C fluxes can be obtained from the balance of ^{12}C fluxes by multiplying each term in Equation 7.4 by the $^{13}C/^{12}C$ ratio associated with the particular flux. In the model the $^{13}C/^{12}C$ ratios associated with Fwc and Fmc are assumed to be identical, as are the $^{13}C/^{12}C$ ratios associated with Fwg and Fmg. Hence the ^{13}C balance analogous to Equation 7.4 takes the form

$$\left(\frac{^{13}C}{^{12}C}\right)_c (F_{wc} + F_{mc}) + \left(\frac{^{13}C}{^{12}C}\right)_g (F_{wg} + F_{mg}) = \left(\frac{^{13}C}{^{12}C}\right)_{bc} F_{bc} + \left(\frac{^{13}C}{^{12}C}\right)_{bc} F_{bg} \tag{7.15}$$

In practice $^{13}C/^{12}C$ ratios are always reported relative to a standard, which happens to be the $^{13}C/^{12}C$ ratio in Cretaceous belemnite from the PeeDee formation in South Carolina, generally designated PDB. The PDB $^{13}C/^{12}C$ ratio is 0.0112372. The relative $^{13}C/^{12}C$ ratio is designated $\delta^{13}C$ and is calculated from the equation

$$\delta^{13}C = \left[\frac{(^{13}C/^{12}C)\ \text{sample}}{(^{13}C/^{12}C)\ \text{standard}} - 1\right] \times 1000 \tag{7.16}$$

The units of $\delta^{13}C$ are per mil, that is, per thousand. With a little algebra it is straightforward to show that Equation 7.15 may be rewritten by replacing each $^{13}C/^{12}C$ ratio with its corresponding $\delta^{13}C$, since due to Equation 7.4 the extra terms arising from the substitutions cancel out. Equation 7.15 then becomes

$$\delta_c(F_{wc} + F_{mc}) + \delta_g(F_{wg} + F_{mg}) = \delta_{bc}F_{bc} + \delta_{bg}F_{bg} \tag{7.17}$$

Equation 7.5 follows directly from Equation 7.17, since $\alpha_c \cong \delta_{bc} - \delta_{bg}$.

Equations 7.12 and 7.13 may similarly be shown to follow from equations describing the rate of change of ^{13}C in the C and G reservoirs, respectively. However, Equations 7.12 and 7.13 are not in a convenient form for our purposes, since it is really the functions δ_c and δ_g that we want to follow over time. We can obtain the derivatives of δ_c and δ_g by applying the rule for the derivative of the product of two functions (Chapter 1) to the left-hand sides of Equations 7.12 and 7.13. We obtain

$$\frac{d(\delta_c C)}{dt} = C\frac{d\delta_c}{dt} + \delta_c\frac{dC}{dt} = \delta_{bc}F_{bc} - \delta_c(F_{wc} + F_{mc}) \tag{7.18}$$

$$\frac{d(\delta_g G)}{dt} = G\frac{d\delta_g}{dt} + \delta_g\frac{dG}{dt} = (\delta_{bc} - \alpha_c)F_{bg} - \delta_g(F_{wg} + F_{mg}) \tag{7.19}$$

Substituting the right-hand sides of Equations 7.10 and 7.11 for dC/dt and

dG/dt, respectively, in Equations 7.18 and 7.19, we obtain

$$\frac{d\delta_c}{dt} = \frac{F_{bc}(\delta_{bc} - \delta_c)}{C} \tag{7.20}$$

$$\frac{d\delta_g}{dt} = \frac{F_{bg}(\delta_{bg} - \delta_g)}{G} \tag{7.21}$$

The remaining equations in the model, Equations 7.6 to 7.9 and 7.14, require some explanation and elaboration. On the time scale of one million years, the rate of burial of carbon as carbonates in sediments, F_{bc}, is assumed to be balanced by the rate of weathering of carbon from silicates, F_{wsi}, and from carbonates, F_{wc}. Hence $F_{wsi} = F_{bc} - F_{wc}$. Both F_{wsi} and F_{wc} are assumed to be functions of the amount of CO_2 in the atmosphere. The functions f_B and f_{BB} account for the effects of atmospheric CO_2 on F_{wsi} and F_{wc} due to the influence of CO_2 on air surface temperature and plant growth. Global circulation models indicate that the mean surface air temperature of the earth over Phanerozoic time can be described by the equation

$$T_t = T_0 + \Gamma \ln(\mathrm{RCO_2}) - W_s(t/570) \tag{7.22}$$

where T_t is the temperature in degrees Celsius at time t millions of years before the present, Γ and W_s are constants, and $\mathrm{RCO_2}$ is the ratio of the level of atmospheric CO_2 in the past at a given time to that at present. The term involving $\ln(\mathrm{RCO_2})$ simulates the greenhouse effect of atmospheric CO_2 on surface air temperature. The term $W_s(t/570)$ simulates the effect of changes in the solar "constant," that is, the output of energy from the sun, which is believed to have increased by about 5% since the beginning of the Cambrian. Temperature influences weathering rates through its effect on dissolution rates (Arrhenius equation) and on rainfall as reflected in river runoff. In the case of the dissolution rate D, the relationship is

$$D(T_t) = D(T_0)\exp[0.09(T_t - T_0)] \tag{7.23}$$

and, in the case of river runoff R,

$$R(T_t) = R(T_0)[1 + 0.038(T_t - T_0)] \tag{7.24}$$

Empirically, one finds that the overall weathering rate is directly proportional to D and proportional to R raised to approximately the 0.65 power. The 0.65 power dependence on R reflects the fact that more runoff implies more dissolution, but empirically one finds that the concentration of dissolved inorganic carbon in rivers draining only silicate rocks varies as R raised to the -0.35. In other words, there is a dilution effect. The flux, which is proportional to R times the inorganic carbon concentration, is therefore proportional to R raised to the $1 - 0.35 = 0.65$ power. After substituting the right-hand side of Equation 7.22 for T_t in Equations 7.23 and 7.24, one concludes that temperature changes caused by changes in the energy output of the sun and the concentration of CO_2 in the atmosphere will cause f_B to depend on time and RCO_2 as follows:

$$f_B = \exp[-0.09W_s(t/570)](RCO_2)^{0.09\Gamma} \times [1 + 0.038\Gamma \ln(RCO_2) - 0.038W_s(t/570)]^{0.65} \quad (7.25)$$

where the first term in the product on the right-hand side of Equation 7.25 simulates the effect of temperature on dissolution rates, and the second term simulates the effect of temperature on river runoff.

To complete the characterization of f_B, we need to know the effect of atmospheric CO_2 on plant growth. The reason is that empirical studies have demonstrated a dramatic increase in silicate mineral weathering rates in vegetated versus nonvegetated regions of the same watershed. For example, field studies on the effects of higher plants on the weathering rate of basalts in Hawaii have demonstrated a large effect of plant rootlets on the initial (centuries to millennia) dissolutive attack of the basalt (Cochran and Berner, 1993). Under present conditions about 35% of plants respond to CO_2 fertilization. GEOCARB assumes that the response is a hyperbolic function of CO_2 concentration and that when plants respond the maximum effect of CO_2 fertilization is a factor of 2 increase in weathering rate. The maximum increase in weathering rate is therefore about a factor of $0.65 + 2(0.35) = 1.35$. There are several possible ways to model this response; GEOCARB assumes that the response takes the form $[2RCO_2/(1 + RCO_2)]^{0.4}$. Given this formulation, the maximum increase in silicate weathering, as $RCO_2 \to \infty$, would be a factor of $2^{0.4} = 1.32$.

The complete equation for f_B *with vascular land plant fertilization* then takes the form

$$f_B = \exp[-0.09W_s(t/570)](RCO_2)^{0.09\Gamma} \times [1 + 0.038\Gamma \ln(RCO_2) - 0.038W_s(t/570)]^{0.65} \times [2RCO_2/(1 + RCO_2)]^{0.4} \quad (7.26)$$

The period from the beginning of the Cambrian to roughly the end of the Devonian (345 million years ago) represents a time before vascular land plants had spread to upland areas and made their influence on weathering felt. During this time weathering was probably mediated by primitive microorganisms, and it is unclear how to formulate the direct effect of CO_2 on the kinetics of the initial dissolution process. If the consumption of CO_2 by weathering processes were very rapid relative to its rate of supply via diffusion or percolation of rainwater, then one would expect the reaction rate to be directly proportional to the concentration of CO_2 in the atmosphere. If the reaction rate were slow compared to the rate of supply of CO_2, then the rate of dissolved CO_2 uptake would have been dependent on reaction kinetics. An analysis of the reaction kinetics indicates that under these conditions the uptake rate would have been proportional to the concentration of atmospheric CO_2 raised to the 0.25 power. GEOCARB takes a position intermediate between these two extremes and assumes that prior to 350 million years ago the rate of dissolved CO_2 uptake associated with silicate mineral weathering was proportional to the square root of the concentration of atmospheric CO_2. Hence *before vascular land plants*

$$f_B = \exp[-0.09W_s(t/570)](\mathrm{RCO_2})^{0.09\Gamma} \times [1 + 0.038\Gamma \ln(\mathrm{RCO_2}) - 0.038W_s(t/570)]^{0.65} \times [\mathrm{RCO_2}]^{0.5} \tag{7.27}$$

The treatment of f_{BB} is similar with two important modifications. First, it is assumed that carbonates saturate the soil or groundwater in contact with them so rapidly that changes in runoff do not result in appreciable dilution. Hence the weathering of carbonates is directly proportional to river runoff. Second, based on the studies of Berner et al. (1983), the combined effects of temperature on the initial dissolution and transport of carbonates can be approximated using the following linear function of temperature:

$$R(T_t)D(T_t) = R(T_0)D(T_0)[1 + 0.087(T_t - T_0)] \tag{7.28}$$

Again substituting the right-hand side of Equation 7.22 for T_t, we conclude that *with vascular land plant fertilization*

$$f_{BB} = [1 + 0.087\Gamma \ln(\mathrm{RCO_2}) - 0.087W_s(t/570)][2\mathrm{RCO_2}/(1 + \mathrm{RCO_2})]^{0.4} \tag{7.29}$$

and *before vascular land plants*

$$f_{BB} = [1 + 0.087\Gamma \ln(\mathrm{RCO_2}) - 0.087W_s(t/570)][\mathrm{RCO_2}]^{0.5} \tag{7.30}$$

where in each case the first term in brackets is $R(T_t)D(T_t)/R(T_0)D(T_0)$ and the second term is the CO_2 fertilization effect.

Equations 7.26 and 7.29 apply from 300 million years ago to the present. Equations 7.27 and 7.30 apply from the beginning of the Phanerozoic to 350 million years ago. Between 350 and 300 million years ago GEOCARB II uses a linear mix of these equations:

$$f_B = \left[\frac{350 - t}{50}\right] f_B(7.26) + \left[\frac{t - 300}{50}\right] f_B(7.27) \tag{7.31}$$

$$f_{BB} = \left[\frac{350 - t}{50}\right] f_{BB}(7.29) + \left[\frac{t - 300}{50}\right] f_{BB}(7.30) \tag{7.32}$$

The other f parameters, that is, f_{LA}, f_D, f_R, f_E, f_G, and f_C, as well as δ_{bc} and α_c are shown in Figures 7.3 to 7.10 and tabulated at 10-million-year intervals in Table 7.1.

The parameter f_{LA} appears only in the equation for F_{wc}, Equation 7.6. It has varied by roughly a factor of 2 during Phanerozoic time, rising in an irregular manner during the Paleozoic, falling during much of the Mesozoic, and rising in an irregular manner to its present value during the last 100 million years. The changes in f_{LA} have been caused in part by the rise and fall of sea level and the concomitant covering and exposure of continental shelves, coastal plains, and interior lowlands. The other factor contributing to the changes in f_{LA} is the variable percentage of land area underlain by carbonates, discussed by Bluth and Kump (1991).

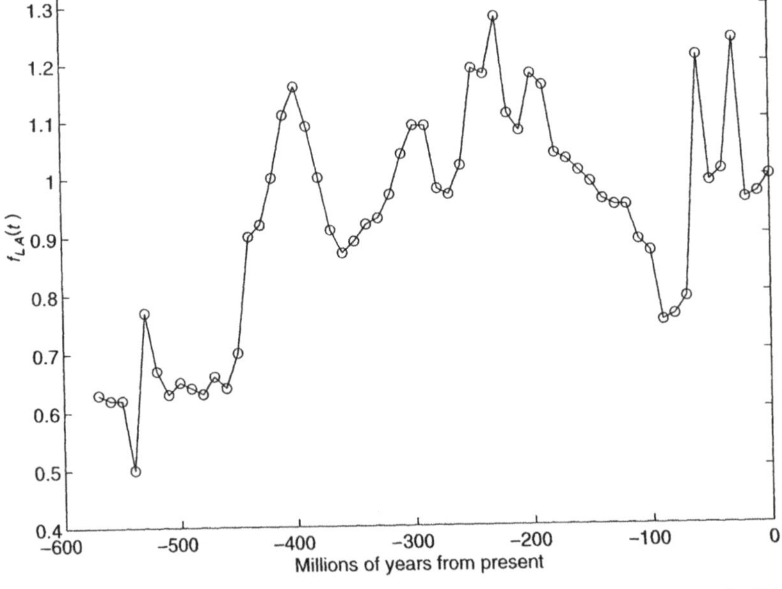

Figure 7.3 Dimensionless parameter f_{LA} versus time from data in Table 7.1.

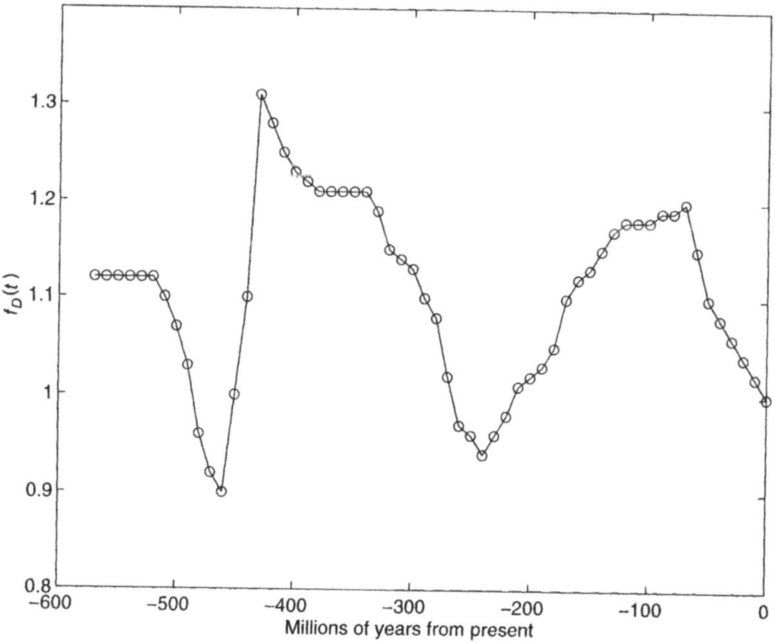

Figure 7.4 Dimensionless parameter f_D versus time from data in Table 7.1.

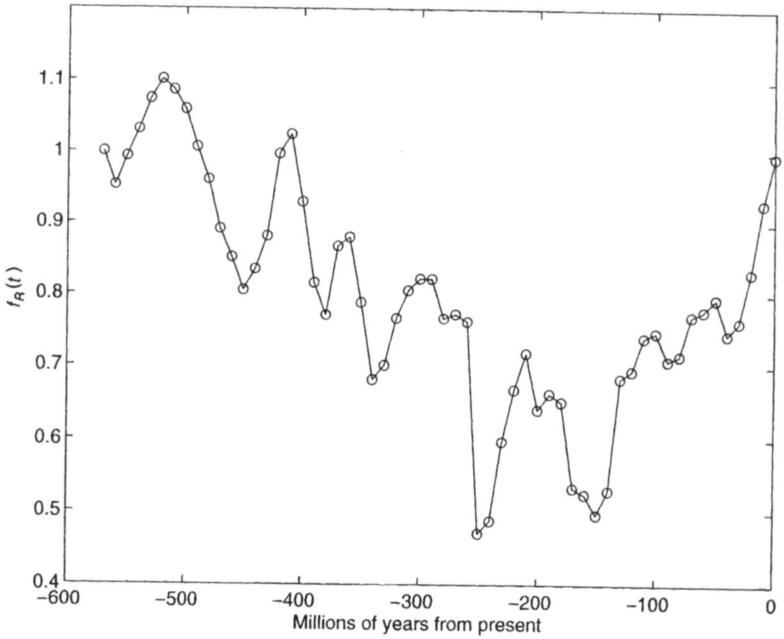

Figure 7.5 Dimensionless parameter f_R versus time from data in Table 7.1.

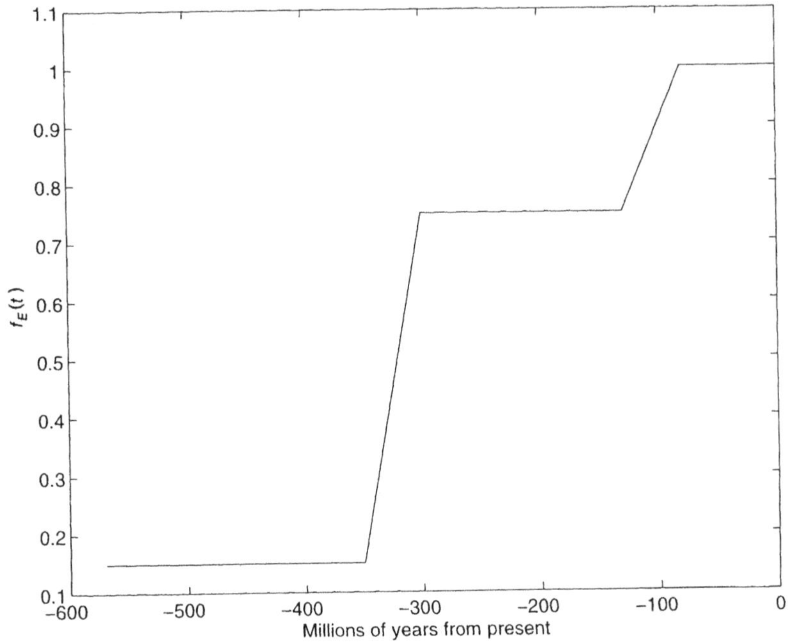

Figure 7.6 Dimensionless parameter f_E versus time from data in Table 7.1.

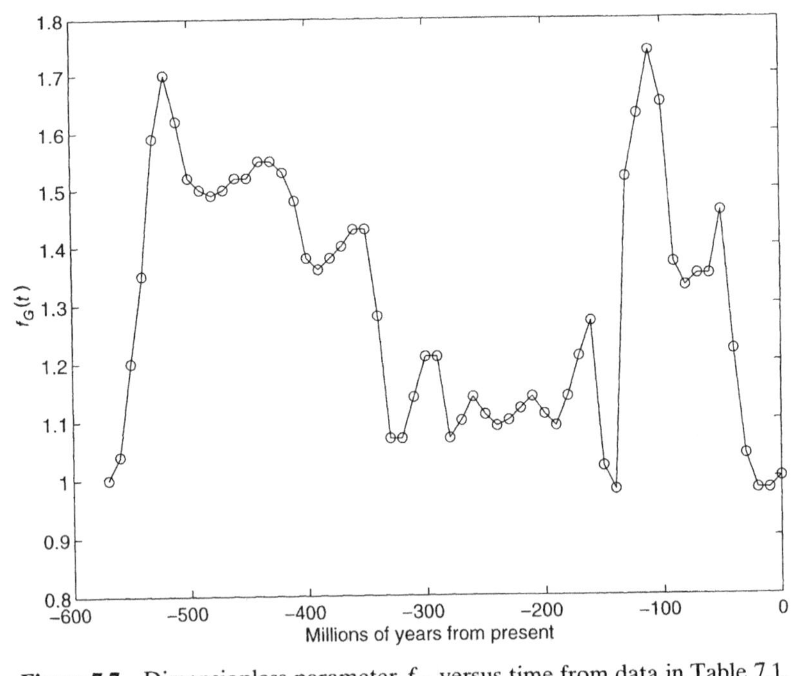

Figure 7.7 Dimensionless parameter f_G versus time from data in Table 7.1.

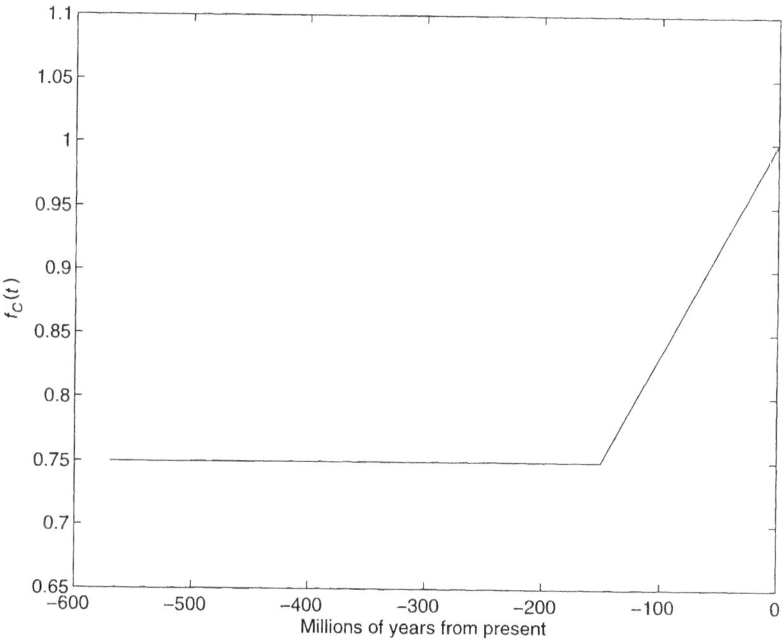

Figure 7.8 Dimensionless parameter f_C versus time from data in Table 7.1.

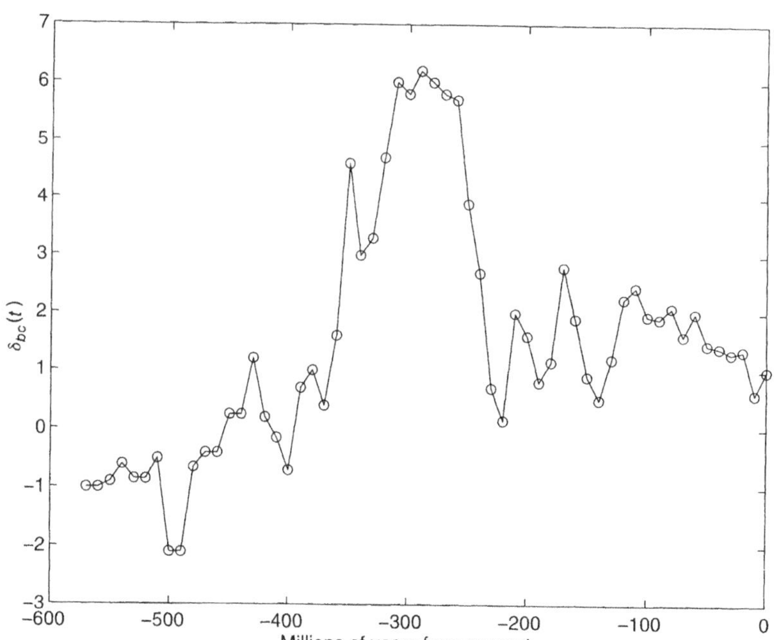

Figure 7.9 δ_{bc} versus time from data in Table 7.1. Units of δ_{bc} are per mil.

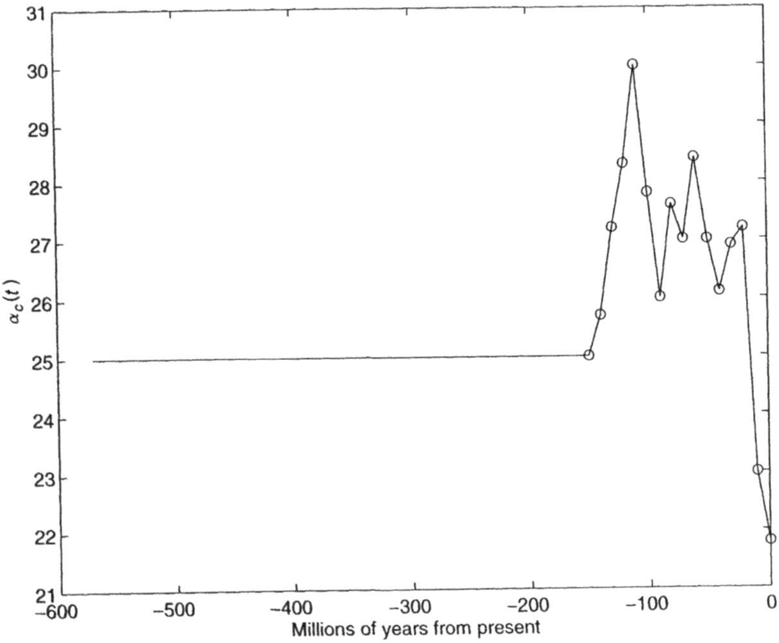

Figure 7.10 α_c versus time from data in Table 7.1 Units of α_c are per mil.

The parameter f_D appears in the equations for all the weathering fluxes, F_{wc}, F_{wg}, and F_{wsi}. It appears in Equation 7.14 raised to the 0.65 power due to the dilution effect discussed previously with respect to f_B. In the case of the weathering of carbonates and organic matter, the dilution effect is assumed to be negligible, and hence F_{wc} and F_{wg} are directly proportional to runoff. f_D has changed by about a factor of 1.5 during Phanerozoic time, falling during the late Cambrian and Ordovician, rising sharply during the Silurian, falling steadily during the remainder of most of the Paleozoic, rising during most of the Mesozoic, and falling to its present value during the Cenozoic. Much of the change in f_D has been caused by plate tectonics as continents have been moved beneath characteristic latitudinal evaporation/precipitation zones. The values of f_D in Table 7.1 are based on a general circulation model of the atmosphere (Otto-Bliesner, 1995) and include, in addition to the effects of evaporation/precipitation zones, factors such as changes in continental size as it affects monsoonality and changes in the distribution of water around continents. The model does not, however, include orographic effects.

The uplift or relief factor f_R appears in the Equations for F_{wg} and F_{wsi}. It does not appear in the equations for the weathering of carbonates, since "relief appears not to be an important factor in carbonate dissolution . . . as

TABLE 7.1 Values of Parameters in GEOCARB II at 10-Million Year Intervals

	Parameters							
Time[a]	f_{LA}	f_D	f_R	f_E	f_G	f_C	δ_{bc}	α_c
0	1.00	1.00	1.00	1.00	1.00	1.00	1.00	21.8
10	0.97	1.02	0.93	1.00	0.98	0.98	0.60	23.0
20	0.96	1.04	0.83	1.00	0.98	0.97	1.35	27.2
30	1.24	1.06	0.76	1.00	1.04	0.95	1.30	26.9
40	1.01	1.08	0.75	1.00	1.22	0.93	1.40	26.1
50	0.99	1.10	0.80	1.00	1.46	0.92	1.45	27.0
60	1.21	1.15	0.78	1.00	1.35	0.90	2.00	28.4
70	0.79	1.20	0.77	1.00	1.35	0.88	1.60	27.0
80	0.76	1.19	0.72	1.00	1.33	0.87	2.10	27.6
90	0.75	1.19	0.71	0.95	1.37	0.85	1.90	26.0
100	0.87	1.18	0.75	0.90	1.65	0.83	1.95	27.8
110	0.89	1.18	0.74	0.85	1.74	0.82	2.45	30.0
120	0.95	1.18	0.69	0.80	1.63	0.80	2.25	28.3
130	0.95	1.17	0.68	0.75	1.52	0.78	1.20	27.2
140	0.96	1.15	0.53	0.75	0.98	0.77	0.50	25.7
150	0.99	1.13	0.50	0.75	1.02	0.75	0.90	25.0
160	1.01	1.12	0.52	0.75	1.27	0.75	1.90	25.5
170	1.03	1.10	0.53	0.75	1.21	0.75	2.80	25.0
180	1.04	1.05	0.65	0.75	1.14	0.75	1.15	25.0
190	1.16	1.03	0.66	0.75	1.09	0.75	0.80	25.0
200	1.18	1.02	0.64	0.75	1.11	0.75	1.60	25.0
210	1.08	1.01	0.72	0.75	1.14	0.75	2.00	25.0
220	1.11	0.98	0.67	0.75	1.12	0.75	0.15	25.0
230	1.28	0.96	0.60	0.75	1.10	0.75	0.70	25.0
240	1.18	0.94	0.49	0.75	1.09	0.75	2.70	25.0
250	1.19	0.96	0.47	0.75	1.11	0.75	3.90	25.0
260	1.02	0.97	0.76	0.75	1.14	0.75	5.70	25.0
270	0.97	1.02	0.77	0.75	1.10	0.75	5.80	25.0
280	0.98	1.08	0.77	0.75	1.07	0.75	6.00	25.0
290	1.09	1.10	0.82	0.75	1.21	0.75	6.20	25.0
300	1.09	1.13	0.82	0.75	1.21	0.75	5.80	25.0
310	1.04	1.14	0.81	0.63	1.14	0.75	6.00	25.0
320	0.97	1.15	0.77	0.51	1.07	0.75	4.70	25.0
330	0.93	1.14	0.70	0.39	1.07	0.75	3.30	25.0
340	0.92	1.21	0.68	0.27	1.28	0.75	3.00	25.0
350	0.89	1.21	0.79	0.15	1.43	0.75	4.60	25.0
360	0.87	1.21	0.88	0.15	1.43	0.75	1.60	25.0
370	0.91	1.21	0.87	0.15	1.40	0.75	0.40	25.0
380	1.00	1.21	0.77	0.15	1.38	0.75	1.00	25.0
390	1.09	1.22	0.82	0.15	1.36	0.75	0.70	25.0
400	1.16	1.23	0.93	0.15	1.38	0.75	−0.70	25.0
410	1.11	1.25	1.03	0.15	1.48	0.75	−0.15	25.0
420	1.00	1.28	1.00	0.15	1.53	0.75	0.20	25.0
430	0.92	1.31	0.88	0.15	1.55	0.75	1.20	25.0
440	0.90	1.10	0.84	0.15	1.55	0.75	0.25	25.0

TABLE 7.1 *Continued*

	Parameters							
Time[a]	f_{LA}	f_D	f_R	f_E	f_G	f_C	δ_{bc}	α_c
450	0.70	1.00	0.81	0.15	1.52	0.75	0.25	25.0
460	0.64	0.90	0.85	0.15	1.52	0.75	0.40	25.0
470	0.66	0.92	0.89	0.15	1.50	0.75	−0.40	25.0
480	0.63	0.96	0.96	0.15	1.49	0.75	−0.65	25.0
490	0.64	1.03	1.01	0.15	1.50	0.75	−2.1	25.0
500	0.65	1.07	1.06	0.15	1.52	0.75	−2.1	25.0
510	0.63	1.10	1.09	0.15	1.62	0.75	−0.50	25.0
520	0.67	1.12	1.10	0.15	1.70	0.75	−0.85	25.0
530	0.77	1.12	1.07	0.15	1.59	0.75	−0.85	25.0
540	0.50	1.12	1.03	0.15	1.35	0.75	−0.60	25.0
550	0.62	1.12	0.99	0.15	1.20	0.75	−0.90	25.0
560	0.62	1.12	0.95	0.15	1.04	0.75	−1.00	25,0
570	0.63	1.12	1.00	0.15	1.00	0.75	−1.00	25.0

[a]Time is millions of years before present.

exemplified by appreciable dissolution in even flat lowland areas such as Florida" (Berner, 1994, p. 61).

Unfortunately, there is a lack of knowledge of paleotopography and paleohypsography, either or both of which would have been useful for estimating f_R. As a result Berner (1994) used a proxy for relief, namely, the $^{87}Sr/^{86}Sr$ ratio of seawater as recorded in limestones. The rationale is that continental silicates have a high $^{87}Sr/^{86}Sr$ ratio compared to carbonates and oceanic crust (basalt). Thus elevated seawater $^{87}Sr/^{86}Sr$ ratios imply an increase in the silicate weathering flux, and presumably in the organic carbon weathering flux as well. There are several caveats to this argument discussed by Berner (1994), but the conclusion is that, "a higher $^{87}Sr/^{86}Sr$ ratio of the source is correlated with enhanced silicate weatherability" (Berner, 1994, p. 62), presumably due to relief effects. The values of f_R estimated by Berner (1994) from $^{87}Sr/^{86}Sr$ data vary by about a factor of 2 during the Phanerozoic, decreasing during the Paleozoic and increasing during the later Mesozoic and Cenozoic.

The parameter f_E appears in the equations for the weathering of both carbonates and silicates, F_{wc} and F_{wsi}. f_E simulates effects on weathering caused by changes in the types and abundance of vegetation, particularly in dry upland areas. Unfortunately, there is rather meager data to quantify the effects of vegetation on weathering rates, but the available data suggest that the effect is quite important. Cawley et al. (1969), for example, found that the HCO_3^- concentration in surface waters draining older vegetated volcanics in Iceland was about three times the concentration of waters draining young unvegetated volcanics. The important transitions considered by Berner (1994) are the rise of vascular land plants in the Silurian–Devonian and the rise of angiosperms in the Cretaceous. The first transition is associated with a linear increase in f_E from 0.15 to 0.75 between 350 and 300 million years ago (Fig. 7.6). The second transition is associated with another linear increase in f_E from 0.75 to 1.00 between 130 and 80 million years ago. The prevascular land plant value of 0.15 for f_E is something of a guesstimate, but is supported by the study of Drever and Zobrist (1992). They found that silicate mineral weathering in a region of the southern Swiss Alps was about seven times higher in vegetated versus unvegetated portions of the same watershed. Angiosperms are believed to bring about more rapid rock weathering than gymnosperms or pteriodophytes, the major vascular land plants before the Cretaceous. The difference has been attributed to less efficient recycling of nutrients by angiosperms (Knoll and James, 1987). The assumption that $f_E = 0.75$ between 300 and 130 million years ago is based on the study of Volk (1989a). Angiosperms had become dominant by 80 million years ago, and f_E is accordingly set equal to 1 for the last 80 million years.

Both f_G and f_C are degassing parameters. The former appears in the equations for F_{mc} and F_{mg}; the latter appears only in the equation for F_{mc}. The temporal dependence of f_G reflects the reasonable assumption that changes in the rate of seafloor subduction will translate into changes in the rate of subduction and breakdown of carbonates and organic matter via metamorphism, volcanism, and diagenesis. Subduction, for example, is a major process whereby sedimentary carbonates are carried downward to higher temperatures and transformed to silicates by metamorphism or volcanism with the consequent release of CO_2 (Berner, 1994). Berner (1994) estimated the rate of seafloor subduction during the past 150 million years based on the study of Engebretson et al. (1992), who calculated oceanic crustal movement relative to a fixed hot spot reference frame. Unfortunately, there is almost no seafloor older than 150 million years, and estimates of f_G from 570 to 150 million years ago had to be made in some other way. Berner (1994) based his estimates on changes in sea level, the assumption being that, "as the rate of generation of seafloor rises, the volume of hot, highstanding young oceanic crust increases, and this additional material brings about the upward displacement of seawater resulting in a rise in sea level" (Berner, 1991, p. 349). In other words, when spreading rates increase along the mid-ocean ridge, sea level is displaced upward, and there is increased subduction and degassing to conserve seafloor area (Berner, 1994). The values of f_G prior to 150 million years ago are based on the analysis of Gaffin (1987), who assumed that sea level changes were due entirely to changes in spreading rates. A caveat to this approach is that it ignores sea level changes due to glaciation. The values of f_G in GEOCARB II almost all exceed 1 and vary by about a factor of 1.7, rising rapidly during the early Cambrian, falling until roughly the beginning of the Cretaceous, rising rapidly until the mid-Cretaceous, and then falling to 1 at the present time.

The parameter f_C accounts for the fact that prior to 150 million years ago virtually all carbonates are believed to have been deposited in shallow water. The evidence on this point is the lack of calcareous planktonic fossils in older sedimentary rocks. During the Jurassic and Cretaceous, however, there evolved calcareous plankton such as foraminifera and coccolithophorids whose skeletal remains could be deposited in deep-sea sediments. Carbonates deposited in deep water may be carried down subduction zones, where they undergo metamorphic breakdown or are dragged downward into the mantle with further heating. Carbonates deposited in shallow water are thermally decomposed only by the diagenesis and metamorphism associated with deep burial in sedimentary rocks (Berner, 1994). Hence the evolution of calcareous plankton would be expected to increase F_{mc}. The assumption that $f_C = 0.75$ prior to 150 million years ago is based on the studies of Wilkinson and Walker (1989) and Volk (1989b). The linear rise in f_C during the last 150 million years is a somewhat arbitrary interpolation.

The temporal behavior of δ_{bc} is based on experimental data reported by Berner (1987, 1989), namely, the $\delta^{13}C$ of calcareous fossils and limestones deposited in shallow seawater. The behavior of α_c during the last 140 million years is based on fractionation data collected by Arthur (1990). From 570 to 150 million years ago Berner (1994) assumed α_c to equal 25 per mil.

Now that we know the source of most of the parameter values, we will consider the problem of integrating the model to follow changes in C, G, δ_c, and δ_g over time. There are a total of 11 dependent variables in the model, namely, C, G, δ_c, δ_g, F_{wc}, F_{mc}, F_{wg}, F_{mg}, F_{bc}, F_{bg}, and RCO_2. Assume that we know the values of all these variables at some point in time, as indeed we must if we are to get the model off the ground. Assume further that the point in time is the present time, $t = 0$. We can then easily determine F_{mc} and F_{mg} from Equations 7.8 and 7.9, respectively, since the right-hand side of each equation involves parameters in Table 7.1 and the initial values of each variable, all of which are given to us. We can easily project C, G, δ_c, and δ_g to time $t + \Delta t$ or $t - \Delta t$ by simple Euler integration using Equations 7.10 and 7.11 and 7.20 and 7.21. Given G at the projected time, we can easily calculate F_{wg} from Equation 7.7. Thus seven of the dependent variables are rather easy to project from one point in time to the next. The remaining four dependent variables are F_{wc}, F_{bc}, F_{bg}, and RCO_2. These must be calculated using Equations 7.4 to 7.6 and 7.14. We can eliminate F_{bg} using Equation 7.4 and then use Equations 7.5 and 7.6 to solve for F_{wc} and F_{bc}. Plugging the solutions for F_{wc} and F_{bc} into Equation 7.14, we obtain the following equation:

$$y_b f_B + y_{bb} f_{BB} + y_c = 0 \tag{7.33}$$

where

$$y_b = \alpha_c f_R f_E F_{wsi}(0) f_D^{0.65} \tag{7.34}$$

$$y_{bb} = f_{LA} f_D f_E k_{wc} C(\delta_{bc} - \delta_c) \tag{7.35}$$

$$y_c = (F_{mc} + F_{wg} + F_{mg})(\delta_{bc} - \alpha_c) - \delta_c F_{mc} - \delta_g(F_{wg} + F_{mg}) \tag{7.36}$$

The coefficients y_b, y_{bb}, and y_c in Equation 7.33 can all be evaluated at our projected point in time, since all quantities on the right-hand sides of Equations 7.34 to 7.36 have been calculated or were given to us. We assume that f_B and f_{BB} are functions of RCO_2 and time, the functional relationships being given by Equations 7.26 and 7.29 for the past 300 million years, by Equations 7.27 and 7.30 from 570 to 350 million years ago, and by Equations 7.31 and 7.32 from 350 to 300 million years ago. Our job is to find the value of RCO_2 that makes the left-hand-side of Equation 7.33 equal to 0. We define $y(R)$ to be $y_b f_B + y_{bb} f_{BB} + y_c$, where R stands for RCO_2. Let Ro be the root of $Y(R)$, that is, the value of R that makes $Y(R)$ equal to 0. Let Re be an estimate of Ro. The Taylor series expansion of y about $R =$ Re gives

$$y(\mathrm{Ro}) = 0 = y(\mathrm{Re}) + \left[\frac{dy}{dR}\right]_{R=\mathrm{Re}} (\mathrm{Ro} - \mathrm{Re}) + \cdots \tag{7.37}$$

If Re is close to Ro, then the additional terms in the Taylor series expansion will hopefully be small, and we can rearrange Equation 7.37 to get an improved estimate of Ro, namely,

$$\mathrm{Ro} = \mathrm{Re} - \frac{y(\mathrm{Re})}{\left[\dfrac{dy}{dR}\right]_{R=\mathrm{Re}}} \tag{7.38}$$

This process may be repeated as many times as we like, the right-hand side of Equation 7.38 in each case giving a hopefully improved estimate of Ro. This particular iterative scheme for finding the root of an equation is called the *Newton–Raphson method*. We have used this approach before to estimate parameters in nonlinear least squares. Let us see how to implement the method in the case of Equation 7.33.

Assuming the method converges if we make a reasonable initial guess of Ro, the only obstacle to implementing Equation 7.38 is calculating the derivatives of f_B and f_{BB} with respect to R, since

$$\frac{dy}{dR} = y_b \frac{df_B}{dR} + y_{bb} \frac{df_{BB}}{dR}$$

The derivatives are a bit messy, but with some work one can show that during the last 300 million years (Eqs. 7.26 and 7.29)

$$\frac{df_B}{dR} = \left(\frac{f_B}{R}\right)\left(0.09\Gamma + \frac{(0.65)(0.038\Gamma)}{1 + 0.038\Gamma \ln(R) - 0.038W_s(t/570)} + \frac{0.4}{1+R}\right) \tag{7.39}$$

$$\frac{df_{BB}}{dR} = \left(\frac{f_{BB}}{R}\right)\left(\frac{0.087\Gamma}{1 + 0.087\Gamma \ln(R) - 0.087W_s(t/570)} + \frac{0.4}{1+R}\right) \tag{7.40}$$

and that from 570 to 350 million years ago (Eqs. 7.27 and 7.30)

$$\frac{df_B}{dR} = \left(\frac{f_B}{R}\right)\left(0.09\Gamma + 0.5 + \frac{(0.65)(0.038\Gamma)}{1 + 0.038\Gamma \ln(R) - 0.038W_s(t/570)}\right) \tag{7.41}$$

$$\frac{df_{BB}}{R} = \left(\frac{f_{BB}}{R}\right)\left(\frac{0.087\Gamma}{1 + 0.087\Gamma \ln(R) - 0.087W_s(t/570)} + 0.5\right) \tag{7.42}$$

Between 350 and 300 million years ago, we must, of course, use Equations 7.31 and 7.32 to calculate f_B and f_{BB}, respectively, and use the corresponding linear combinations of Equations 7.39 to 7.42 to calculate $(df_B/dR$ and df_{BB}/dR. For the purposes of integrating the GEOCARB II equations, it is sufficient to iterate Equation 7.38 until

$$\left|\frac{y(\mathrm{Re})}{\left[\dfrac{dy}{dR}\right]_{R=\mathrm{Re}}}\right| < 0.001$$

Having obtained RCO_2 by Newton–Raphson iteration, we can calculate f_B and f_{BB}. We then calculate F_{wc} from Equation 7.6, F_{bc} from Equation 7.14, and F_{bg} from either Equation 7.4 or 7.5. We have now determined all 11 dependent variables at time $t + \Delta t$ or $t - \Delta t$ and can begin the process again to make the next projection.

Table 7.2 gives the values of constant parameters in the GEOCARB II model as well as the values of the dependent variables needed to characterize the system at the present time. Given this information, it is straightforward to calculate the present values of the other five dependent variables in the model. The present value of RCO_2 is, of course, 1, and from Equations 7.26 and 7.29 the present values of f_B and f_{BB} are also 1. Since the values of all the f parameters in Table 7.1 are, by definition, 1 at the present time, it is easy to calculate the present values of F_{wc} and F_{wg} from Equations 7.6 and 7.7. The present values of F_{mg} and F_{mc} may then be obtained from Equations 7.4 and 7.5 as follows:

$$F_{mg} = \frac{\delta_{bc}F_{bc} + (\delta_{bc} - \alpha_c)F_{bg} - \delta_c(F_{bc} + F_{bg})}{\delta_g - \delta_c} - F_{wg} \tag{7.43}$$

$$F_{mc} = F_{bc} + F_{bg} - F_{wc} - F_{wg} - F_{mg} \tag{7.44}$$

where all quantities on the right-hand sides of Equations 7.43 and 7.44 are

TABLE 7.2 Values of Constant Parameters and Certain Dependent Variables at the Present Time for the GEOCARB II Model

Parameter	Value
k_{wc}	0.00267 per million year
k_{wg}	0.0030 per million year
W_s	12.9°C
Γ	6°C

Dependent Variable	Value
C	5000×10^{18} mol
G	1250×10^{18} mol
δ_c	1.5 per mil
δ_g	−23.5 per mil
F_{bg}	5×10^{18} mol per million year
F_{bc}	20×10^{18} mol per million year

to be evaluated at the present time. Since we do not know the values of the dependent variables 570 million years ago, we will integrate the GEOCARB II equations backwards in time, using a Δt of -1 million years. Since the parameters in Table 7.1 are given only at 10-million-year intervals, it will be necessary to interpolate these parameters. A simple linear interpolation will suffice:

$$f(t-x) = \left(1 - \frac{x}{10}\right)f(t) + \frac{x}{10}f(t-10)$$

The initial guess of RCO_2 for the Newton–Raphson iteration is always taken to be the value of RCO_2 at time $t - 1$ million years ago. Thus, for example, the initial guess of RCO_2 at time 1 million years ago is 1, the present value of RCO_2.

Figure 7.11 shows RCO_2 during Phanerozoic time as calculated from the GEOCARB II model. The model indicates that during most of Phanerozoic time the concentration of CO_2 in the atmosphere has been substantially higher than at present, particularly during the first half of the Paleozoic era. These higher CO_2 concentrations would have caused a significant warming of the lower atmosphere and hence counteracted the cooling effect associated with the "pale sun." We will have a chance to explore these issues more in the Exercises and to examine the impact of various assumptions in the model on the behavior of RCO_2. Suffice it to say for now that the predictions of the model are consistent with what we think we know about the paleoclimate of the earth and the cycling of carbon on a geological time scale.

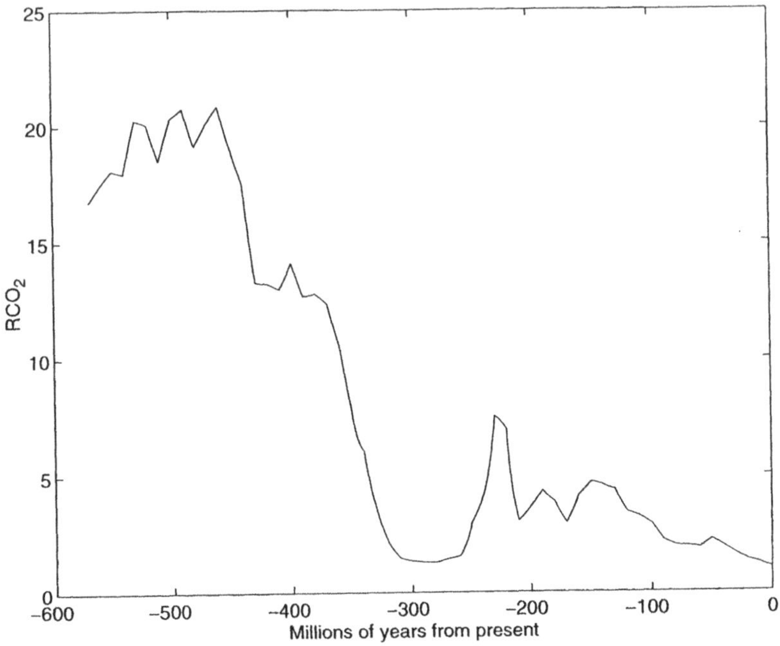

Figure 7.11 RCO_2 during Phanerozoic time as calculated from the GEOCARB II model.

PELAGIC FOOD CHAIN

Our second box model example is intended to simulate an open-ocean grazing pelagic food chain. Conceptually, the model is similar to the box model in Figure 7.1, but without specific inclusion of the detritivore trophic level. The model assumes that detritus produced by the grazing food chain either sinks out of the system or is broken down and converted into inorganic nutrients. This scenario is consistent with both theoretical and experimental studies which have indicated that the principal role of the microbial loop is to recycle nutrients. Only a small percentage of the organic matter excreted as detritus by the grazing food chain finds its way back into the grazing food chain via the microbial loop (Ducklow et al., 1986).

The symbols used in the pelagic food chain model are indicted in Table 7.3. We assume the system to be nutrient limited, and for the purposes of this discussion we will assume nitrogen to be the limiting nutrient. All concentrations are therefore expressed in terms of N. The flow of N through the system is indicated in Figure 7.12.

For the time being we will consider the simplest possible representation of the model and will not consider the details of the functional form of the f_m. We consider a one-carnivore model, although later we learn how to extend

TABLE 7.3 Symbols Used in Pelagic Food Chain Model

Symbol	Meaning and Units
X_1	Concentration of inorganic N (μg m^{-3})
X_2	Phytoplankton concentration as N (μg m^{-3})
X_3	Herbivore concentration as N (μg m^{-3})
X_4	Carnivore concentration as N (μg m^{-3})
X_5	Concentration of detritus as N (μg m^{-3})
A_m	Maximum fractional ingestion rate of N by X_m (d^{-1})
F	Loss rate as N of top-level carnivore due to natural mortality (μg m^{-3} d^{-1})
f_m	Rate of uptake or ingestion of N by X_m (μg m^{-3} d^{-1})
M	Fractional loss rate due to natural mortality of top-level carnivore (d^{-1})
P_m	Concentration of X_m as N below which uptake or ingestion of X_m ceases (μg m^{-3})
Q_m	Fraction of ingested N converted to biomass by X_m
R_m	Fraction of ingested N respired or excreted by X_m
S_m	Fraction of ingested N egested as fecal material by X_m
D	Fractional loss rate of detritus due to sinking (d^{-1})
k	Fractional regeneration rate of inorganic N from detritus (d^{-1})
L	External loading rate of inorganic N (μg m^{-3} d^{-1})

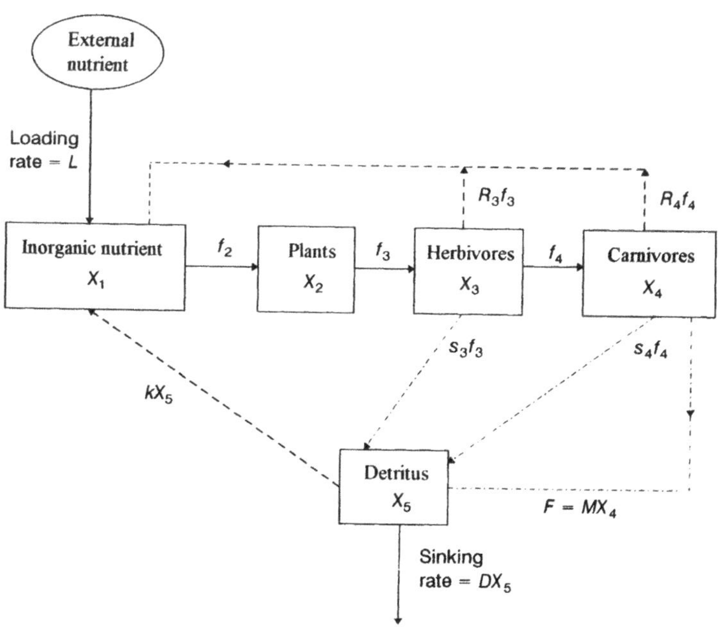

Figure 7.12 Flow of N through the pelagic food chain model.

the model to an arbitrary number of carnivore trophic levels. The differential equations governing the concentrations in this five-box model are as follows:

$$\frac{dX_1}{dt} = R_3f_3 + R_4f_4 - f_2 + kX_5 + L \tag{7.45}$$

$$\frac{dX_2}{dt} = f_2 - f_3 \tag{7.46}$$

$$\frac{dX_3}{dt} = Q_3f_3 - f_4 \tag{7.47}$$

$$\frac{dX_4}{dt} = Q_4f_4 - F \tag{7.48}$$

$$\frac{dX_5}{dt} = S_3f_3 + S_4f_4 + F - (D + k)X_5 \tag{7.49}$$

We will use this simple representation of the system to determine how the f ratio, the ratio of primary production supported by external nutrient inputs to total primary production, depends on the characteristics of the system we have modeled at steady state. Note that all external nutrient that enters the system is assumed to be taken up by the phytoplankton. The only N export occurs via the sinking of detritus, DX_5. Note also that phytoplankton are assumed to incorporate all the N they take up into biomass, that is, $Q_2 = 1$. The f ratio in this system will be L/f_2. At steady state we can see from Equation 7.45 that

$$L = f_2 - R_3f_3 - R_4f_4 - kX_5 \tag{7.50}$$

However, we can see from Equations 7.46 and 7.47 that at steady state $f_3 = f_2$ and that $f_4 = Q_3f_3 = Q_3f_2$. Furthermore, from Equation 7.49 we have $X_5 = (S_3f_3 + S_4f_4 + F)/(D + k)$. Hence Equation 7.50 becomes

$$L = f_2(1 - R_3 - R_4Q_3) - \frac{k}{D + k}(S_3f_3 + S_4f_4 + F) \tag{7.51}$$

Substituting f_2 for f_3 and Q_3f_2 for f_4 and noting from Equation 7.48 that at steady state $F = Q_4f_4 = Q_4Q_3f_2$, we find that Equation 7.51 becomes

$$L = f_2\left[1 - R_3 - R_4Q_3 - \frac{k}{D + k}(S_3 + S_4Q_3 + Q_4Q_3)\right] \tag{7.52}$$

By definition, however, $Q_m + R_m + S_m = 1$ for all m. Substituting $1 - Q_3 - R_3$ for S_3 and $1 - Q_4 - R_4$ for S_4, we find that $S_3 + S_4Q_3 + Q_4Q_3 = 1 - R_3 - R_4Q_3$. Equation 7.52 therefore becomes

$$L = f_2\left(1 - \frac{k}{D + k}\right)(1 - R_3 - R_4Q_3) = f_2\left(\frac{D}{D + k}\right)(1 - R_3 - R_4Q_3) \tag{7.53}$$

Hence the f ratio is given by the equation

$$f \text{ ratio} = \frac{L}{f_2} = \left(\frac{D}{D+k}\right)(1 - R_3 - R_4 Q_3) \qquad (7.54)$$

The f ratio in this system is therefore given by the product of two terms, the first of which, $D/(D + k)$, is the fraction of detrital losses accounted for by sinking, and the second of which, $1 - R_3 - R_4 Q_3$, is the fraction of the N taken up by the phytoplankton that is not respired by the grazing food chain. This realization allows us to tentatively extend our f ratio equation to a system with any number of carnivore trophic levels. We defer a rigorous proof of this extension to the Exercises. For now, we hypothesize that, for example, in a system with two carnivore trophic levels, the f ratio would be given by the equation

$$f \text{ ratio} = \left(\frac{D}{D+k}\right)(1 - R_3 - R_4 Q_3 - R_5 Q_4 Q_3) \qquad (7.55)$$

It is now of interest to make some estimate of the values of the parameters in the f ratio equation and to examine how the f ratio might depend on the length of the grazing food chain. Feeding studies with zooplankton indicate that for herbivorous zooplankton $S \cong 0.35$, $R \cong 0.4$, and $Q \cong 0.25$. For carnivorous zooplankton $S \cong 0.15$, $R \cong 0.5$, and $Q \cong 0.35$ (Raymont, 1983; Lalli and Parsons, 1993). In our model system with one carnivore trophic level, we have $1 - R_3 - R_4 Q_3 = 1 - 0.4 - (0.5)(0.25) = 0.485$. In our two-carnivore-trophic-level model, we have $1 - R_3 - R_4 Q_3 - R_5 Q_4 Q_3 = 1 - 0.4 - (0.25)(0.5)(1 + 0.35) = 0.43$. If we extend the food chain to include an infinite number of carnivore trophic levels, each with the same Q, R, and S values, the f ratio becomes $1 - 0.4 - (0.25)(0.5)(1 + 0.35 + 0.35^2 + 0.35^3 + \cdots) = 0.41$. Thus, according to the model, the range of maximum possible f ratios, when $D \gg k$, is roughly 0.4 to 0.5. This is consistent with estimates of f ratios from field work (Eppley and Peterson, 1979). It is apparent from this analysis that while lengthening the food chain does tend to reduce the f ratio, long food chains cannot account for the f ratios of 0.1 to 0.2 reported in many open-ocean systems (Eppley and Peterson, 1979). Most of the recycling of nutrients occurs near the base of food chains, and hence adding more trophic levels to the top of the chain has little effect on the extent of nutrient recycling. f ratios of 0.1 to 0.2 imply that $D < k$, and f ratios of 0.4 to 0.5 imply that $D \gg k$. Food chains in which the herbivores and primary carnivores are mainly protozoans tend to be characterized by low f ratios, because the waste material produced by protozoans takes the form of either dissolved organics or small particles, both of which have little or no tendency to sink and are rapidly broken down by microorganisms. Hence $k > D$ in such systems. Food chains in which the herbivores and primary carnivores are mainly macrozooplankton which produce encapsulated fecal pellets are likely to be characterized by f ratios of 0.4 to 0.5. Such pellets sink rapidly and, because of their encapsulation, are broken down slowly. $D \gg k$ in such systems.

In the next step of our analysis, we consider the functional form of the f_m and derive expressions for the concentration of N in each box of our model at equilibrium. The assumed functional form is

$$f_m = \frac{A_m(X_{m-1} - P_{m-1})X_m}{X_{m-1}} \tag{7.56}$$

f_m is therefore assumed to be directly proportional to X_m and a hyperbolic function of prey or substrate concentration X_{m-1}. f_m is 0 if $X_{m-1} \leq P_{m-1}$ and equals $0.5A_mX_m$ when $X_{m-1} = 2P_{m-1}$. We will also assume that $F = MX_4$, that is, the natural mortality rate of the top-level carnivore is directly proportional to the concentration of the top-level carnivore. In order to determine the equilibrium concentration of N in each box of our model, we note that by adding Equations 7.45 to 7.49 and recalling that $Q_m + R_m + S_m = 1$ it follows that

$$\frac{d(X_1 + X_2 + X_3 + X_4 + X_5)}{dt} = -DX_5 + L \tag{7.57}$$

Hence at equilibrium

$$X_5 = \frac{L}{D} \tag{7.58}$$

To find the equilibrium concentration of the carnivore in the system, we note that at equilibrium the rate of production of carnivore biomass equals $Q_4f_4 = Q_3Q_4f_3 = Q_3Q_4f_2 = Q_3Q_4L/(f \text{ ratio})$. This must be balanced by the rate of loss of carnivore biomass, which equals MX_4. Hence at equilibrium $MX_4 = Q_3Q_4L/(f \text{ ratio})$, or

$$X_4 = \frac{LQ_3Q_4}{M(f \text{ ratio})} \tag{7.59}$$

To obtain the equilibrium concentration of the herbivore, we similarly note that the rate of herbivore production equals $Q_3f_3 = Q_3L/(f \text{ ratio})$. At equilibrium this must be balanced by the grazing loss, which equals $[A_4(X_3 - P_3)X_4]/X_3$. Hence at equilibrium

$$\frac{Q_3L}{f \text{ ratio}} = \frac{A_4(X_3 - P_3)X_4}{X_3} \tag{7.60}$$

Solving Equation 7.60 for X_3 gives

$$X_3 = \frac{P_3}{1 - \dfrac{LQ_3}{A_4 X_4 (f \text{ ratio})}} \tag{7.61}$$

To obtain the equilibrium concentration of the phytoplankton, we note that the rate of phytoplankton production equals $f_2 = L/(f \text{ ratio})$. This must be balanced by the grazing loss, which equals $[A_3(X_2 - P_2)X_3]/X_2$. Hence at equilibrium

$$\frac{L}{f \text{ ratio}} = \frac{A_3(X_2 - P_2)X_3}{X_2} \tag{7.62}$$

Solving this equation for X_2 gives

$$X_2 = \frac{P_2}{1 - \dfrac{L}{A_3 X_3 (f \text{ ratio})}} \tag{7.63}$$

To determine X_1, we use the fact that

$$\frac{L}{f \text{ ratio}} = f_2 = \frac{A_2(X_1 - P_1)X_2}{X_1} \tag{7.64}$$

Solving this equation for X_1 gives

$$X_1 = \frac{P_1}{1 - \dfrac{L}{A_2 X_2 (f \text{ ratio})}} \tag{7.65}$$

The pattern should by now be clear. For all trophic levels below the highest in the grazing food chain, the equilibrium concentration of N is related to the equilibrium concentration of N in the next-higher trophic level by the equation

$$X_m = \frac{P_m}{1 - \dfrac{L \prod_{j=1}^{m} Q_j}{A_{m+1} X_{m+1} (f \text{ ratio})}} \tag{7.66}$$

where we define $Q_1 = Q_2 = 1$. This follows directly from Equation 7.56 and the fact that

$$f_{m+1} = f_2 \prod_{j=1}^{m} Q_j = \frac{L}{(f \text{ ratio})} \prod_{j=1}^{m} Q_j$$

We now have a simple means for calculating the equilibrium concentration of N for each box in our model no matter how many carnivore trophic levels we include in the grazing food chain. The concentration of detritus, X_n, is given by L/D (Eq. 7.58), the concentration of N in the top trophic level of the grazing food chain, X_{n-1}, is given by

$$X_{n-1} = \left(\frac{L}{M(f \text{ ratio})}\right) \prod_{j=1}^{n-1} Q_j \tag{7.67}$$

and the concentration of N in successively lower trophic levels is given by Equation 7.66. The f ratio is given by

$$f \text{ ratio} = \left(\frac{D}{D+k}\right)\left(1 - \sum_{j=3}^{n-1} R_j \prod_{k=2}^{j-1} Q_k\right) \tag{7.68}$$

Now that we know how to calculate the equilibrium concentrations of N in our box model, we will begin assigning values to the parameters in the model. We will use field data collected primarily from the euphotic zone of the North Pacific subtropical gyre, which we assume to be 150 m deep (Laws et al., 1990). Table 7.4 summarizes what we think we know about this system. The derived parameters were obtained from the equations in the model and the parameters determined from field or laboratory data. As a reference case we assume that all organisms in the grazing food chain are growing at 50% of their maximum possible rate. We will examine the implications of changing this assumption in the Exercises. One crude test of the implications of the model is the reasonableness of the derived parameters. The maximum growth rate of phytoplankton at temperatures of 20 to 25°C, approximately the temperature of the system we are modeling, is about 1.0 to 1.4 d^{-1} with a 12 h photoperiod (Eppley, 1972). Since our model averages the growth rate of the phytoplankton from the surface to the 0.1% light level (150 m), it is reasonable to assume that the maximum growth rate of the phytoplankton in the euphotic zone would be less than 1.0 to 1.4 d^{-1}. The average light intensity from the surface to the 0.1% light level is about 14% of the irradiance just below the surface. Photosynthetic rates in the North Pacific subtropical gyre are light saturated at irradiances above about 30% of near-surface values and roughly proportional to light intensity at lower irradiances (Laws et al., 1990). It is therefore reasonable to expect that the maximum growth rate of phytoplankton in our model system would be approximately 0.5 to 0.7 d^{-1}. The derived A_2 of 0.545 d^{-1} is therefore

TABLE 7.4 Characteristics of Euphotic Zone of North Pacific Subtropical Gyre Used to Parameterize One-Carnivore Trophic Level Food Chain Model

Parameter	Value	Source
	Parameters Obtained from Field or Laboratory Data	
X_1	420 μg N m^{-3}	Hawaiian Ocean Time Series data reports
X_2	1560 μg N m^{-3}	Williams (1984)
X_3	1625 μg N m^{-3}	Williams (1984)
X_4	385 μg N m^{-3}	Williams (1984)
X_5	890 μg N m^{-3}	Laws et al. and Williams (1984)
f ratio	0.2	Laws (1991)
L	85 μg N m^{-3}d^{-1}	Laws (1991)
Q_3	0.25	Raymont (1983) and Lalli and Parsons (1993)
Q_4	0.35	Raymont (1983) and Lalli and Parsons (1993)
S_3	0.35	Raymont (1983) and Lalli and Parsons (1993)
S_4	0.15	Raymont (1983) and Lalli and Parsons (1993)
	Derived Parameters	
D	0.0955 d^{-1}	Equation 7.58
k	0.1313 d^{-1}	Equation 7.54
M	0.0966 d^{-1}	Equation 7.59
P_1	210 μg N m^{-3} (= 0.5X_1)	Equation 7.56 assuming phytoplankton growing at 50% of maximum rate
P_2	780 μg N m^{-3} (= 0.5X_2)	Equation 7.56 assuming herbivores growing at 50% of maximum rate
P_3	812.5 μg N m^{-3} (= 0.5 X_3)	Equation 7.56 assuming carnivores growing at 50% of maximum rate
A_2	0.545 d^{-1}	Equation 7.64
A_3	0.523 d^{-1}	Equation 7.62
A_4	0.553 d^{-1}	Equation 7.60

reasonable. The maximum growth rates of the herbivores and carnivores in the system, Q_3A_3 and Q_4A_4, are 0.13 and 0.19 d^{-1}, respectively. For comparison, Huntley and Lopez's (1992) analysis of physiological rate measurements on marine copepods indicates maximum growth rates of 0.15 to 0.23 at temperatures of 20 to 25°C. Thus the derived parameters do seem reasonable.

STABILITY ANALYSIS AND PERTURBATION THEORY

While the food chain model seems to give a reasonable description of the system of interest, we have as yet no guarantee that the equilibrium distribution of N in the model represents a stable situation. What would happen, for example, if our system, initially at equilibrium, were perturbed by a change in the amount of N in one or more of the boxes? There are several possibilities. The system could return to its equilibrium configuration, in which case we say that the system is stable. The other possibility, of course, is that the system is unstable. In this case the perturbation might produce one of two types of results. First, the concentrations of N in the boxes might begin to oscillate about their equilibrium values, never settling down again but never going to zero or increasing indefinitely. Second, the concentration of N in one or more of the boxes might go to 0 or begin to increase indefinitely. How the system responds may depend on the size of the perturbation. A system that is stable to small perturbations may be unstable to large perturbations. In the former case we say that the system has *local stability* about its equilibrium point. If the system is stable to virtually any perturbation from its equilibrium point, we say that the system has *global stability*. Obviously, a system which lacks local stability also lacks global stability. To be stable in any sense, a system must have local stability. We will therefore begin our analysis of system stability by considering the issue of local stability.

We begin by considering a very simple system. We imagine a one-box model in which the rate of change of X is described by the equation

$$\frac{dX}{dt} = F(X) \tag{7.69}$$

Let X^* be the equilibrium value of X, that is, the value of X such that $F(X^*) = 0$. Now assume that X is perturbed from its equilibrium value by a small amount ΔX. Since ΔX is small, we should be able to approximate $F(X^* + \Delta X)$ by expanding $F(X)$ in its Taylor series about $X = X^*$. To first order we obtain

$$F(X) = F(X^* + \Delta X) = F(X^*) + \left[\frac{dF}{dX}\right]_{X=X^*} \Delta X = \left[\frac{dF}{dX}\right]_{X=X^*} \Delta X \tag{7.70}$$

However, since X^* is a constant,

$$\frac{dX}{dt} = \frac{d(X^* + \Delta X)}{dt} = \frac{d(\Delta X)}{dt} \tag{7.71}$$

We conclude that, in the limit as $\Delta X \to 0$,

$$\frac{d(\Delta X)}{dt} = \left[\frac{dF}{dX}\right]_{X=X^*} \Delta X$$

Now define $[dF/dX]_{X=X^*}$ to be α. Then, in the limit as $\Delta X \to 0$,

$$\frac{d(\Delta X)}{dt} = \alpha \Delta X \tag{7.72}$$

The solution of Equation 7.72 is $\Delta X = \Delta X_0 e^{\alpha t}$, where ΔX_0 is the original perturbation. Clearly, the perturbation will decay away if $\alpha < 0$ and will increase if $\alpha > 0$. The system is therefore stable to small perturbations if $\alpha < 0$ and unstable if $\alpha > 0$. If $\alpha > 0$ and the perturbation continues to grow, Equation 7.72 will eventually no longer apply, because higher-order terms in the Taylor series expansion of $F(X^* + \Delta X)$ will become important. In other words, the fact that $\alpha > 0$ does not necessarily imply that the perturbation will grow exponentially forever. Higher-order terms in the Taylor series may limit the magnitude of ΔX and cause X to oscillate about X^*. The important point is that if $\alpha > 0$ there is no way for the system to settle back down to its equilibrium state once it has been perturbed.

As an example of the application of perturbation theory, consider the logistic equation, which we first encountered in Exercise 6.5. According to the logistic equation,

$$\frac{dX}{dt} = F(X) = rX\left(1 - \frac{X}{K}\right) \tag{7.73}$$

We can easily see that

$$\frac{dF}{dX} = r\left(1 - \frac{X}{K}\right) - r\frac{X}{K}$$

The equilibrium points are $X = 0$ and $X = K$. At $X = 0$, $dF/dX = r$. At $X = K$, $dF/dX = -r$. Thus the equilibrium at $X = 0$ is unstable, because $\alpha = r > 0$. The equilibrium at $X = K$ is stable, because $\alpha = -r < 0$. In this case at least we could have determined the stability of the equilibria by graphical methods. The function $F(X)$ for the logistic equation contains two terms, one of which, $rX(= rK(X/K))$, causes X to increase, and the other of which, $-rX^2/K(= -rK(X/K)^2)$, causes X to decrease. If we plot X/K and $(X/K)^2$ versus X, the equilibria will be the points where the two functions are equal. Such a graph is shown in Figure 7.13. At $X = K$ an increase in X to $K + \Delta X$ causes rX^2/K to exceed rX. Hence X decreases and returns toward its equilibrium value, K. Similarly, if X decreases to $K - \Delta X$, rX exceeds rX^2/K. Hence X increases and returns toward K. At $X = 0$ an increase in X to ΔX causes rX to exceed rX^2/K. Hence X becomes even bigger and moves farther away from the equilibrium point at $X = 0$. In fact, it should be obvious from Figure 7.13 that no matter what positive initial value we may choose to give X, the rate of change of X will always move X toward the equilibrium point at $X = K$. Thus we could say that for all $X > 0$ the equilibrium point at $X = K$ is globally stable.

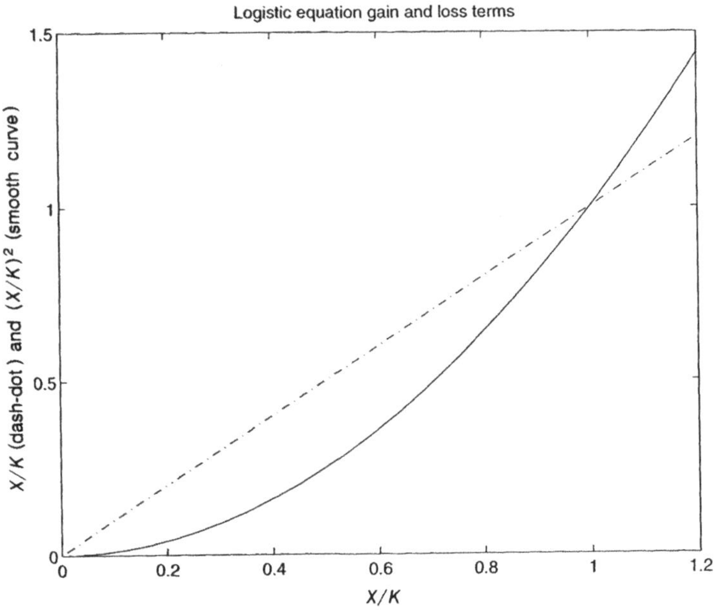

Figure 7.13 Plot of (X/K) and $(X/K)^2$ versus X/K. The equilibria of the logistic equation correspond to the intersection of these two functions.

Now we will consider a little more difficult problem. In Chapter 6 we examined a simple competing species model, that is, Equations 6.44 and 6.45. We now examine the stability characteristics of the equilibrium points. The competing species model takes the form

$$\frac{dX}{dt} = F_1(X,Y) = (a - bX - cY)X \tag{7.74}$$

$$\frac{dY}{dt} = F_2(X,Y) = (e - fX - gY)Y \tag{7.75}$$

There are actually four equilibrium points to the model. Three of them are $(X,Y) = (0,0)$, $(X,Y) = (0, e/g)$, and $(X,Y) = (a/b, 0)$. If either $X = 0$ or $Y = 0$, we are back to the logistic equation for the dynamics of the nonzero species. The equilibrium point $(0,0)$ is not very interesting, since both species have been eliminated. The interesting equilibrium point is the point at which both X and Y are not automatically 0. It is easy to show that at this point

$$X = \frac{ag - ce}{bg - cf} \tag{7.76}$$

$$Y = \frac{be - af}{bg - cf} \tag{7.77}$$

We will call the right-hand sides of Equations 7.76 and 7.77 X^* and Y^*, respectively. Is this equilibrium point stable? To answer this question, we let $X = X^* + \Delta X$ and $Y = Y^* + \Delta Y$. We then expand F_1 and F_2 to first order using the Taylor series expansion of both functions about (X^*, Y^*):

$$F_1(X,Y) = F_1(X^* \;\; Y^*) + \left[\frac{\partial F_1}{\partial X}\right]_{(X^*,Y^*)} \Delta X + \left[\frac{\partial F_1}{\partial Y}\right]_{(X^*,Y^*)} \Delta Y \quad (7.78)$$

$$F_2(X,Y) = F_2(X^*,Y^*) + \left[\frac{\partial F_2}{\partial X}\right]_{(X^*,Y^*)} \Delta X + \left[\frac{\partial F_2}{\partial Y}\right]_{(X^*,Y^*)} \Delta Y \quad (7.79)$$

We now define $\alpha_{1,1} = [\partial F_1/\partial X]_{(X^*,Y^*)}$, $\alpha_{1,2} = [\partial F_1/\partial Y]_{(X^*,Y^*)}$, $\alpha_{2,1} = [\partial F_2/\partial X]_{(X^*,Y^*)}$, and $\alpha_{2,2} = [\partial F/\partial Y]_{(X^*,Y^*)}$. Since, by definition, $F_1(X^*,Y^*) = F_2(X^*,Y^*) = 0$, we have

$$\frac{dX}{dt} = \frac{d(\Delta X)}{dt} = F_1(X,Y) = \alpha_{1,1}\Delta X + \alpha_{1,2}\Delta Y \quad (7.80)$$

$$\frac{dY}{dt} = \frac{d(\Delta Y)}{dt} = F_2(X,Y) = \alpha_{2,1}\Delta X + \alpha_{2,2}\Delta Y \quad (7.81)$$

Equations 7.80 and 7.81 can be written more compactly in matrix notation:

$$\frac{d(\mathbf{\Delta Z})}{dt} = \boldsymbol{\alpha}\mathbf{\Delta Z} \quad (7.82)$$

where

$$\boldsymbol{\alpha} = \begin{bmatrix} \alpha_{1,1} & \alpha_{1,2} \\ \alpha_{2,1} & \alpha_{2,2} \end{bmatrix} \quad \text{and} \quad \mathbf{\Delta Z} = \begin{bmatrix} \Delta X \\ \Delta Y \end{bmatrix}$$

The matrix $\boldsymbol{\alpha}$ is called the *community matrix*. We now claim, without proof, that there are certain combinations of ΔX and ΔY such that $\boldsymbol{\alpha}\mathbf{\Delta Z} = \lambda\mathbf{\Delta Z}$, where λ is a scalar. The column vectors $\mathbf{\Delta Z}$ which satisfy this condition are called the eigenvectors of $\boldsymbol{\alpha}$, and the λ are the corresponding eigenvalues. We also claim, again without proof, that any 2×1 column vector $\mathbf{\Delta Z}$ can be expressed as a linear combination of the eigenvectors of the 2×2 matrix $\boldsymbol{\alpha}$.

Now suppose that we have identified an eigenvector of $\boldsymbol{\alpha}$. Call this eigenvector $\mathbf{\Delta E}$. Since $\mathbf{\Delta E}$ is an eigenvector we know that

$$\frac{d(\mathbf{\Delta E})}{dt} = \lambda\mathbf{\Delta E} \quad (7.83)$$

The solution of Equation 7.83, analogous of the solution to Equation 7.72, is

$$\mathbf{\Delta E} = \mathbf{\Delta E}_0 e^{\lambda t} \tag{7.84}$$

Thus the perturbation will decay if $\lambda < 0$ and will grow if $\lambda > 0$. The system is locally stable about the equilibrium point if all eigenvalues of the community matrix are negative. This follows from the fact that any perturbation can be expressed as a linear combination of the eigenfunctions of the community matrix. If even one of the eigenvalues is positive, the equilibrium point is unstable, because there exists a perturbation which will not decay away over time.

A caveat to this simple explanation is that the eigenvalues of the community matrix may be complex numbers. In this case the eigenvalues always occur in complex conjugate pairs, as do their associated complex eigenvectors. It is straightforward to show that the perturbations associated with these complex eigenvalues and eigenvectors will decay away, although not in a simple exponential manner, if the real parts of the eigenvalues are negative. The mode of decay will be an oscillation with an exponentially damped amplitude, the frequency of the oscillation being determined by the imaginary part of the eigenvalue and the rate of decay determined by the real part of the eigenvalue. Thus the criterion for a stable equilibrium point is that the real parts of all eigenvalues, $\mathrm{Re}(\lambda)$, be negative.

Now return to Equations 7.74 and 7.75 and determine the elements of the 2×2 community matrix for our competing species model, remembering that we are to evaluate the partial derivatives of F_1 and F_2 using the values of X and Y that make $a - bX - cY$ and $e - fX - gY = 0$. We conclude that

$$\alpha_{1,1} = -bX^* \tag{7.85}$$

$$\alpha_{1,2} = -cX^* \tag{7.86}$$

$$\alpha_{2,1} = -fY^* \tag{7.87}$$

$$\alpha_{2,2} = -gY^* \tag{7.88}$$

The eigenfunctions must satisfy the equation $\boldsymbol{\alpha}\mathbf{\Delta E} = \lambda\mathbf{\Delta E}$, or $(\mathbf{a} - \lambda\mathbf{I})\mathbf{\Delta E} = \mathbf{0}$. Solving this matrix equation for ΔX and ΔY gives

$$\Delta X = \Delta Y = \frac{0}{(\alpha_{1,1} - \lambda)(\alpha_{2,2} - \lambda) - \alpha_{1,2}\alpha_{2,1}} \tag{7.89}$$

Needless to say this is not a very interesting solution, unless, of course, the denominator on the right-hand side happens to be 0. In that case, as we will see in the Exercises, ΔX and ΔY can be determined to within a constant multiple, normally chosen so that $\mathbf{\Delta E}'\mathbf{\Delta E} = 1$. For the denominator to be 0, we must have

$$\lambda^2 - (\alpha_{1,1} + \alpha_{2,2})\lambda + \alpha_{1,1}\alpha_{2,2} - \alpha_{1,2}\alpha_{2,1} = 0 \tag{7.90}$$

The two values of λ that satisfy this equation are the two eigenvalues of the community matrix. They are given by

$$\lambda = \frac{1}{2}\left[\alpha_{1,1} + \alpha_{2,2} \pm \sqrt{(\alpha_{1,1} + \alpha_{2,2})^2 + 4(\alpha_{1,2}\alpha_{2,1} - \alpha_{1,1}\alpha_{2,2})}\right] \tag{7.91}$$

We can see from Equations 7.85 to 7.88 that if X^* and Y^* are positive, all the elements of α will be negative. With a little thought, it should then be obvious that both eigenvalues will be negative as long as $\alpha_{1,2}\alpha_{2,1} - \alpha_{1,1}\alpha_{2,2} < 0$. From Equations 7.85 to 7.88, this condition implies that $bg - cf > 0$. However, if X^* and Y^* are both positive and $bg - cf > 0$, then Equations 7.76 and 7.77 imply that $ag > ce$ and $be > af$. Taken together, however, these latter two conditions automatically ensure that $bg - cf > 0$. We conclude that X^* and Y^* will both be positive and the equilibrium at (X^*, Y^*) will be locally stable as long as $ag > ce$ and $be > af$.

Now we will return to our food chain model. In order to determine whether the equilibrium N concentrations in our five-box model are stable to small perturbations, we need to determine the eigenvalues of the community matrix. The community matrix in this case will be a 5×5 matrix and will have five eigenvalues. All five eigenvalues must have negative real parts for the equilibrium to be stable. In order to evaluate the elements of the community matrix, we will go back to Equations 7.45 to 7.49. We must take the partial derivative of the right-hand side of each of these equations with respect to X_1, X_2, X_3, X_4, and X_5. Referring to Equation 7.56, it is easy to see that

$$\frac{\partial f_m}{\partial X_m} = \frac{f_m}{X_m} \tag{7.92}$$

$$\frac{\partial f_m}{\partial X_{m-1}} = \frac{f_m P_{m-1}}{(X_{m-1} - P_{m-1})X_{m-1}} \tag{7.93}$$

All derivatives are, of course, to be evaluated at the equilibrium point when calculating the elements of the community matrix. This means, for example, that in the case of Equation 7.48, the only nonzero partial derivative will be the derivative with respect to X_3. The eigenvalues of the community matrix in the standard case, with X_2, X_3, and X_4 growing at 50% of their maximum rates, are as follows:

$$\lambda_1 = -0.7565$$

$$\lambda_2 = -0.3625$$

$$\lambda_3 = -0.0414 + 0.0781i$$

$$\lambda_4 = -0.0414 - 0.0781i$$

$$\lambda_5 = -0.0369$$

Three of the eigenvalues are real and two are complex, but all five have negative real parts. We conclude that the equilibrium in the standard case is stable to small perturbations. In the Exercises you will have a chance to verify this calculation and to explore the conditions which lead to stability and instability in this model system.

STABILITY OF FINITE-DIFFERENCE MODELS

The foregoing analysis is appropriate for systems of differential equations but must be modified when the equations are expressed in finite-difference form. The analog of Equation 7.69 in finite-difference form, for example, is

$$\frac{X(t + \Delta t) - X(t)}{\Delta t} = F(X) \tag{7.94}$$

The equilibrium point is still defined by $F(X^*) = 0$. Let $X = X^* + \Delta X$. Then

$$\begin{aligned} X(t + \Delta t) - X(t) &= X^* + \Delta X(t + \Delta t) - X^* - \Delta X(t) \\ &= \Delta X(t + \Delta t) - \Delta X(t) \end{aligned} \tag{7.95}$$

If we expand $F(X(t))$ in a Taylor series about $X = X^*$, we obtain

$$\begin{aligned} F(X^* + \Delta X(t)) &= F(X^*) + \left[\frac{\partial F}{\partial X}\right]_{X=X^*} \Delta X(t) + \cdots \\ &= \alpha \Delta X(t) + \cdots \end{aligned} \tag{7.96}$$

where $\alpha = [\partial F / \partial X]_{X=X^*}$. Substituting the right-hand side of Equation 7.95 for $X(t + \Delta t) - X(t)$ and the right-hand side of Equation 7.96 for $F(X)$ in Equation 7.94 and rearranging, we obtain

$$\Delta X(t + \Delta t) = (1 + \alpha \Delta t) \Delta X(t) \tag{7.97}$$

In order for the equilibrium point to be stable, $|\Delta X(t+\Delta t)/\Delta X(t)| < 1$. The criterion for stability is therefore that $|1+\alpha\Delta t| < 1$. Clearly, this condition requires that the real part of α be negative, which should come as no surprise. If α is real, we now also require that $\alpha\Delta t > -2$. In the limit as $\Delta t \to 0$, this second condition is automatically satisfied. It is not, however, automatically satisfied for finite Δt. When α is complex, the condition that $|1+\alpha\Delta t| < 1$ implies that

$$0 < \Delta t < \frac{-2\,\mathrm{Re}(\alpha)}{|\alpha|^2} \tag{7.98}$$

In the more general case of two or more finite-difference equations, the conditions for a stable equilibrium are that

$$0 < \Delta t < \frac{-2\,\mathrm{Re}(\lambda)}{|\lambda|^2} \tag{7.99}$$

for all eigenvalues of the community matrix.

We should now be able to appreciate the results we obtained in Exercise 6.6. The differential equations in that case were

$$\frac{dX}{dt} = (a - cY)X \tag{7.100}$$

$$\frac{dY}{dt} = (-e + fX)Y \tag{7.101}$$

The community matrix for this problem is

$$\boldsymbol{\alpha} = \begin{bmatrix} 0 & -\dfrac{ce}{f} \\ \dfrac{af}{c} & 0 \end{bmatrix}$$

The eigenvalues of this matrix are $\lambda = \pm\sqrt{-ae}$. Hence $\mathrm{Re}(\lambda) = 0$. The system is neutrally stable and will oscillate indefinitely if displaced from the equilibrium point. If the equations are integrated numerically by the Euler method, however, the amplitude of the oscillations will grow in an exponential manner. We were able to suppress this behavior by using a fourth-order Runge–Kutta integration method. The lesson to be learned here is that integrating a system of differential equations numerically is not a foolproof method of telling whether the equilibrium points are stable to small perturbations. The eigenvalues of the community matrix are needed to answer that question.

EXERCISES

7.1 Why would you expect river runoff to be positively correlated with the temperature of the lower atmosphere (Eq. 7.24)?

7.2 To characterize the effect of atmospheric CO_2 on plant growth, GEOCARB II assumes a functional dependence of the form $[2RCO_2/(1 + RCO_2)]^{0.4}$. This formulation is based on the assumption that about 35% of plants respond to CO_2 fertilization and that when plants respond the maximum effect is a factor of 2 increase in weathering rate. Can you suggest an alternative formulation of the response function which does not involve exponentiating the function $RCO_2/(1 + RCO_2)$?

7.3 Derive Equations 7.33 to 7.36 and 7.43.

7.4 Write a computer program to numerically integrate the GEOCARB II model backward in time to the beginning of the Cambrian. Plot RCO_2, C, G, δ_c, and δ_g versus time.

7.5 Using the program you wrote in Exercise 7.4, plot $T_t - T_0$ versus time. On the same graph, plot $T_t - T_0$ if $RCO_2 = 1$ at all times.

7.6 Consider the following variations on the GEOCARB II model and plot RCO_2 versus time, in each case using the computer program you wrote in Exercise 7.4:

a. The solar constant has really been constant ($W_s = 0$).

b. $f_E = 0.5$ prior to 350 million years ago, that is, the effect of land plants on weathering is much less than assumed in the standard GEOCARB II model.

c. $f_R = f_G = 1$ for all times, that is, there is no change in continental relief and no change in degassing due to tectonics.

7.7 Use induction to prove Equation 7.68. *Hint*: We have already seen that the theorem is true for a one-carnivore food chain. Let the top-carnivore trophic level be trophic level m. If the theorem is true for $m = n - 1$, show that it is true for $m = n$. Note that, for $m = n$,

$$L = f_2 - \sum_{j=3}^{n-1} R_j f_j - R_n f_n - kX_{n+1}$$

$$X_{n+1} = \frac{1}{D + k}\left(\sum_{j=3}^{n-1} S_j f_j + S_n f_n + F\right)$$

$$F = f_2 \prod_{i=2}^{n} Q_i = f_2 Q_n \prod_{i=2}^{n-1} Q_i = f_2(Q_n - 1 + 1) \prod_{i=2}^{n-1} Q_i$$

7.8 Why is the average light intensity from the surface to the 0.1% light level equal to about 14% of the irradiance just below the surface? *Hint*: Assume that $I = I_0 e^{-kz}$, where z is depth. What is the value of kz when $I = 0.001 I_0$?

7.9 We concluded from an analysis of Figure 7.13 that the logistic equation equilibrium point at $X = K$ was globally stable for $X > 0$. Would this conclusion also be true if we allowed X to be negative as well as positive?

7.10 Consider the predator–prey model:

$$\frac{dX}{dt} = (6 - 2X - Y)X$$

$$\frac{dY}{dt} = (8 + 4X - 3Y)Y$$

Find the nonzero equilibrium values of X and Y. Find the elements of the community matrix and the eigenvalues of the community matrix. Find an eigenvector associated with each eigenvalue.

Now suppose dX/dt and dY/dt are replaced by the following finite-difference expressions:

$$\frac{X(t + \Delta t) - X(t)}{\Delta t} = (6 - 2X - Y)X$$

$$\frac{Y(t + \Delta t) - Y(t)}{\Delta t} = (8 + 4X - 3Y)Y$$

What condition must now be satisfied by the eigenvalues of the community matrix to ensure that the equilibrium point you identified in the first part of this exercise is stable to small perturbations?

Numerically integrate the differential Equations from $t = 0$ to $t = 1.2$ using the Euler method starting with initial values $X = 2$ and $Y = 3$. Try the following values of Δt: 0.1, 0.15, and 0.20. Graph X and Y versus time. Based on your stability analysis, how would you account for the pattern of results you obtained with the three values of Δt?

Now try the same numerical integration using the fourth-order Runge–Kutta method you learned in Chapter 6, again with $\Delta t = 0.1$, 0.15, and 0.20. Plot X and Y against time. How would you account for any differences you see in the results of the Euler and Runge–Kutta integrations?

7.11 Set up the community matrix and find the eigenvalues for the standard five-box food chain model assuming that X_2, X_3, and X_4 are growing at 50% of their maximum rates (see Table 7.4).

7.12 Consider the possibility that the growth rates in our five-box food chain model are different from 50% of the maximum rates. Let frac_m be the growth rate of trophic level m expressed as a fraction of its maximum rate. Show that the equilibrium biomass values for the model will remain unchanged if

$$A_m = (L)\prod_{k=1}^{m-1} Q_k/[(X_m)(f\ \text{ratio})(\text{frac}_m)]$$

$$P_m = X_m(1 - \text{frac}_{m+1})$$

7.13 Repeat Exercise 7.11 with the fractional growth rates set to 0.6 and the values of A_m and P_m determined by the equations in Exercise 7.12. Repeat with fractional growth rates of 0.7 and again with fractional growth rates of 0.8. Now try fractional growth rates of 0.4 and 0.3. Make a plot of the least negative or most positive eigenvalue of the community matrix versus fractional growth rate. Is there any systematic correlation between the stability of the equilibrium point and fractional growth rate?

7.14 Numerically integrate Equations 7.45 to 7.49 using a Δt of 1 and the initial values of all the X_m set equal to their equilibrium values except for X_2, which is initially set equal to 1500. Integrate from time 0 to time 300. Set the fractional growth rates equal to 0.5. Repeat this exercise for $\Delta t = 2$ and again for $\Delta t = 3$. How would you account for the difference in the results you obtain?

7.15 There is nothing sacred about the hyperbola $(X_m - P_m)/X_m$ to describe nutrient-limited or food-limited growth. This particular hyperbola implies that the growth rate will be 50% of the maximum value when $X_m = 2P_m$, but why should we think that this is the case? Let us make the model more general to allow for other possibilities. We will make the hyperbola $((X_m - P_m)/X_m)^n$. If $n < 1$ the hyperbola rises more steeply than in the standard case. If $n > 1$ the rise is more gradual. Show that with this form of the hyperbola the equations in Exercise 7.12 become

$$A_m = (L)\prod_{k=1}^{m-1} Q_k/[(X_m)(f\ \text{ratio})(\text{frac}_m)]$$

$$P_m = X_m\left(1 - \sqrt[n]{\text{frac}_{m+1}}\right)$$

Show furthermore that Equation 7.93 becomes

$$\frac{\partial f_m}{\partial X_{m-1}} = \frac{n f_m P_{m-1}}{(X_{m-1} - P_{m-1}) X_{m-1}} \tag{7.102}$$

Repeat Excrcise 7.13 with $n = 0.5$ and again with $n = 2$. Which change makes the equilibrium points more stable?

7.16 Generalize the program you wrote in Exercise 7.11 so that it can handle an arbitrary number of carnivore trophic levels. Test the model using a two-carnivore food chain with an f ratio of 0.2. Assume that X_5 (secondary carnivore N) = 385 and that $Q_5 = 0.35$ and $S_5 = 0.15$. The fractional growth rates of all plant and animal trophic levels are 0.5, the same as the standard model, and the biomass of X_1 (dissolved N), X_6 (detrital N), and all other plant and animal biomass are as given in Table 7.4. Let n (Exercise 7.15) equal 1. The eigenvalues of the community matrix in this case should be

$$\lambda_1 = -0.7652$$
$$\lambda_2 = -0.3286$$
$$\lambda_3 = -0.0262 + 0.0928i$$
$$\lambda_4 = -0.0262 - 0.0928i$$
$$\lambda_5 = -0.0513$$
$$\lambda_6 = -0.0202$$

REFERENCES

Arthur, M. A. 1990. Personal communication to R. A. Berner.

Berner, R. A. 1987. Models for carbon and sulfur cycles and atmospheric oxygen: application to Paleozoic geologic history. *Amer. J. Sci.* **287** 177–196.

Berner, R. A. 1989. Biogeochemical cycles of carbon and sulfur and their effect on atmospheric oxygen over Phanerozoic time. *Global and Planetary Change* **1** 97–122.

Berner, R. A. 1991. A model for atmospheric CO_2 over Phanerozoic time. *Amer. J. Sci.* **291** 339–376.

Berner, R. A. 1994. Geocarb II: a revised model of atmospheric CO_2 over Phanerozoic time. *Amer. J. Sci.* **294** 56–91.

Berner, R. A., A. C. Lasaga, and R. M. Garrels. 1983. The carbonate–silicate geochemical cycle and its effect on atmospheric carbon dioxide over the past 100 million years. *Amer. J. Sci.* **283** 641–683.

Bluth, G. J. S., and L. R. Kump. 1991. Phanerozoic paleogeology. *Amer. J. Sci.* **291** 284–308.

Cawley, J. L., R. C. Burruss, and H. D. Holland. 1969. Chemical weathering in central Iceland: an analog of pre-Silurian weathering. *Science* **165** 391–392.

Cochran, M. F., and R. A. Berner. 1993. Enhancement of silicate weathering rates by vascular land plants: quantifying the effect. *Chem. Geo.* **107** 213–215.

Drever, J. I., and J. Zobrist. 1992. Chemical weathering of silicate rocks as a function of elevation in the southern Swiss Alps. *Geochim. Cosmochim. Acta* **56** 3209–3216.

Ducklow, H. W., D. A. Purdie, P. J. LeB. Williams, and J. M. Davies. 1986. Bacterioplankton: a sink for carbon in a coastal marine plankton community. *Science* **232** 863–867.

Engebretson, D. C., K. P. Kelley, H. J. Cashman, and M. A. Richards. 1992. 180 million years of subduction. *GSA Today* **2** 93–95, 100.

Eppley, R. W. 1972. Temperature and phytoplankton growth in the sea. *Fish. Bull.* **70** 1063–1085.

Eppley, R. W., and B. J. Peterson. 1979. Particulate organic matter flux and planktonic new production in the deep ocean. *Nature* **282** 677–680.

Frost, B. W., and N. C. Franzen. 1992. Grazing and iron limitation in the control of phytoplankton stock and nutrient concentration: a chemostat analogue of the Pacific equatorial upwelling zone. *Mar. Ecol. Prog. Ser.* **83** 291–303.

Gaffin, S. 1987. Ridge volume dependence on seafloor generation rate and inversion using long term sea level change. *Amer. J. Sci.* **287** 596–611.

Huntley, M. E., and M. D. G. Lopez. 1992. Temperature-dependent production of marine copepods: a global synthesis. *Amer. Nat.* **140** 201–242,

Knoll, M. A., and W. C. James. 1987. Effect of the advent and diversification of vascular land plants on mineral weathering through geologic time. *Geology* **15** 1099–1102.

Lalli, C. M., and T. R. Parsons. 1993. *Biological Oceanography: An Introduction.* Pergamon, New York.

Laws, E. A. 1984. Analytic models of NH_4^+ uptake and regeneration experiments. *Limnol. Oceanogr.* **30** 1340–1350.

Laws, E. A. 1991. Photosynthetic quotients, new production, and net community production in the open ocean. *Deep-Sea Res.* **38** 143–167.

Laws, E. A., G. R. DiTullio, P. R. Betzer, and S. Hawes. 1990. Primary production in the deep blue sea. *Deep-Sea Res.* **37** 715–730.

Laws, E. A., D. G. Redalje, L. W. Haas, P. K. Bienfang, R. W. Eppley, W. G. Harrison, D. M. Karl, and J. Marra. 1984. High phytoplankton growth and production rates in oligotrophic Hawaiian coastal waters. *Limnol. Oceanogr.* **29** 1161–1169.

Longhurst, A. R. 1991. Role of the marine biosphere in the global carbon cycle. *Limnol. Oceanogr.* **36** 1507–1526.

Otto-Bliesner, B. 1995. Continental drift, runoff and weathering feedbacks: Implications from climate model experiments. *J. Geophys. Res.* **100** 11537–11548.

Pauly, D., J. Ingles, and R. Neal. 1984. Application to shrimp stocks of objective methods for the estimation of growth, mortality and recruitment-related parameters from length–frequency data (ELEFAN I and II). In J. A. Gulland and B. J.

Rothschild (eds), *Penaeid Shrimps—Their Biology and Management*, Fishing News Books, Farnham, UK.

Raymont, J. E. G. 1983. *Plankton and Productivity in the Oceans. Volume 2. Zooplankton., 2nd ed.* Pergamon, New York.

Urey, H. C. 1952. *The Planets: Their Origin and Development.* Yale University Press, New Haven.

Volk, T. 1989a. Rise of angiosperms as a factor in long-term climatic cooling. *Geolog* **17** 107–110.

Volk, T. 1989b. Sensitivity of climate and atmospheric CO_2 to deep-ocean and shallow-ocean carbonate burial. *Nature* **337** 637–640.

Wilkinson, B. H., and J. C. G. Walker. 1989. Phanerozoic cycling of sedimentary carbonate. *Amer. J. Sci.* **289** 525–548.

Williams, P. J. LeB. 1984. Bacterial production in the marine food chain: the emperor's new suit of clothes? In M. J. R. Fasham (ed.), *Flows of Energy and Materials in Marine Ecosystems. Theory and Practice.* Plenum, New York.

8

TIME SERIES ANALYSIS

A time series has been rather glibly described as "one damned thing after another" (quote attributed to R. A. Fisher). More correctly, a time series is any set of ordered observations of a variable for which either time or distance provides the ordering metric. Figures 8.1 to 8.3 provide three examples of time series. Figure 8.1 shows the monthly mean concentrations of CO_2 in the atmosphere measured at a station atop the Mauna Loa volcano on the island of Hawaii from 1958 to 1993. Figure 8.2 shows annual recruitment to the southern stock of silver hake off the northeastern coast of the United States from 1955 to 1984. Figure 8.3 is a time series of 12-month running means of the Southern Oscillation Index (SOI), the difference in sea level pressure between Easter Island and Darwin, Australia.

The principal goal of time series analysis is to develop quantitative methods which allow one to characterize time series, in particular, to say quantitatively how two series differ or how they are related (Percival and Walden, 1993). Some of the practical applications of time series analysis are the following:

1. *Testing of theories*. For example, can the pattern of the ice ages be explained by variations in the characteristics of the orbit of the earth around the sun (Milankovitch theory)?
2. *Investigating data*. In geophysics, for example, time series analysis has been used to study the attenuation of compressional waves and intraplate volcanism in the Pacific Ocean basin. In oceanography, time series analysis has been used to study tidal cycles and the effect of pressure and wind stress on the ocean.

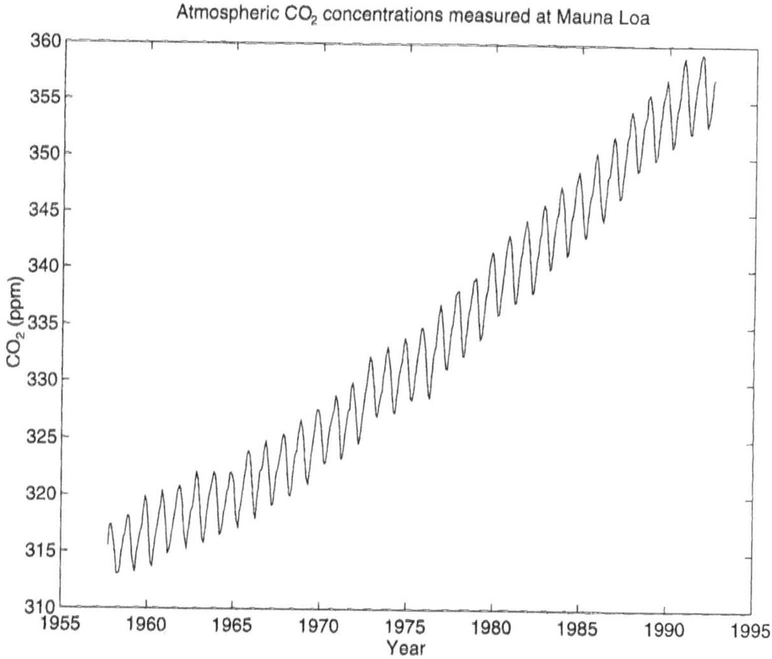

Figure 8.1 Monthly mean concentrations of CO_2 in the atmosphere measured at a station atop the Mauna Loa volcano on the island of Hawaii.

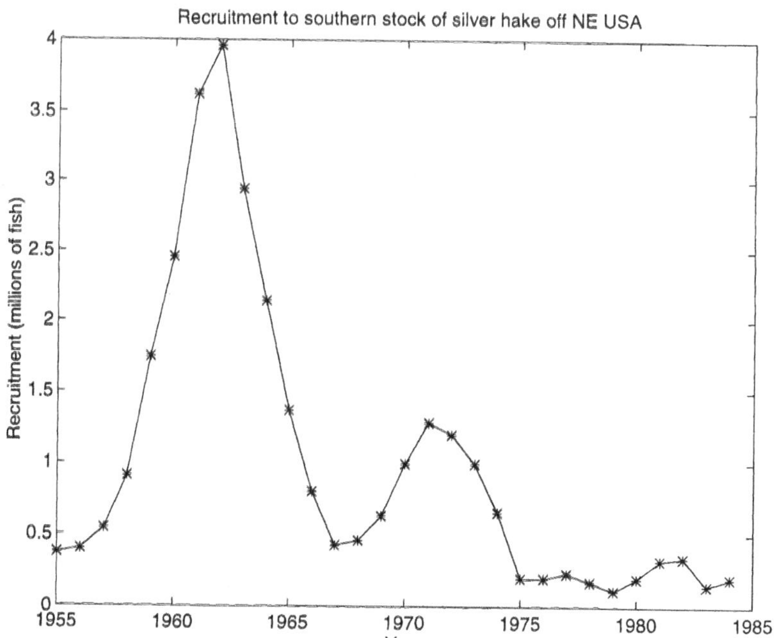

Figure 8.2 Annual recruitment to the southern stock of silver hake off the northeastern coast of the United States from 1955 to 1982 (Sissenwine et al., 1988).

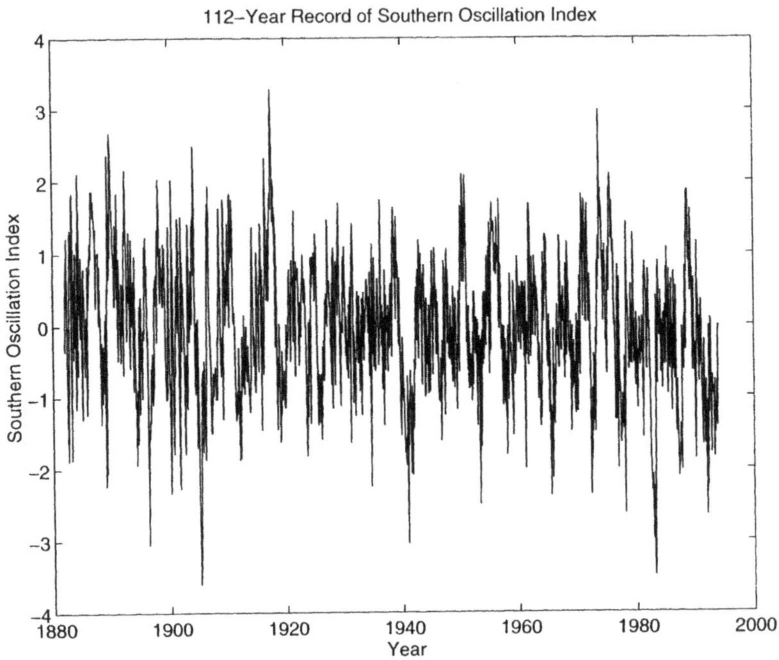

Figure 8.3 Twelve-month running mean of the Southern Oscillation Index.

3. *Discriminating data*. What are the differences, for example, in the records of sea level at locations throughout the Pacific Ocean?
4. *Performing diagnostic tests*. If we fit the data in Figure 8.1, for example, to a simple exponential function or low-order polynomial, the residuals would form a second time series. If the errors in the data are random and the function we chose accounted for the essential features of the data, the residuals would form a series of random numbers. Time series analysis could be used to determine whether the pattern of the residuals was consistent with the results expected for a series of random numbers. In the case of Figure 8.1, it should be obvious that the deviations from a simple exponential function or low-order polynomial would not form a random time series.
5. *Assessment of predictability*. One very practical application of time series analysis is prediction. Predicting the tides is one obvious example. Could we use data such as those shown in Figure 8.3 to predict the occurrence of the next El Niño? With what confidence could we predict the onset of the next ice age?

Of the time series shown in Figures 8.1 to 8.3, the Mauna Loa CO_2 data are probably the easiest to explain in terms of mechanisms. Even a casual examination of the data reveals that there is an annual cycle, with CO_2 concentrations falling in the Northern Hemisphere in the summer and rising in the winter. It does not take much imagination to realize that this pattern reflects the photosynthetic uptake of CO_2 by plants during the summer and the release of CO_2 by respiration during the winter when photosynthetic rates are minimal. Superimposed on this annual cycle is a long-term rise in CO_2, which amounted to about 1.2 ppm yr^{-1} from 1958 to 1993. This annual increase largely reflects the burning of fossil fuels by humans.

Figure 8.2 is probably the most challenging of the three figures from the standpoint of analysis. It seems fair to say, even from a casual examination of the ordered data, that the pattern of recruitment from one year to the next is not random, but the nature of the pattern is not as easy to describe as in the case of the Mauna Loa CO_2 data. The SOI time series is intermediate in terms of difficulty of analysis. There is certainly a pattern, but the amplitude and frequency of the oscillations are much more irregular than is the case for the CO_2 data.

CLASSIFICATION OF TIME SERIES DATA

Before discussing techniques for analyzing time series data, it will be useful to consider in general terms the types of data which we may be dealing with. How we approach the problem of analysis will depend on the nature of the data. Time series data are considered to fall into two general categories, deterministic data and random data. *Deterministic data* are data which can be described or predicted by an explicit mathematical relationship. The relationship may be derived either from theory or from observation. The distance between the earth and the sun is an example of a time series which can be described by an explicit mathematical relationship. *Random data* are data which cannot be described or predicted by an explicit mathematical relationship, because each series of observations is unique. An example of a time series of random data would be the daily rainfall at a particular location. Analysis of many years of such data might allow us to characterize the data in a statistical sense. For example, we might discover that there was a rainy season and a dry season. We could not, however, predict the amount of rainfall on a particular day with an explicit mathematical relationship.

In truth the data in many time series lie somewhere between strictly deterministic and random. There is always a limit on how accurately we can predict a phenomenon with mathematical equations. If the errors are insignificant, we may consider the phenomenon deterministic, even though we know there are small errors in our predictions. On the other hand, our inability to describe or predict certain phenomena may simply reflect out lack of understanding of the laws of nature or our inability to express their effects with a mathematical equation. In other words, the phenomenon may well be deterministic, but mathematical limitations and/or our lack of understanding prevent us from describing or predicting the phenomenon with an explicit mathematical relationship. Note that random data as defined here by no means necessarily imply disorderly data. The pattern of rainy days and dry days at a particular location in a given year may follow a very nonrandom pattern, with long periods of dry weather followed by several days of rain when a low-pressure system moves over the area. Such data, however, constitute a random time series, because we cannot predict whether it will rain or not on a given day in a given year with an explicit mathematical function.

DETERMINISTIC DATA

Deterministic data are commonly classified as indicated in Figure 8.4. *Periodic data* are data that can be described by a function $f(t)$ which satisfies the equation

$$f(t) = f(t \pm n\tau) \tag{8.1}$$

where τ is the period of the function and n is any integer. The frequency of the phenomenon, or the number of complete cycles per unit time, is τ^{-1}. The number of daylight hours at a particular location on the earth is an example of a periodic function with a period τ of one year. *Sinusoidal data* are periodic data which can be described by a function $f(t)$ of the form

$$f(t) = A \sin\left(2\pi\frac{t}{\tau} + \theta\right) \tag{8.2}$$

where A is the amplitude of the function and θ is the phase angle. *Complex periodic data* are data which, with few exceptions in practice, can be described

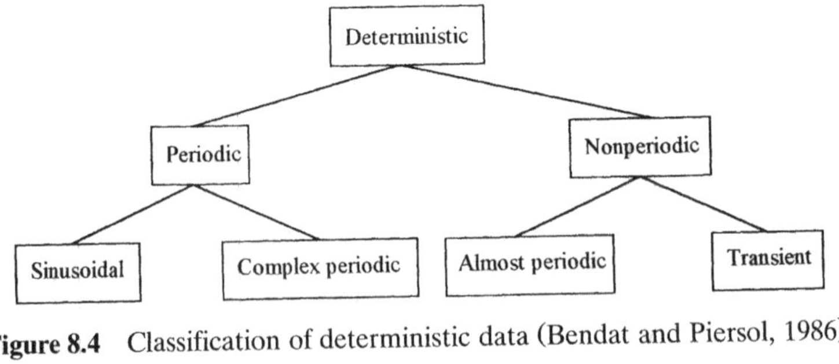

Figure 8.4 Classification of deterministic data (Bendat and Piersol, 1986).

by a function of the form

$$f(t) = \frac{A_0}{2} + \sum_{n=1}^{\infty}\left[A_n \cos\left(2\pi n\frac{t}{\tau}\right) + B_n \sin\left(2\pi n\frac{t}{\tau}\right)\right] \tag{8.3}$$

where

$$A_n = \frac{2}{\tau}\int_{-\tau/2}^{\tau/2} f(t)\cos\left(\frac{2\pi nt}{\tau}\right) dt \tag{8.4}$$

$$B_n = \frac{2}{\tau}\int_{-\tau/2}^{\tau/2} f(t)\sin\left(\frac{2\pi nt}{\tau}\right) dt \tag{8.5}$$

When A_n and B_n are given by Equations 8.4 and 8.5, respectively, the right-hand side of Equation 8.3 is called the *Fourier series* expansion of the function $f(t)$. The *fundamental period* of the function is τ, and τ^{-1} is referred to as the *fundamental frequency*, that is, the frequency at which the process repeats itself. For example, the function

$$f(t) = A_1 \sin\left(\frac{2\pi t}{3}\right) + A_2 \sin\left(\frac{2\pi t}{5}\right) \tag{8.6}$$

is a complex periodic function with a fundamental period of 15. The number of hours of daylight at a particular location on the earth is a good example of a complex periodic function.

Nonperiodic data are divided into two categories, almost periodic and transient. An *almost periodic* function $f(t)$ can be described by an equation of the form

$$f(t) = \sum_{n=1}^{\infty}\left[A_n \sin\left(\frac{2\pi t}{\tau_n} + \theta_n\right)\right] \tag{8.7}$$

where the ratio τ_m/τ_n is not equal to a rational number for at least one pair of periods. For example, the function

$$f(t) = A_1 \sin\left(\frac{2\pi t}{3} + \theta_1\right) + A_2 \sin\left(\frac{2\pi t}{5} + \theta_2\right) + A_3 \sin\left(\frac{2\pi t}{\sqrt{37}} + \theta_3\right) \tag{8.8}$$

is an almost periodic function, because $\sqrt{37}/3$ and $\sqrt{37}/5$ are not rational numbers. If $\sqrt{37}$ were replaced by $\sqrt{36} = 6$, the function would be periodic with a fundamental period of 30. In general, the sum of two or more periodic functions will be periodic only if the ratio τ_m/τ_n is a rational number for all possible pairs of τ_m and τ_n. Almost periodic functions are functions that can be expressed as a sum of periodic functions, but they are not themselves periodic.

Transient functions in time series analysis are functions which can be expressed as an explicit function of time but do not fall into one of the three categories just discussed. A good example of a transient function would be the biomass or numbers of an organism whose population dynamics are described by the logistic equation and which is displaced at time $t = 0$ from its equilibrium value of $X(t) = K$ by a small amount of ΔX_0. As we have seen in Chapter 7, the function describing $X(t)$ will be

$$X(t) = K + \Delta X_0 e^{-rt} \tag{8.9}$$

This is certainly not a periodic function, nor is it a sum of periodic functions, but it is a function which can be expressed by an explicit mathematical relationship and is hence deterministic.

RANDOM DATA

Random data are commonly classified as indicated in Figure 8.5. Since the data cannot be described or predicted by an explicit mathematical relationship, they are characterized by statistics. A single time history of a random or *stochastic process* $X(t)$ is called a *sample function*, and a finite portion of a time history is a *sample record*. The collection of all possible sample functions that a random process $X(t)$ might have produced is called the *ensemble* of $X(t)$ and is denoted $\{X(t)\}$. The mean value of the process at time t is the expectation value of the process at time t and is calculated by taking the average of all possible sample functions at time t:

$$\mu_x(t) = E[X(t)] = \lim_{N \to \infty} \frac{1}{N} \sum_{j=1}^{N} X_j(t) \tag{8.10}$$

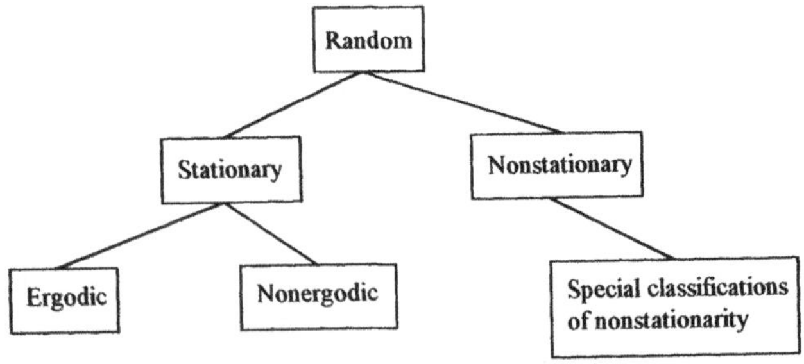

Figure 8.5 Classification of random data (Bendat and Piersol, 1986).

where $\mu_x(t)$ is the average value of process X at time t, $X_j(t)$ denotes the value of the jth sample function at time t, and we have assumed that each sample function is equally probable. The mean value so defined is also referred to as the *first moment* of the random process (see Chapter 6). The *autocorrelation function* of the random process is the expectation value or ensemble average of the product of sample functions at times separated by an interval Δt:

$$R_{xx}(t, \Delta t) = E[X(t)X(t + \Delta t)] = \lim_{N \to \infty} \frac{1}{N} \sum_{j=1}^{N} X_j(t)X_j(t + \Delta t) \quad (8.11)$$

R_{xx} is an example of a *joint moment* and is a measure of the correlation between the values of the process at two times. One may similarly define the *covariance function* C_{xx} by

$$C_{xx}(t, \Delta t) = E\{[X(t) - \mu_x(t)][X(t + \Delta t) - \mu_x(t + \Delta t)]\}$$
$$= \lim_{N \to \infty} \frac{1}{N} \sum_{j=1}^{N} \{[X_j(t) - \mu_x(t)][X_j(t + \Delta t) - \mu_x(t + \Delta t)]\} \quad (8.12)$$

If either μ_x or R_{xx} varies with time or, equivalently, if either μ_x or C_{xx} varies with time, the process is considered to be *nonstationary*. If μ_x is constant, then

$$C_{xx}(t, \Delta t) = R_{xx}(t, \Delta t) - \mu_x^2 \quad (8.13)$$

Hence if μ_x is constant, then C_{xx} is independent of t and dependent only on Δt if and only if R_{xx} is independent of t and dependent only on Δt. If μ_x is constant and R_{xx} or, equivalently, C_{xx} is dependent only on Δt, then the process is considered to be *weakly stationary*. If all higher-order moments and joint moments are also time invariant, then the process is said to be *strongly stationary*. Since there are an infinite number of possible higher-order moments and joint moments, demonstration or proof that all moments and joint moments are time invariant is usually impossible. For practical purposes, however, demonstration that a process is weakly stationary will often justify the assumption that the process is strongly stationary. When $\mu_x \neq 0$ it is more convenient to work with C_{xx} than R_{xx} (Bendat and Piersol, 1986).

The definitions of R_{xx} and C_{xx} in Equations 8.11 and 8.12 are consistent with the engineering literature, but the definition of R_{xx} in particular is somewhat inconsistent with the concept of correlation coefficient in statistics. Not surprisingly, some time series analysts prefer to define an autocorrelation function $\rho_{xx}(t, \Delta t)$ as

$$\rho_{xx}(t, \Delta t) = \frac{E\{[X(t) - \mu_x(t)][X(t + \Delta t) - \mu_x(t + \Delta t)]\}}{\sqrt{E\{[X(t) - \mu_x(t)]^2\}E\{[X(t + \Delta t) - \mu_x(t + \Delta t)]^2\}}}$$

$$= \frac{C_{xx}(t, \Delta t)}{\sqrt{C_{xx}(t,0)C_{xx}(t + \Delta t, 0)}} \tag{8.14}$$

One can then, for example, define a weakly stationary random process as one for which μ_x is independent of time and ρ_{xx} is dependent only on Δt. It is not our intention to take sides on this issue, but rather to warn the reader that in the literature on time series analysis there is disagreement on how to define the autocorrelation function. Being aware of this fact can help to avoid confusion. In the remainder of this chapter, we will refer to ρ_{xx} as the *autocorrelation coefficient function* to distinguish it from R_{xx}.

It is quite common for a scientist to have available only one or a few sample records with which to estimate the statistical properties of a random process. Obviously, this situation precludes calculating ensemble averages using equations such as Equations 8.10 to 8.12. Ensemble averages may, however, be estimated from a single sample function for stationary random processes if the random process is ergodic. An *ergodic random process* is a stationary random process for which the time average of any property such as μ_x, R_{xx}, or C_{xx} is the same for each sample function and equal to the ensemble average of the corresponding property. For sample function X_j the time-averaged values of μ_x, R_{xx}, C_{xx}, and ρ_{xx} are calculated as follows:

$$\mu_x(j) = \lim_{T\to\infty} \frac{1}{T}\int_0^T X_j(t)\, dt \tag{8.15}$$

$$R_{xx}(\Delta t, j) = \lim_{T\to\infty} \frac{1}{T}\int_0^T X_j(t)X_j(t + \Delta t)\, dt \tag{8.16}$$

$$C_{xx}(\Delta t, j) = \lim_{T\to\infty} \frac{1}{T}\int_0^T \{[X_j(t) - \mu_x(t)][X_j(t + \Delta t) - \mu_x(t + \Delta t)]\}\, dt \tag{8.17}$$

$$\rho_{xx}(\Delta t, j) = \frac{C_{xx}(\Delta t, j)}{C_{xx}(0, j)} \tag{8.18}$$

If the stationary random process is ergodic, then these averages are independent of j. Hence $\mu_x(j) = \mu_x$, $R_{xx}(\Delta t, j) = R_{xx}(\Delta t)$, $C_{xx}(\Delta t, j) = C_{xx}(\Delta t)$, and $\rho_{xx}(\Delta t, j) = \rho_{xx}(\Delta t)$, where μ_x, $R_{xx}(\Delta t)$, $C_{xx}(\Delta t)$, and $\rho_{xx}(\Delta t)$ are ensemble averages. In practice, stationary random processes are often ergodic, in which case their statistical characteristics may be estimated from the analysis of a single time series.

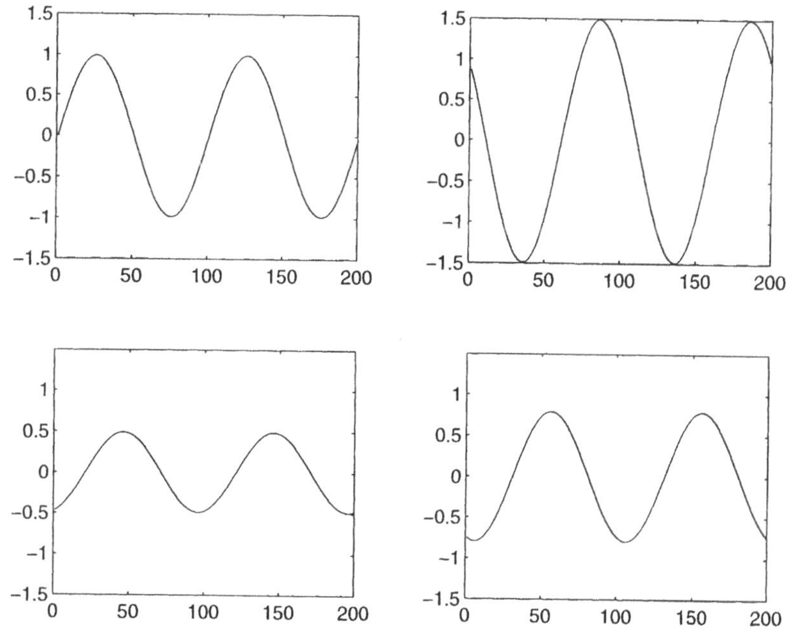

Figure 8.6 Sample records from sample functions of a nonergodic random process. The functions are sine waves with a period of 100 but with random amplitudes and phase angles.

Figure 8.6 is an example of four sample records drawn from sample functions of a stationary but nonergodic random process. The random process is a sine wave with a random amplitude and phase angle and a period of 100. Because the phase angle is random, the expectation values of all moments and joint moments are independent of time. In this case $\mu_x(j) = 0$ for all j, but because the amplitudes are random, $R_{xx}(\Delta t, j)$ and $C_{xx}(\Delta t, j)$, which are equal in this example, will depend on j.

Nonstationary random processes may prove difficult to analyze, because it is often not feasible or practical to obtain enough sample records to carry out ensemble averaging. Analysis of a single sample record may, however, be sufficient in special cases, particularly when the nonstationary process involves a time-dependent deterministic component and a stationary random component. Consider, for example, the function

$$f_j(t) = \alpha + \beta t + \varepsilon_j \tag{8.19}$$

where ε_j is a random error term with an expectation value of 0. The function $f_j(t)$ should seem like an old friend to us, but it is a sample function from a nonstationary random process. However, since $f_j(t)$ is the sum of a deterministic function and a sample function from an ergodic random process, the characteristics of $\{f(t)\}$, in a statistical sense, are easily calculated from a single sample function. Consider also the function

$$g_j(t) = f_j(t)u(t) \tag{8.20}$$

where $f_j(t)$ is given by the right-hand side of Equation 8.19 and $u(t)$ is a deterministic multiplication factor. The function $g_j(t)$ is again a sample function from a nonstationary random process. However, because of the special form of the function, the statistical characteristics of $\{g(t)\}$ can be calculated from analysis of a single sample function.

STATIONARY SAMPLE RECORDS

When we say that a random process is stationary, we mean that the ensemble-averaged characteristics of the process are time invariant. The term stationary is also used with respect to sample records, and in that case has a different meaning. When we say that a sample record is stationary, we mean that the mean value and the autocorrelation function calculated over short time intervals do not vary significantly from one time interval to the next. The differences are judged to be significant if they are greater than would be expected due to normal statistical sampling variations. If the width of the time interval is T, then $\mu_x(t_0)$, $R_{xx}(t_0, \Delta t)$, $C_{xx}(t_0, \Delta t)$, and $\rho_{xx}(t_0, \Delta t)$ are estimated from the jth sample function from the following equations:

$$\mu_x(t_0, j) = \frac{1}{T}\int_{t_0}^{t_0+T} X_j(t)\,dt \tag{8.21}$$

$$R_{xx}(t_0, \Delta t, j) = \frac{1}{T}\int_{t_0}^{t_0+T} X_j(t)X_j(t+\Delta t)\,dt \tag{8.22}$$

$$C_{xx}(t_0, \Delta t, j) = \frac{1}{T}\int_{t_0}^{t_0+T} \{[X_j(t) - \mu_x(t)][X_j(t+\Delta t) - \mu_x(t+\Delta t)]\}\,dt \tag{8.23}$$

$$\rho_{xx}(t_0, \Delta t, j) = \frac{C_{xx}(t_0, \Delta t, j)}{\sqrt{C_{xx}(t_0, 0, j)C_{xx}(t_0+\Delta t, 0, j)}} \tag{8.24}$$

If the random process is ergodic, then a sample record from a sample function of the process will be stationary; that is, the differences between $\mu_x(t_0, j)$, $R_{xx}(t_0, \Delta t, j)$, $C_{xx}(t_0, \Delta t, j)$, and $\rho_{xx}(t_0, \Delta t, j)$ for different values of t_0 will be within the range of variability expected due to normal statistical sampling variations. It is also true that sample records from most scientifically interesting nonstationary random processes will be nonstationary. Hence if it is reasonable to assume that a random process is ergodic if it is stationary, then verification that a sample record from a sample function of the process is stationarity justifies the assumption of stationarity and ergodicity for the random process.

TIME DOMAIN TECHNIQUES

We have noted that the goal of time series analysis is to develop quantitative methods which allow one to characterize time series. There are two general classes of such characterizations, time domain techniques and frequency domain techniques. Time domain techniques are not as powerful as frequency domain methods, but they are easier to understand, and the study of time domain methods lays some groundwork for the study of frequency domain techniques. It is therefore worth spending a little time in the time domain before moving on to the frequency domain. We will use the sample records in Figures 8.1 and 8.3 for illustrative purposes.

The first step in analyzing a sample record is to look at the data. In other words, look at a plot of $X(t)$ versus t. A visual examination of the data can serve two purposes. First, it can serve to identify obviously nonstationary data. One might argue, for example, that the data in Figure 8.1 came from a nonstationary random process. When an obvious temporal trend is apparent, the conclusion of nonstationarity is justified if (1) it is reasonable to assume that the sample record reflects the temporal characteristics of the random process and (2) the sample record is long compared to the longest period associated with random fluctuations in the data. In the case of Figure 8.1, this second caveat is nontrivial. Figure 7.11 makes it clear that the roughly 10% increase in atmospheric CO_2 that has occurred during the last 35 years is small compared to changes that have occurred over geologic time scales in the past. When a temporal trend is apparent in a sample record, it is impossible without additional information to say whether the trend is the result of a nonstationary process or a long-period component of a stationary random process.

The second purpose of a visual examination is to identify bad data. Data points which lie far above or below adjacent points in the time series should be regarded as suspect. They may be real, but they may also be the result of equipment failure or human error. If there is good reason to believe that they are bad data and it is not obvious how to correct the error, the data should be thrown out.

Bad Data

When there are many data in a time series and visual inspection indicates that one or more of the data points may be bad, it is often desirable to let a computer find the bad data and discard them. In fact, it may be desirable to

TABLE 8.1 Intervals Within Which All Data Would Be Expected to Fall 95% of the Time if the Data Are Normally Distributed and μ and σ Are Estimated by $\bar{X}$ and s

Sample Size	Interval
100	$\bar{X} \pm 3.60s$
1,000	$\bar{X} \pm 4.07s$
10,000	$\bar{X} \pm 4.56s$
100,000	$\bar{X} \pm 5.02s$

run a check with a computer even if visual inspection does not determine that there are bad data. If the sample record appears to be stationary, the usual method of checking for bad data is to calculate the mean and standard deviation of the entire record and then to assume that the data are normally distributed. One should be conservative in throwing out data, the null hypothesis being that all the data are good. The criterion for throwing out data is that the probability of making a type I error, that is, incorrectly rejecting the null hypothesis, is no more than 5%. Assume, for example, that there are 1000 data points in our sample record. We assume that the data are normally distributed, but we do not know the true mean and standard deviation. We estimate μ with $\bar{X}$ and σ with s. Based on the t distribution function, there is a 95% probability that all 1000 of the data points will lie within $\pm 4.07s$ of $\bar{X}$. We therefore discard points that lie outside of $\bar{X} \pm 4.07s$. Table 8.1 gives the corresponding intervals for sample records having 100 to 10^5 data points. It is a good idea to repeat this process if some of the original data lay well outside the intervals indicated in Table 8.1, since bad data distort $\bar{X}$ and s.

Figure 8.7 shows the results of applying this test to the Mauna Loa CO_2 and Southern Oscillation Index sample records. In both cases all the data fell within the 95% confidence limit bounds, in the former case even though the sample record shows an obvious temporal trend. There appear to be no bad data, and hence all data were included in the subsequent calculations.

Test for Stationarity

The next test is for stationarity. The null hypothesis is that the sample record is stationary. The criterion for a stationary record is that $\mu_x(t_0, j)$ and either $R_{xx}(t_0, \Delta t, j)$, $C_{xx}(t_0, \Delta t, j)$, or $\rho_{xx}(t_0, \Delta t, j)$ calculated from the right-hand sides of Equations 8.21 to 8.24 show no temporal trend. There are several ways to test this assumption, and none is sacred. A common procedure is to divide the sample record into subintervals containing equal or nearly equal numbers of data points. The subintervals should be long enough so that the data in each subinterval can be considered independent, and there should be

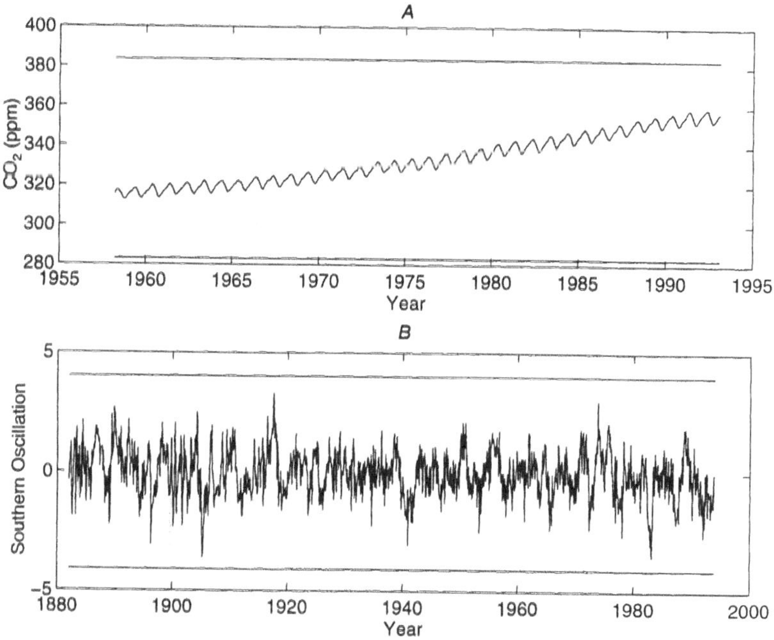

Figure 8.7 Mauna Loa CO_2 data (A) and Southern Oscillation Index data (B) with estimates of the upper and lower bounds within which all the data would be expected to fall 95% of the time if the data were normality distributed.

a minimum of roughly seven to eight subintervals. One then calculates $\overline{X}$ and either $\overline{X^2}$ or $\overline{(X-\overline{X})^2}$ for each subinterval. A simple test for trends is then to rank the $\overline{X}$ and $\overline{X^2}$ or $\overline{(X-\overline{X})^2}$ statistics and calculate either the Spearman rank or the Kendall rank correlation coefficient for both statistics. We reject the null hypothesis that the sample record is stationary using the critical values for the correlation coefficients in Appendices D and E. The conclusion that $\overline{X^2}$ or $\overline{(X-\overline{X})^2}$ is stationary obviously does not guarantee that R_{xx} or C_{xx} is stationary for all values of Δt. It is uncommon, however, for R_{xx} or C_{xx} to be time dependent for $\Delta t \neq 0$ and time invariant for $\Delta t = 0$. Thus acceptance of the hypothesis that there is no temporal trend in $\overline{X^2}$ or $\overline{(X-\overline{X})^2}$ is generally considered sufficient to accept the hypothesis that there is no temporal trend in R_{xx} or C_{xx}. Note that $\rho_{xx}(t_0, \Delta t, j)$ is, by definition, independent of t_0 when $\Delta t = 0$.

An alternative to a rank correlation test on $\overline{X}$ and $\overline{X^2}$ or $\overline{(X-\overline{X})^2}$ is a runs test. A run is an uninterrupted sequence of either positive or negative results. Suppose, for example, that we perform a series of experiments, the results of the experiments being recorded as either positive or negative. Assume further that there are N_1 positive results and N_2 negative results. If the N_1 positive results and N_2 negative results are randomly distributed among the $N = N_1 + N_1$ experiments, the average number of runs, that is, the average number of sequences of all positive or all negative experimental results, will be $(2N_1N_2)/N + 1$. The minimum number of runs is clearly 2 as long as N_1 and N_2 are both greater than 0. If $N_1 = N_2$ the maximum possible number of runs is N. If $N_1 < N_2$ the maximum possible number of runs is $2N_1 + 1$. As N_1 and N_2 both become large, the probability that the number of runs in a random sequence will be close to the maximum or minimum possible value becomes small. Hence an unusually small or unusually large number of runs can be taken as evidence that the sequence of positive and negative results was probably not drawn from a random sample.

Perhaps the most common use of the runs test is to compare a series of numbers to their median value. In this case $N_1 = N_2$, values above the median being given a positive score and values below a negative score. If there is a temporal trend in a time series, one expects the number of runs above and below the median to be lower than usual. Hence the test of runs above and below the median, when applied to a sequence of $\overline{X}$, $\overline{X^2}$, or $\overline{(X - \overline{X})^2}$ values to determine whether there is a temporal trend, is a one-sided test. An unusually large number of runs would imply that the value of T in Equations 8.21 to 8.23 is too small; that is, the data in successive subintervals are not independent. Appendix F lists the critical number of runs above and below the median for $N_1 = N_2 \leq 100$.

Table 8.2 gives the results of applying the Kendall and Spearman rank correlation tests to the Southern Oscillation Index data in Figure 8.3. In this case the sample record spans 112 years, and the data were divided into 14 subsets of exactly 8 years duration. At $p = 0.05$ the critical number of runs for $N_1 = N_2 = 4$ is 2, and none of the time series of $\overline{X}$, $\overline{X^2}$, or $\overline{(X - \overline{X})^2}$ contained less than 7 runs. The critical values at $p = 0.05$ for the magnitude of the Kendall and Spearman correlation coefficients in this case are 0.407 and 0.539. None of the calculated coefficients exceeds the respective critical value. Hence we accept the null hypothesis that the Southern Oscillation Index sample record is stationary. At the risk of belaboring the obvious, the 35-year Mauna Loa CO_2 sample record was divided into seven subsets of

TABLE 8.2 Kendall and Spearman Rank Correlation Coefficients for Southern Oscillation Index Time Series (Fig. 8.3) Divided into 14 Subsets with a Duration T of 8 Years Each

Kendall			Spearman		
$\overline{X}$	$\overline{X^2}$	$\overline{(X - \overline{X})^2}$	$\overline{X}$	$\overline{X^2}$	$\overline{(X - \overline{X})^2}$
−0.363	−0.187	−0.077	−0.459	−0.235	0.077

exactly 5 years duration. The Kendall and Spearman correlation coefficients for $\overline{X}$, $\overline{X^2}$, or $\overline{(X - \overline{X})^2}$ were all 1.0. On the basis of this test, we reject the null hypothesis that the Mauna Loa CO_2 data come from a stationary random process.

Presence of Periodicities

We have now concluded that our sample records contain no bad data and that one is stationary and the other nonstationary. What other differences or similarities can we find in the sample records? We notice that both sample records seem to oscillate, and since it is reasonable to assume that what we are seeing may have something to do with meteorological or biological phenomena, we might ask whether there is a 12-month cycle in the sample records. To check this hypothesis, we might plot $X(t)$ versus $X(t + 6)$ or $X(t)$ versus $X(t + 12)$, where t is time in months. If there is an annual cycle, we might expect $X(t)$ and $X(t + 6)$ to be negatively correlated and $X(t)$ and $X(t + 12)$ to be positively correlated.

Figure 8.8*A* and *B* shows plots for the Mauna Loa CO_2 data in Figure 8.1. Both plots reveal a high degree of correlation between the variables being plotted, and Figure 8.8*B* does show the expected positive correlation. Figure 8.8*A*, however, also shows positive correlation, and if we plot $\rho_{xx}(\Delta t)$ for this sample record, we find that, although it does show an obvious cyclical pattern (Fig. 8.8*C*), the values of $\rho_{xx}(\Delta t)$ all lie between 0.94 and 1.00 for $0 < \Delta t < 60$ months. What is the explanation for this peculiar behavior?

The answer, of course, is that we are not dealing with a stationary sample record. In fact, the sample record has a very obvious trend toward higher CO_2 concentrations with increasing time over the entire period of the record. Therefore CO_2 concentrations which are above (below) the mean of the variable on the ordinate tend to be paired with CO_2 concentrations which are similarly above (below) the mean of the variable on the abscissa when we calculate $C_{xx}(\Delta t)$. The result is a positive correlation as long as Δt is a small fraction of the sample record length.

This example illustrates rather well why we need to detrend nonstationary sample records for purposes of time series analysis. In a case such as this, the detrending operation may be as simple as drawing a smooth curve through the sample record. Figure 8.9*A* shows a smooth curve, in this case a third-order polynomial fit by least squares, drawn through the data in Figure 8.1. The residuals, that is, the deviations between the curve and the data, are shown in Figure 8.9*B*. There is no visual evidence of a trend in the residual plot, and when we compare residual CO_2 concentrations separated by either 6 or 12 months, we obtain the expected results. When the lag time is 6 months, there is a negative correlation (Fig. 8.9*C*), and when the lag time is 12 months, the correlation is positive (Fig. 8.9*D*). When we now plot $\rho_{xx}(\Delta t)$, we find that it is close to 1 when Δt is an integer multiple of 12 months and

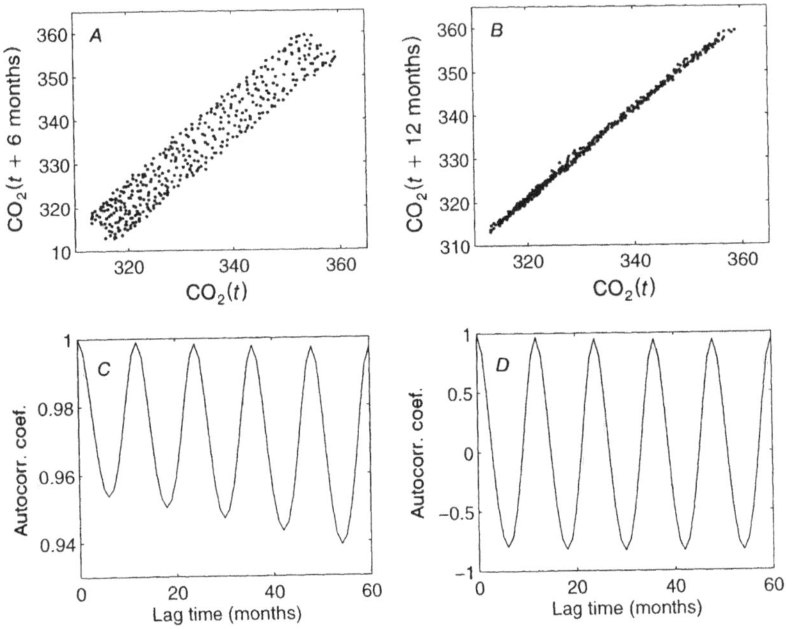

Figure 8.8 (*A*) Mauna Loa CO_2 concentrations versus concentrations 6 months later. (*B*) Mauna Loa CO_2 concentrations versus concentrations 12 months later. (*C*) Autocorrelation coefficient function for data in Figure 8.1. (*D*) Autocorrelation coefficient function for detrended data in Figure 8.9*B*.

close to -1 when Δt is a half-integer multiple of 12 months (Fig. 8.8*D*). This behavior is consistent with our hypothesis that there is an annual cycle in the data.

The Southern Oscillation Index (SOI) data provide an interesting comparison. The sample record, as already confirmed, appears stationary. Hence there is no need to detrend anything. The lag time plots of monthly mean SOI values separated by 6 or 12 months, Figures 8.10*A* and *B*, show very little correlation. This is due to the fact that the data are 12-month running means. The result is that the annual cycle of the SOI has been effectively filtered out. When we plot $\rho_{xx}(\Delta t)$ for the SOI data (Fig. 8.10*C*), we find that there is very poor correlation at time lags greater than 1 year. We do see a peak at a time lag of about 60 months (5 years) and a trough at a time lag of about 120 months (10 years), but the associated ρ_{xx} values are only about 0.1 and -0.1, respectively. The data are clearly much noisier than the Mauna Loa CO_2 time series. These observations are consistent with two characteristics of the El Niño cycle. The cycle has a periodicity that averages about 5 years, and the interval between El Niño events is irregular.

The rather small magnitude of $\rho_{xx}(\Delta t)$ in the case of the SOI sample record raises the question of whether any of the values are significant in a

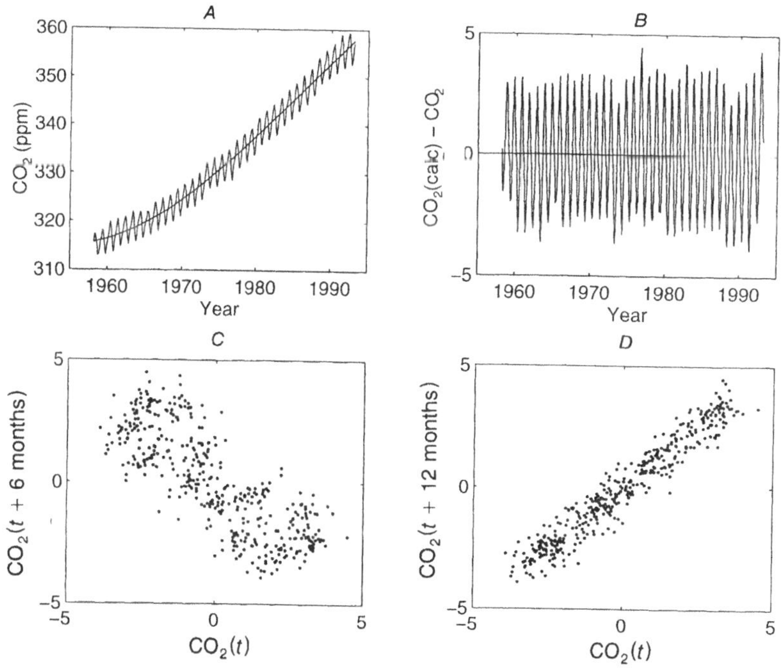

Figure 8.9 (*A*) Mauna Loa CO_2 data with cubic polynomial trend curve. (*B*) Detrended Mauna Loa CO_2 data, that is, the difference between the original data and the trend curve. (*C*) Detrended Mauna Loa CO_2 concentrations versus concentrations 6 months later. (*D*) Detrended Mauna Loa CO_2 concentrations versus concentrations 12 months later.

statistical sense. When calculating $\rho_{xx}(\Delta t)$ two rules of thumb should be followed. First, the number of data points in the sample record should be large, and second, the lag time Δt should be small compared to the length of the record. Large in this context means greater than about 50, and small means less than about 25% of the record length (Davis, 1986). In the case of the SOI data, for example, the sample record spans 112 years and contains 1344 data points, so the first rule of thumb is easily satisfied. According to the second rule of thumb, the maximum Δt should be about 28 years. These rules of thumb arise from the fact that as Δt increases, $\rho_{xx}(\Delta t)$ is based on fewer and fewer observations. This causes an increase in the variance of $\rho_{xx}(\Delta t)$ and results in an increasing violation of Equations 8.17 and 8.18, which state that $\rho_{xx}(\Delta t)$ was calculated from an infinitely long time series. The variance of $\rho_{xx}(\Delta t)$ is given by the equation (Davis, 1986, p. 223)

$$V[\rho_{xx}(\Delta t)] = \frac{1}{N\left(1 - \frac{\Delta t}{T}\right) + 3} \tag{8.25}$$

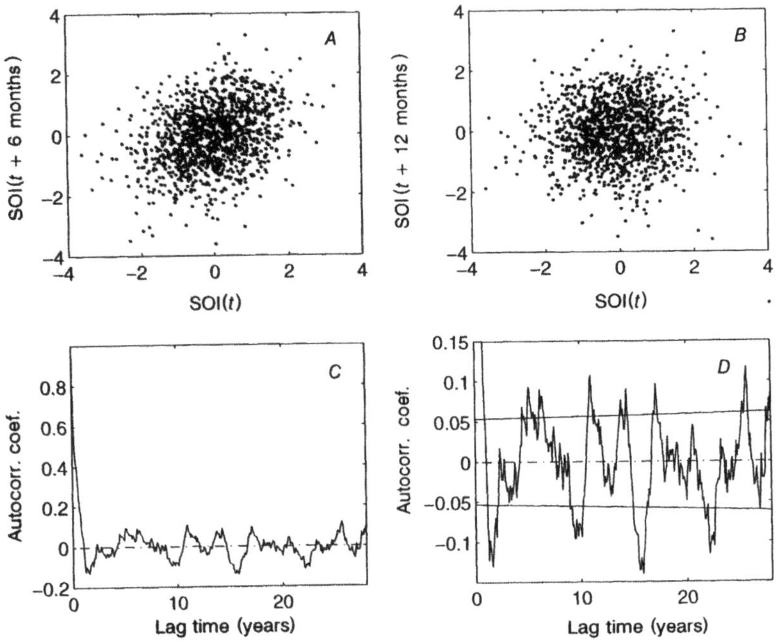

Figure 8.10 (*A*) Monthly mean Southern Oscillation Index versus monthly mean Southern Oscillation Index 6 months later. (*B*) Monthly mean Southern Oscillation Index versus monthly mean Southern Oscillation Index 12 months later. (*C*) Autocorrelation coefficient function for monthly mean Southern Oscillation Index data. (*D*) Autocorrelation coefficient function for 12-month running mean Southern Oscillation Index data in Figure 8.3.

where N is the number of data points in the sample record and T is the duration of the record. Applying this equation to the SOI $\rho_{xx}(\Delta t)$ results (Fig. 8.10*D*), we can determine whether any of the autocorrelation coefficients are significant. Δt in this case is the time lag in months, and N and T equal 1344. Hence $V[\rho_{xx}(\Delta t)] = 1/(1347 - \Delta t)$. Assuming a normal distribution of errors, the 95% confidence interval for the autocorrelation coefficients is $\rho_{xx}(\Delta t) \pm 1.96/\sqrt{1347 - \Delta t}$. This interval is shown in Figure 8.10*D* along with $\rho_{xx}(\Delta t)$ for $-0.15 \leq \rho_{xx}(\Delta t) \leq 0.15$. We discover that there are several significant peaks and troughs in the autocorrelation coefficient function at time lags of one year or more.

Test for Normality

A random process is considered to be normal or Gaussian if at each point in time the $\{X_k(t)\}$ follow a multidimensional normal (Gaussian) distribution function, the multidimensional analog of the bivariate normal distribution function discussed in Chapter 3. If the process is ergodic and normal, then the probability density function (pdf) for any sample record will become identical to the normal pdf as $T \to \infty$. This conclusion follows from the fact that for an ergodic process the statistical characteristics of the sample record as $T \to \infty$ become identical to those of the corresponding ensemble averages at a fixed time. An important characteristic of the normal pdf is that all moments and joint moments can be derived from μ_x and R_{xx}. Hence for a Gaussian random process weak stationarity implies strong stationarity. A common test in time series analysis is therefore to determine whether the pdf for a sample record is consistent with the assumption of normality.

We begin by assuming that the sample record came from an ergodic normal random process. A common test of this hypothesis is the χ^2 goodness-of-fit test. We rank the N observations in the sample record and then divide the ranked observations into K intervals, called *class intervals*. The number of observations in the jth interval is called the *observed frequency* and will here be designated fo_j. We then calculate the mean and variance of all the data in the sample record. These become our estimates of μ and σ^2 for the sample function. Given these estimates of μ and σ^2, we calculate the expected number of observations in the jth class interval, assuming a normal pdf. The expected number of observations is designated fe_j. We then calculate the statistic

$$\chi^2 = \sum_{j=1}^{K} \frac{(fo_j - fe_j)^2}{fe_j} \tag{8.26}$$

It can be shown that the distribution of χ^2 defined by the right-hand side of Equation 8.26 is approximately the same as the distribution of χ^2 discussed in Chapter 2 with $n = K - 3$ degrees of freedom. We subtract 3 from K to determine the df because only $K - 1$ of the fo_j are independent, the Kth being equal to N minus the sum of the first $K - 1$ observed frequencies, and because we have estimated two parameters, μ and σ^2. We reject the null hypothesis if the value of χ^2 calculated from the right-hand side of Equation 8.26 is greater than the critical value of the χ^2 statistic tabulated in Appendix G. In making the χ^2 test, it is common to choose class intervals of equal width and to choose the width to be approximately $0.4s$, where s is the standard deviation of the sample record. A more important consideration is that the fe_j be large enough that the right-hand side of Equation 8.26 is an acceptable approximation to the χ^2 statistic with $K - 3$ degrees of freedom. The usual recommendation is that $fe_j > 3$ for all j. In other words, if setting the interval width equal to $0.4s$ results in some $fe_j \leq 3$, then the interval width should be increased until all expected frequencies are greater than 3.

As a simple example of the application of this test, consider the case where the sample record is a sine function and T is an integer number of periods. A histogram of the observed frequencies for $N = 1000$ and the

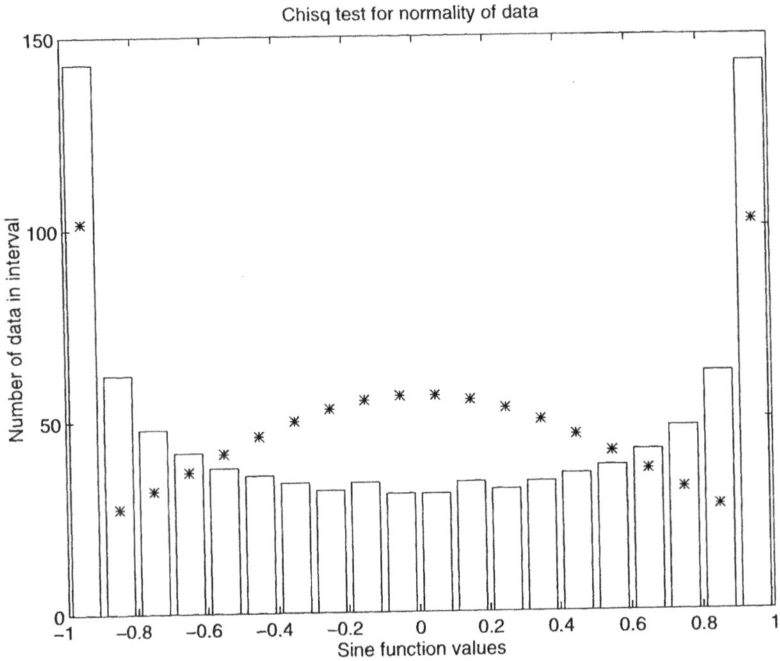

Figure 8.11 Histogram of observed frequencies of function values for a sample record of a sine function with 1000 equally spaced observations over a time equal to one period. The asterisks indicate the expected frequencies if the pdf for the sine function were a Gaussian. In the case of the first and last class interval, the Gaussian expected frequencies include the frequencies of observations of values less than -1 and greater than 1, respectively, as well as the expected frequencies within the indicated interval.

expected frequencies if the pdf were a Gaussian is shown in Figure 8.11. The observed values of the sine function were grouped in 20 class intervals. The χ^2 statistic in this case is 208, which is significant at $p < 0.005$ (Appendix G). We can therefore safely reject the hypothesis that the pdf for a sine function is a Gaussian.

This result should not be very surprising. The range of the sine function is $[-1, 1]$, and its variance over an integer number of periods is 0.5. The range of the function therefore corresponds to $\pm 1/\sqrt{0.5} = \pm\sqrt{2}$ standard deviations from the mean value of 0. Reference to Appendix A shows that for a normally distributed function, almost 8% of the function values will lie above this range and almost 8% below. In Figure 8.11 we considered the first and last class intervals to extend from $-\infty$ to -0.9 and from 0.9 to ∞. Had we truncated these intervals at -1 and 1, respectively, the χ^2 statistic would have been even greater, since the expected frequencies in the truncated intervals for a Gaussian pdf would have been only a little over 20.

What is the pdf for a sine function? The glib answer to this question is to refer the reader to Figure 8.11 and say, "You are looking at it." Let us see if we can be more quantitative.

We will let $y = \sin(\theta)$. By definition, the probability that y lies between y and $y + dy$ is $p(y)\,dy$ in the limit as $dy \to 0$. Looking at Figure 8.11, we can see that the width of each class interval is $\Delta \sin(\theta) = \Delta y$. The height of the bars is ΔN, the number of times the value of $\sin(\theta)$ falls within the interval $[y, y + \Delta y]$. Clearly, the sum of the heights of all the bars is N, and the fraction of the time that $\sin(\theta)$ falls within the interval $[y, y + \Delta y]$ is therefore $\Delta N/N$. In generating the data for Figure 8.11, the argument θ of the sine function was set equal to $\pi n/N$, where $n = 1, 2, 3, \ldots, 1000$. The reason for considering the interval $[0, \pi]$ rather than the interval $[0, 2\pi]$ is that the sine function on the interval $[\pi, 2\pi]$ is a mirror image of the sine function on the interval $[0, \pi]$. Hence we learn all we need to know about the probability density function for the sine function by considering its behavior on the interval $[0, \pi]$. If $\theta = \pi n/N$, then $\Delta\theta = \pi \Delta N/N$. Without loss of generality we can therefore equate $\Delta N/N$ to $\Delta\theta/\pi$, where θ is the argument of the sine function in radians. The fraction of the time that $\sin(\theta)$ falls between y and $y + \Delta y$ is therefore $\Delta\theta/\pi$. Taking the limit as $\Delta\theta \to 0$, we conclude that $p(y)\,dy = d\theta/\pi$. Hence $\pi p(y)(dy/d\theta) = 1$. Since $y = \sin(\theta)$, $dy/d\theta = \cos(\theta) = \sqrt{1 - y^2}$. We conclude that the pdf for $y = \sin(\theta)$ is $1/\pi\sqrt{1 - y^2}$. We can now understand why the histogram in Figure 8.11 is a minimum for the intervals closest to 0 and is a maximum for the intervals closest to ± 1.

One might infer from this example that the pdf for any periodic or nearly periodic function would not remotely resemble a Gaussian pdf. This inference, however, would be false. The pdf for both complex periodic and nearly periodic functions may closely resemble a Gaussian pdf. Figure 8.12*A*, for example, shows a sample record of 400 data points from a nearly periodic function consisting of a sum of six sine functions. The amplitudes and phase angles of all six functions are identical; only the periods are different. When a histogram of the 400 data points in Figure 8.12*A* is constructed, it rather closely resembles the histogram expected for a Gaussian pdf (Fig. 8.12*B*). The χ^2 statistic in this case is 15.4, which is less than the critical value of 16.9 for 9 df at $p = 0.05$. The fact is, however, that the pdf for this nearly periodic function is not a Gaussian, and if we increase the frequency of sampling, we can show beyond reasonable doubt that the two pdf's are different. Figure 8.12*C* shows the same portion of the sample function, this time sampled at 800 points in time. The histogram of the observed frequencies again closely resembles the results expected for a Gaussian pdf (Fig. 8.12*D*), but this time the χ^2 statistic is 21.6, and we could reject the null

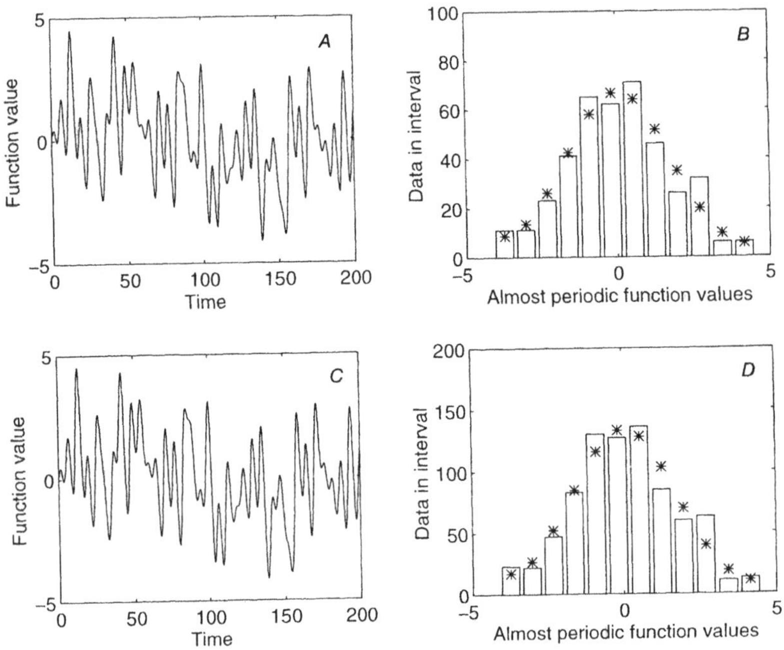

Figure 8.12 (A) Sample record consisting of 1000 data points from a nearly periodic process. (B) Histogram of the 400 data points from (A) and the expected frequencies assuming a Gaussian pdf. In this case the χ^2 test does not reveal a difference significant at $p = 0.05$. (C) The same sample record as in (A), but this time sampled at 2000 points in time. (D) Histogram of the 800 data points from (C) and the expected frequencies assuming a Gaussian pdf. This time the χ^2 test reveals a difference that is significant at $p < 0.05$.

hypothesis that the pdf for our sample record is a Gaussian at almost the $p = 0.01$ confidence level (Appendix G).

The fact is that Gaussian random processes are quite prevalent in physical problems, and, as we can see from Figure 8.12, the statistical characteristics of some deterministic phenomena may closely resemble those of a Gaussian random process. All moments and joint moments of the sample function which produced the sample record in Figure 8.12 are, in fact, time invariant. Hence acceptance of the hypothesis of strong stationarity based on acceptance of the hypothesis of weak stationarity and the fact that the pdf estimated from the sample record does not seem to differ significantly from a Gaussian (Fig. 8.12B) would not be misleading. Furthermore, the process which produced the nearly periodic sample record would be ergodic if each sample function in the ensemble were characterized by a random phase angle uniformly distributed on the interval $[0, 2\pi]$ but otherwise consisted of a sum of sine functions with amplitudes and frequencies identical to those in the example.

Now we will take a look at the results of χ^2 tests for normality of the SOI data (Fig. 8.3) and the detrended Mauna Loa CO_2 data (Fig. 8.9*B*). The histogram of values for the SOI data, Figure 8.13, is close to expectations for a Gaussian pdf, and the χ^2 statistic of 9.40 is less than the critical value of 11.07 for 5 degrees of freedom. Therefore we accept the null hypothesis that the pdf for the SOI data is a Gaussian. The detrended CO_2 histogram, Figure 8.14, bears little resemblance to a Gaussian pdf, a fact that is confirmed by the χ^2 test. The χ^2 statistic of 92.3 allows us to reject the null hypothesis at $p < 0.005$. Interestingly, the detrended CO_2 histogram does not look much like the histogram for a sine function, either. Given the highly periodic nature of the detrended CO_2 data (Fig. 8.8*D*), one is inclined to think that this sample record may have come from a complex periodic process.

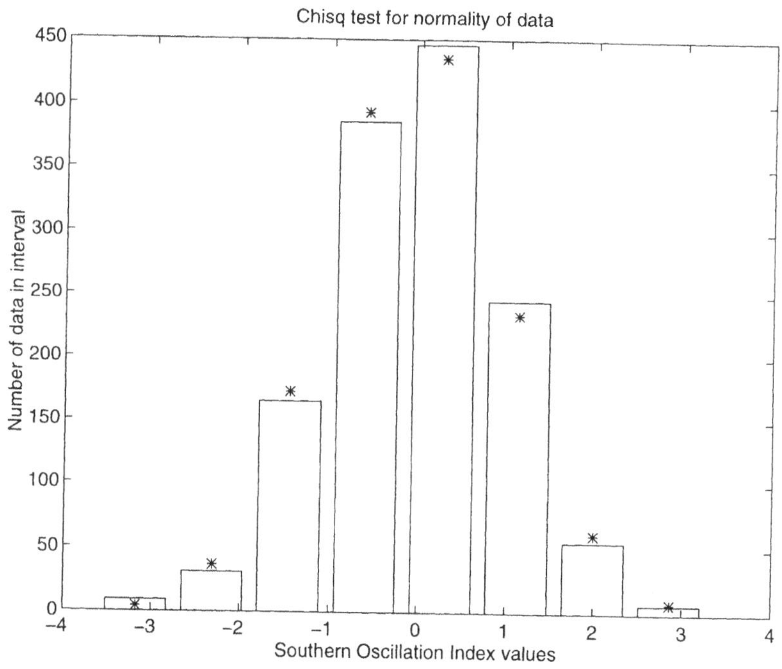

Figure 8.13 Histogram of SOI data in Figure 8.3 and the expected frequencies assuming a Gaussian pdf. $\chi^2 = 9.40$, df = 5. Since the χ^2 statistic for 5 degrees of freedom is 11.07 at $p = 0.05$, we accept the null hypothesis that the pdf for these data is a Gaussian.

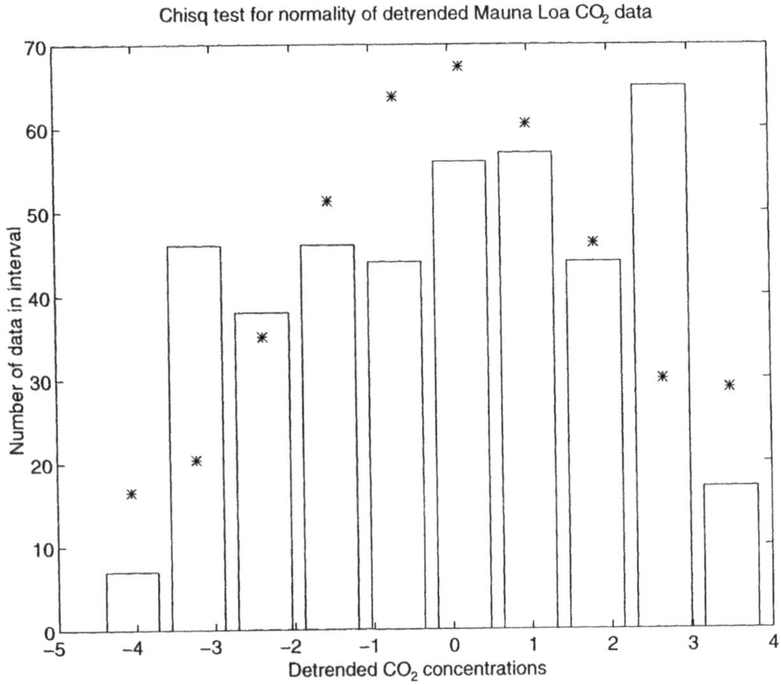

Figure 8.14 Histogram of detrended CO_2 data in Figure 8.9*B* and the expected frequencies assuming a Gaussian pdf. $\chi^2 = 92.3$, df $= 7$. Hence we reject at $p < 0.005$ the hypothesis that the pdf for the detrended CO_2 data is a Gaussian.

FREQUENCY DOMAIN TECHNIQUES

We have now proceeded through four steps which should begin any time series analysis:

1. Examination of the time series for bad data
2. Test of stationarity
3. Test of periodicities
4. Test for normality of the pdf

If we were content to work entirely in the time domain, the analyses conducted in carrying out these four steps might complete our work. If we concluded, for example, that the sample record we were examining came from an ergodic process and the pdf closely approximated a Gaussian, then knowledge of μ_x and $C_{xx}(\Delta t)$ would completely characterize the process.

There are good reasons, however, for extending our analysis beyond the time domain:

1. The parameters of a model describing a process should make it easy for us to visualize typical sample functions that might be produced by the process. This is sometimes very difficult to do with statistics derived from a strictly time domain analysis, for example, μ_x and $C_{xx}(\Delta t)$.
2. In most cases $\rho_{xx}(\Delta t_1)$ and $\rho_{xx}(\Delta t_2)$ are highly correlated when Δt_1 is comparable to Δt_2. Furthermore, as we have seen from Equation 8.25, the variance of $\rho_{xx}(\Delta t)$ is not constant. The correlation between nearby estimators of $\rho_{xx}(\Delta t)$ and the lack of a constant variance confound the interpretation of a plot of $\rho_{xx}(\Delta t)$ and make it difficult to devise good ways to test various hypotheses of interest based on estimates of $\rho_{xx}(\Delta t)$ (Percival and Walden, 1993).
3. Even in cases where a model based entirely on a time domain analysis is adequate, it is often useful to have a second characterization as a complementary way of understanding the behavior of our data.

Spectral analysis is the prime example of a frequency domain technique. It is essentially an analysis of variance procedure based on representing a stationary time series as a sum of sines and cosines of different frequencies and amplitudes. The variance of the time series is partitioned into a number of components, each of which is associated with a particular frequency. The ensemble of components is called the *spectrum*.

Fourier Analysis

The basic tool of spectral characterization of a process is Fourier analysis. We will begin with a simple example. Consider a periodic process $g(t)$ with a fundamental period τ. Recall from Equation 8.3 that this process can be described by an equation of the form

$$g(t) = \frac{A_0}{2} + \sum_{n=1}^{\infty} \left[A_n \cos\left(\frac{2\pi nt}{\tau}\right) + B_n \sin\left(\frac{2\pi nt}{\tau}\right) \right] \tag{8.27}$$

By making use of the fact that $\exp(i\theta) = \cos(\theta) + i\sin(\theta)$, where $i = \sqrt{-1}$, we can rewrite Equation 8.27 as

$$g(t) = \sum_{n=-\infty}^{\infty} G_n \exp\left(\frac{2\pi int}{\tau}\right) \tag{8.28}$$

where $G_n = (A_{|n|} + iB_{|n|})/2$ for $n < 0$ $G_n = (A_n + iB_n)/2$ for $n > 0$, and, $G_0 = A_0/2$. Given $g(t)$, how do we determine the G_n? Assume that we know the value of $g(t)$ over one complete period. We then multiply $g(t)$ by

$\exp(-2\pi imt/\tau)$ and integrate over one period:

$$\int_{-\tau/2}^{\tau/2} g(t)\exp\left(-\frac{2\pi imt}{\tau}\right) dt = \int_{-\tau/2}^{\tau/2} \sum_{n=-\infty}^{\infty} G_n \exp\left(-\frac{2\pi i(n-m)t}{\tau}\right) dt \tag{8.29}$$

We now interchange the order of integration and summation on the right-hand side of Equation 8.29 and take advantage of the fact that

$$\int_{-\tau/2}^{\tau/2} \exp\left[\frac{2\pi i(n-m)t}{\tau}\right] dt = \delta_{mn}\tau \tag{8.30}$$

where $\delta_{mn} = 1$ if $m = n$ and $\delta_{mn} = 0$ if $m \neq n$. Equation 8.30 follows from the fact that

$$\int_{-\tau/2}^{\tau/2} \sin\left(\frac{2\pi nt}{\tau}\right)\sin\left(\frac{2\pi mt}{\tau}\right) dt = \int_{-\tau/2}^{\tau/2} \cos\left(\frac{2\pi nt}{\tau}\right)\cos\left(\frac{2\pi mt}{\tau}\right) dt = \frac{\tau}{2}\delta_{mn} \tag{8.31}$$

and

$$\int_{-\tau/2}^{\tau/2} \sin\left(\frac{2\pi nt}{\tau}\right)\cos\left(\frac{2\pi mt}{\tau}\right) dt = 0 \tag{8.32}$$

Given Equation 8.30, we can see from Equation 8.29 that

$$G_n = \frac{1}{\tau}\int_{-\tau/2}^{\tau/2} g(t)\exp\left(-\frac{2\pi int}{\tau}\right) dt \tag{8.33}$$

The right-hand side of equation 8.28 is called the *Fourier series representation* of a periodic function, and G_n is referred to as the nth *Fourier coefficient*.

It follows from the definition of A_n in Equation 8.23 that

$$E[g(t)] = \frac{1}{\tau}\int_{-\tau/2}^{\tau/2} g(t)\, dt = \frac{A_0}{2} = G_0 \tag{8.34}$$

Hence

$$g(t) - E[g(t)] = \sum_{n=1}^{\infty}\left[A_n \cos\left(\frac{2\pi nt}{\tau}\right) + B_n \sin\left(\frac{2\pi nt}{\tau}\right)\right]$$

and

$$V(g) = \frac{1}{\tau}\int_{-\tau/2}^{\tau/2} \sum_{n=1}^{\infty} \left[A_n \cos\left(\frac{2\pi nt}{\tau}\right) + B_n \sin\left(\frac{2\pi nt}{\tau}\right)\right]$$
$$\times \sum_{m=1}^{\infty} \left[A_m \cos\left(\frac{2\pi mt}{\tau}\right) + B_m \sin\left(\frac{2\pi mt}{\tau}\right)\right]$$

Taking advantage of Equations 8.31 and 8.32, we can see that this equation reduces to

$$V(g) = \frac{1}{2}\sum_{n=1}^{\infty} (A_n^2 + B_n^2) = 2\sum_{n=1}^{\infty} |G_n|^2 \tag{8.35}$$

Thus the variance of a periodic function can be partitioned into a sum of components, each of which is associated with a sinusoidal function having a period equal to the fundamental period of the function or an integral multiple thereof. The orthogonality relationship, Equation 8.31, ensures that these components are independent of one another.

Substituting the right-hand side of Equation 8.33 for G_n in Equation 8.28, we can see that

$$g(t) = \frac{1}{\tau}\sum_{n=-\infty}^{\infty} \exp\left(\frac{2\pi int}{\tau}\right)\int_{-\tau/2}^{\tau/2} g(x)\exp\left(-\frac{2\pi inx}{\tau}\right) dx \tag{8.36}$$

We now notice that consecutive terms in this summation are associated with sinusoidal functions having frequencies f_n and f_{n+1} equal to n/τ and $(n+1)/\tau$, respectively. The difference in consecutive frequencies is therefore $1/\tau$, which we will call Δf. Equation 8.36 can therefore be written as

$$g(t) = \sum_{n=-\infty}^{\infty} G(f_n)\exp(2\pi if_n t)\,\Delta f \tag{8.37}$$

where

$$G(f_n) = \int_{-\tau/2}^{\tau/2} g(x)\exp(-2\pi if_n x)\, dx \tag{8.38}$$

If we now take the limit as $\tau \to \infty$, we realize from our discussion of integrals in Chapter 6 that

$$g(t) \to \int_{-\infty}^{\infty} G(f)\exp(2\pi ift)\, df \tag{8.39}$$

where

$$G(f) = \int_{-\infty}^{\infty} g(x)\exp(-2\pi ifx)\, dx \qquad (8.40)$$

The right-hand side of Equation 8.39 is called the *Fourier integral representation* of $g(t)$. G and g are referred to as a *Fourier transform pair.* $G(f)$ is the *Fourier transform* of g, and $g(t)$ is the *inverse Fourier transform* of G.

Equations 8.39 and 8.40 allow us, at least theoretically, to deal with functions $g(t)$ whose period is infinite. Almost periodic functions fall into this category. In reality, however, we must always work with sample records of finite length. As we will see, this fact imposes some constraints on what we can learn about a process from spectral analysis.

Convolution

A very important tool in spectral analysis is convolution. It is often regarded as a smoothing or filtering operation, that is, a means of filtering out wiggles within a certain frequency range. This can sometimes be a great help when one is trying to distinguish the forest from the trees.

Suppose g and h are two functions of time. The convolution of g and h, designated $g * h(t)$, is defined by

$$g * h(t) \equiv \int_{-\infty}^{\infty} g(x)h(t-x)\, dx \qquad (8.41)$$

where the symbol $\equiv$ means equals by definition. Note that if we change the dummy integration variable to $y = t - x$, then the right-hand side of Equation 8.41 becomes

$$\int_{-\infty}^{\infty} g(t-y)h(y)\, dy \equiv h * g(t)$$

Hence $g * h(t) = h * g(t)$.

Using Equation 8.40, we can see that the Fourier transform of $g * h(t)$, which we will call $G * H(f)$, is given by

$$G * H(f) = \int_{-\infty}^{\infty} g * h(x)\exp(-2\pi ifx)\, dx$$

$$= \int_{-\infty}^{\infty}\int_{-\infty}^{\infty} g(y)h(x-y)\exp(-2\pi ifx)\, dy\, dx \qquad (8.42)$$

$$= \int_{-\infty}^{\infty} g(y)\exp(-2\pi ify)\, dy \int_{-\infty}^{\infty} h(x-y)\exp[-2\pi if(x-y)]\, dx \qquad (8.43)$$

Substituting $z = x - y$ for the dummy variable in the second integrand, we obtain

$$G * H(f) = \int_{-\infty}^{\infty} g(y)\exp(-2\pi i f y)\,dy \int_{-\infty}^{\infty} h(z)\exp(-2\pi i f z)\,dz$$
$$= G(f)H(f) \tag{8.44}$$

where $G(f)$ and $H(f)$ are the Fourier transforms of $g(t)$ and $h(t)$, respectively. We thus see that the Fourier transform of the convolution of two functions is the product of the Fourier transforms of the functions. This fact is sometimes expressed as $g * h \subset GH$, where the symbol $\subset$ means has the Fourier transform. It is also straightforward to show (see Exercise 8.5) that the converse of this theorem is also true, that is, $gh \subset \int_{-\infty}^{\infty} G(s)H(f-s)\,ds$. In plain English:

1. The Fourier transform of the convolution of two functions is the product of their Fourier transforms.
2. The Fourier transform of the product of two functions is the convolution of their Fourier transforms.

Now we will consider a couple of examples to illustrate some of the properties of convolution. First consider a function Box defined by

$$\text{Box}(t) \equiv \begin{cases} 1 & \text{for } -\dfrac{1}{2} \le t \le \dfrac{1}{2} \\ 0 & \text{otherwise} \end{cases} \tag{8.45}$$

The function defined by Equation 8.45 is often referred to as a *boxcar* or *gate function* and is illustrated in Figure 8.15*A*. It is a square with a base and height of 1 with the center of its base at the origin. To generalize this function, we will define a normalized boxcar function Π by

$$\Pi(t, \tau) \equiv \frac{1}{\tau}\text{Box}\left(\frac{t}{\tau}\right) \tag{8.46}$$

$\Pi(t, \tau)$ is a rectangle with an area of 1 and a width of τ. Now consider the effect of substituting $\Pi(t, \tau)$ for $h(t)$ in Equation 8.41. The result is

$$g * \Pi(t, \tau) = \int_{-\infty}^{\infty} g(x)\frac{1}{\tau}\text{Box}\left(\frac{t-x}{\tau}\right) dx$$
$$= \frac{1}{\tau}\int_{t-\tau/2}^{t+\tau/2} g(x)\,dx \tag{8.47}$$

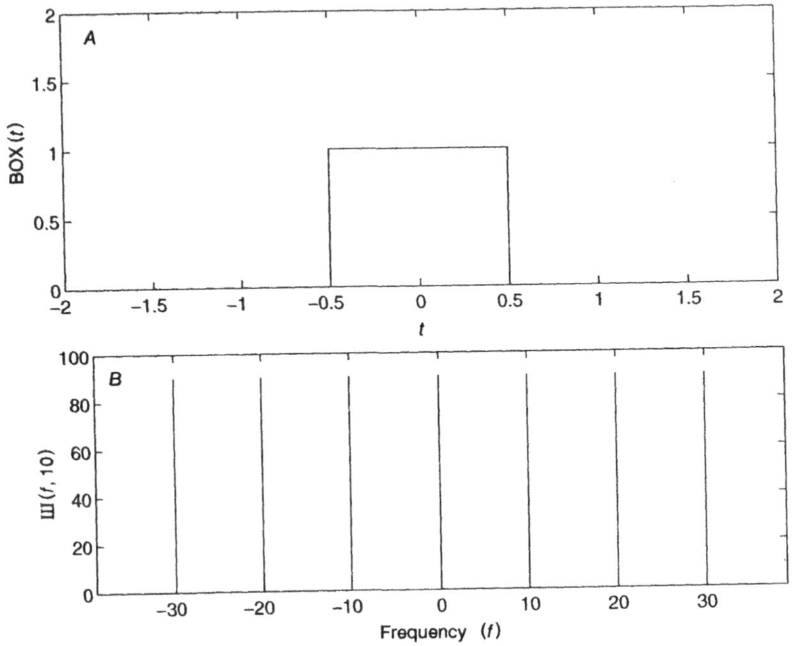

Figure 8.15 (A) The boxcar function defined by Equation 8.45. (B) The Ш function defined by Equation 8.48.

Thus the function $g(t) * \Pi(t, \tau)$ is a running average of $g(t)$ over the time interval $t - \tau/2$ to $t + \tau/2$. This example illustrates why convolution is often regarded as a smoothing operation. When convolution is used for this purpose, the function $h(t)$ in Equation 8.41 is called the *smoothing kernel*, and g is the function to be smoothed.

Now we will consider another example. Consider the function $\delta(t - a)$ defined by

$$\int_{-\infty}^{\infty} \delta(t - a) g(t)\, dt = g(a) \tag{8.48}$$

for any function $g(t)$. Note that $\delta(t - a)$, commonly called a Dirac delta function,[1] is not a function in the usual sense of the word, but it may be thought of, for example, as $\lim_{\tau \to 0} \Pi(t - a, \tau)$. Now assume that we are working in the frequency domain and consider the function, $Ш(f, \Delta f)$ defined by

$$Ш(f, \Delta f) \equiv \sum_{n=-\infty}^{\infty} \delta(f - n\Delta f) \tag{8.49}$$

[1]Named after the theoretical physicist P. M. Dirac, who first made use of the function.

The function defined by Equation 8.49 is referred to as a *comb* or *shah function*[2] and is illustrated in Figure 8.15*B*. It consists of an infinite series of spikes, each having unit area, and centered at $f = n\Delta f$. When we substitute $Ш(f, \Delta f)$ for h in Equation 8.41, we obtain

$$
\begin{aligned}
g * Ш(f, \Delta f) &= \int_{-\infty}^{\infty} g(x) Ш(f - x, \Delta f)\, dx \\
&= \int_{-\infty}^{\infty} g(x) \sum_{n=-\infty}^{\infty} \delta(f - x, n\Delta f)\, dx \\
&= \sum_{n=-\infty}^{\infty} \int_{-\infty}^{\infty} g(x)\, \delta(f - x, n\Delta f)\, dx \\
&= \sum_{n=-\infty}^{\infty} g(f - n\Delta f) \qquad (8.50)
\end{aligned}
$$

The effect of convolution with $Ш(f, \Delta f)$ is therefore to give us the sum of $g(f)$ and all values of the function for which the argument differs from f by an integral multiple of Δf.

Examples of the effect of convolution with the Π and $Ш$ functions are shown in Figures 8.16 and 8.17. In the first case we have used the Π function with $\tau = 12$ months to remove the annual cycle in the Mauna Lao CO_2 time series. This would be a useful exercise if we wanted to closely examine the nature of the longer-term trend in the data. The Southern Oscillation Index data (Fig. 8.3) are an example of a time series where the longer-term trends are generally of more interest than the annual cycle, and for this reason are typically smoothed by convoluting with a filter such as $\Pi(t, 12 \text{ months})$.

In the second example we consider a simple time series whose Fourier transform is shown graphically in Figure 8.17*A*. The spectrum contains no sinusoidal components with frequencies greater than 50, and in this simple case we assume that $G(f)$ declines linearly from a peak at $f = 0$ to 0 at $f = \pm 50$. The convolution of this function with $Ш$ gives Figure 8.17*B* when $\Delta f = 200$, with the caveat that the convoluted Fourier transform repeats itself ad infinitum at intervals of $\Delta f = 200$. The convoluted Fourier transform consists of an infinite number of triangular peaks whose centers are separated by a distance Δf of 200. Because the width of the base of each peak is 100, there is a gap of 100 frequency units between the bases of adjacent peaks.

In Figure 8.17*C*, Δf has been reduced to 100, and the gap between the bases of adjacent peaks has shrunk to 0. Each peak, however, is still distinct from its neighbors. In Figure 8.17*D*, Δf has been reduced to 75, and 25% of

[2]The symbol Ш is pronounced *shah* after the Cyrillic character Ш, which is believed to have been derived from the Hebrew letter *shin*, which in turn may have been derived from an Egyptian hieroglyph depicting papyrus plants on the Nile.

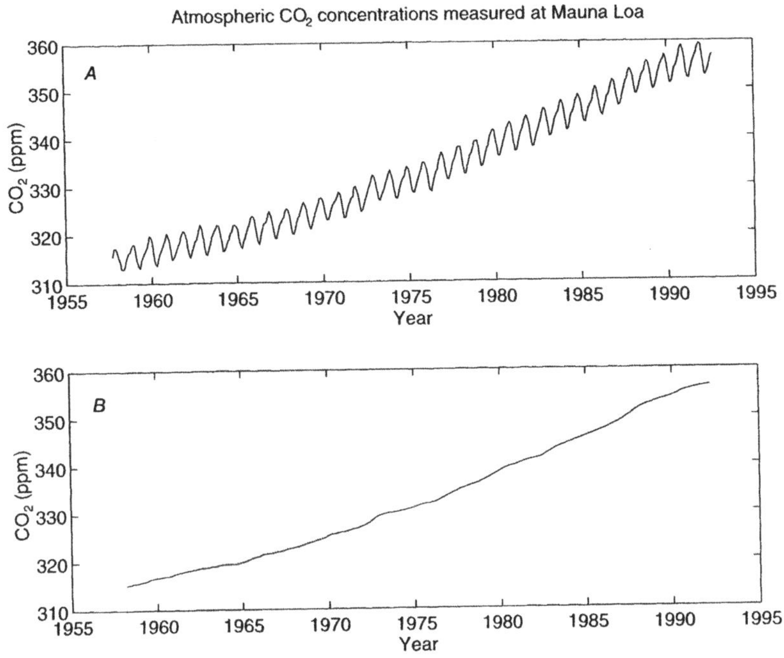

Figure 8.16 (A) The monthly mean CO_2 concentrations atop Mauna Loa shown in Figure 8.1 and (B) the convolution of the time series in Figure 8.13A with $\Pi(t, 12)$, that is, the 12-month running means of the Mauna Loa CO_2 concentrations.

each peak is overlapped by its neighbor on the right and 25% by its neighbor on the left. In Figure 8.17E, Δf has been reduced to 50. Now 50% of each peak is overlapped by an identical percentage of the peak on the right and 50% by the peak on the left. Because of the triangular shape of the peaks, the overlap produces a constant function at all frequencies which certainly bears no resemblance to the original Fourier transform. Finally, in Figure 8.17F, Δf has been reduced to 35. In this case not only is each peak overlapped by its neighbors to the right and left, but the neighbors themselves overlap over a bandwidth of 30 frequency units in the center of each peak. The result is a sawtooth pattern with many small peaks atop a constant background.

What, we may ask, does Figure 8.17 have to do with time series analysis? The answer is that in practice the sample records which we analyze always consist of data collected at discrete points in time rather than continuously. Assume that the interval between discrete time points is constant and equal to Δt and that we have sampled $g(t)$ at an infinite number of such discrete time points. If we know the value of the function only at points in time separated by an interval Δt, then our best estimate of the value of the function during the time interval $[n\Delta t - \Delta t/2, n\Delta t + \Delta t/2]$ is just $g(n\Delta t)$,

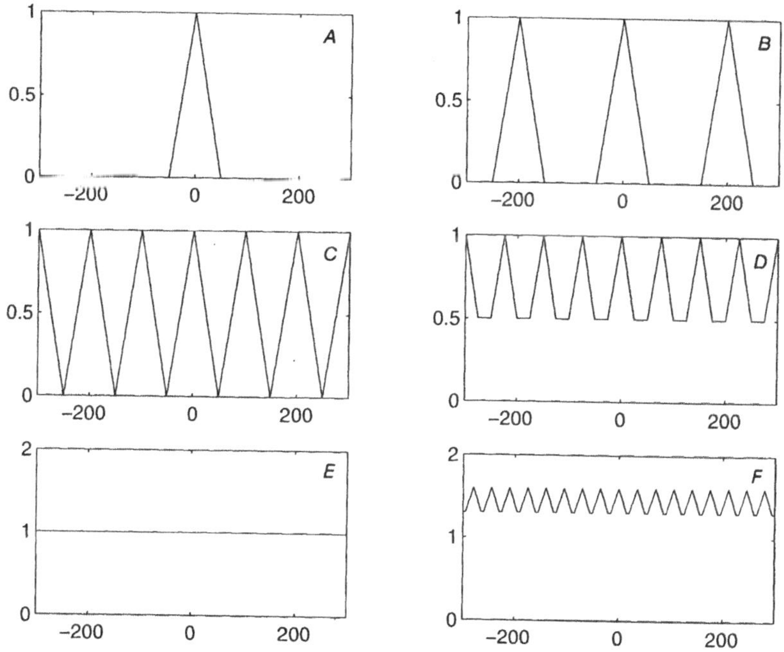

Figure 8.17 (A) A simple triangular function $g(f)$ which is nonzero on the interval $[-50, 50]$ and the results of convoluting $g(f)$ with $\text{III}(f, \Delta f)$ for (B) $\Delta f = 200$, (C) $\Delta f = 100$, (D) $\Delta f = 75$, (E) $\Delta f = 50$, and (F) $\Delta f = 35$.

and our best estimate of the integral of the function over this same time interval is therefore $g(n\Delta t)\,\Delta t$. A mathematical representation of the discretely sampled function which satisfies this condition is $g(t)\,\Delta t\,\delta(t - n\Delta t)$, because

$$\int_{n\Delta t-\Delta t/2}^{n\Delta t+\Delta t/2} g(t)\,\Delta t\,\delta(t - n\Delta t)\,dt = g(n\Delta t)\,\Delta t \tag{8.51}$$

It thus seems reasonable to say that a mathematical representation of the discretely sampled function, which we will call $\hat{g}(t)$, would be

$$\begin{aligned} \hat{g}(t) &= g(t)\,\Delta t \sum_{n=-\infty}^{\infty} \delta(t - n\Delta t) \\ &= g(t)\,\Delta t\,\text{III}(t, \Delta t) \end{aligned} \tag{8.52}$$

Since the Fourier transform of the product of two functions is equal to the convolution of the Fourier transforms of the individual functions, it follows that the Fourier transform of $\hat{g}(t)$, which we will call $\hat{G}(f)$, is equal to the convolution of $G(f)$ and the Fourier transform of $\Delta t\,\text{III}(t, \Delta t)$. The Fourier transform of $\Delta t\,\text{III}(t, \Delta t)$ is straightforward enough to calculate. Since $\Delta t\,\text{III}(t, \Delta t)$ is clearly a complex periodic function with a fundamental period at Δt, it follows that the Fourier series expansion of $\Delta t\,\text{III}(t, \Delta t)$ will contain sinusoidal functions with frequencies of $1/\Delta t$ and integral multiples thereof. Since the Fourier transform of $\delta(t - n\Delta t)$ is $\exp(2\pi i f n \Delta t)$, it follows from Equation 8.37 that the coefficient of $(1/\Delta t)\exp(2\pi i m t/\Delta t)$ in the Fourier series expansion of $\delta(t - n\Delta t)$ is $\exp(2\pi i m n \Delta t/\delta t) = \exp(2\pi i m n) = 1$.[3] It follows that the Fourier series expansion of $\Delta t\,\text{III}(t, \Delta t)$ equals $\sum_{m=-\infty}^{\infty} \exp(2\pi i m t/\Delta t)$. Since $\exp(ix) = \cos(x) + i\sin(x)$ and since $\cos(x)$ and $\sin(x)$ are even and odd functions of x, respectively, it follows that

$$\Delta t\,\text{III}(t, \Delta t) = \sum_{m=-\infty}^{\infty} \cos\left(\frac{2\pi m t}{\Delta t}\right) \tag{8.53}$$

A truncated version of the right-hand side of Equation 8.53 is shown in Figure 8.18 for $\Delta t = 10$. The Fourier series, even when truncated at $m = 100$, clearly gives a very good approximation to the sum of δ functions. It follows from Equations 8.40 and 8.53 that the Fourier transform of $\Delta t\,\text{III}(t, \Delta t)$ is equal to

$$\sum_{m=-\infty}^{\infty} \int_{-\infty}^{\infty} \exp\left[-2\pi i t\left(\frac{n-m}{\Delta t}\right)\right] dt = \sum_{m=-\infty}^{\infty} \delta\left(\frac{n-m}{\Delta t}\right) = \text{III}\left(f_n, \frac{1}{\Delta t}\right) \tag{8.54}$$

where $f_n = n/\Delta t$ and we have taken advantage of the fact (see Exercise 8.4) that $\int_{-\infty}^{\infty} \exp(2\pi i f t)\, df = \delta(t)$.

We conclude that the Fourier transform of $\hat{g}(t)$ is equal to $G * \text{III}(f_n, 1/\Delta t)$. The important implication of this conclusion is that when we sample a continuous function at discrete time points separated by an interval Δt and Fourier transform the results, the Fourier transform of the discretely sampled function will consist of discrete frequencies f_n from the Fourier transform of the continuous function $G(f)$ along with an infinite number of replicates offset by $\pm m\Delta f = \pm m/\Delta t$. If $G(f)$ is 0 for $|f| > f_{\max}$, then $G(f)$ is said to be *band limited*, and we can cleanly separate $\hat{G}(f)$ from its neighbors by choosing $\Delta t < 1/(2f_{\max})$ (e.g., Fig. 8.17*B*). In this case the Fourier transform of $\hat{g}(t)$ is exactly $\hat{G}(f)$ in the frequency interval $[-1/(2\Delta t), 1/(2\Delta t)]$. Furthermore, no matter how good our peripheral vision is, if $f_{\max}$ is finite we can move the side peaks out of sight by choosing Δt sufficiently small. Problems occur when $\Delta t > 1/(2f_{\max})$. In this case the

[3]Note that Δf in Equation 8.37 $= 1/\Delta t$.

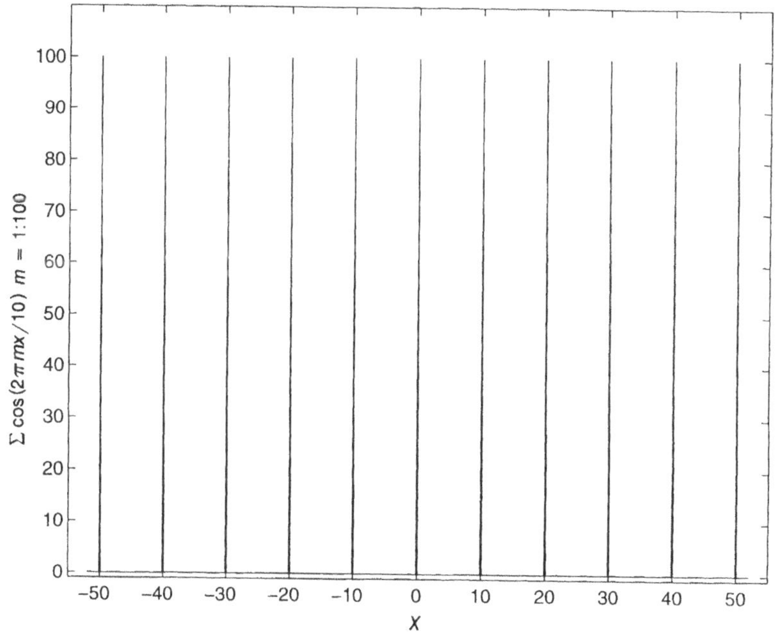

Figure 8.18 The right-hand side of Equation 8.52 evaluated through $m = \pm 100$ with $\Delta t = 10$.

side peaks began to overlap with the central peak (e.g., Fig. 8.17*D*–*F*). This problem is known as *aliasing*, and we will have more to say about it shortly.

Finite Sample Records with Sampling at Discrete Points

Equation 8.40 is a theoretically convenient expression for calculating the Fourier transform of a function, but in practice sample records never extend over an infinite time interval, and data are invariably taken at discrete time points rather than continuously. The former consideration leads to a problem called *leakage* and the latter to aliasing. Before becoming embroiled in a discussion of leakage and aliasing, we will consider the practical problem of carrying out a Fourier analysis of discretely sampled data over a finite time interval.

Assume that the data were collected at N evenly spaced points in time with the time interval Δt between points being τ/N, where τ is the duration of the sample record. For the sake of simplicity, we will assume that N is odd. We will let the first point in the time series be $t = -((N - 1)/2)\,\Delta t$. The last point is therefore $t = ((N - 1)/2)\,\Delta t$. The first and last points therefore correspond to times $(-1/2 + 1/(2N))\tau$ and $(1/2 - 1/(2N))\tau$, respectively. By analogy with Equation 8.28, we might imagine expressing a stationary function $g(t)$ sampled at these points in time by the equation

$$g(n\Delta t) = \sum_{k=-(N-1)/2}^{(N-1)/2} G_k \exp\left(\frac{2\pi ikn\Delta t}{\tau}\right) = \sum_{k=-(N-1)/2}^{(N-1)/2} G_k \exp\left(\frac{2\pi ikn}{N}\right) \tag{8.55}$$

Equation 8.55 can be written in the equivalent form

$$g(n\Delta t) = \frac{A_0}{2} + \sum_{k=1}^{(N-1)/2} A_k \cos\left(\frac{2\pi kn}{N}\right) + B_k \sin\left(\frac{2\pi kn}{N}\right) \tag{8.56}$$

The coefficients A_k and B_k are related to G_k by

$$A_0 = 2G_0 \tag{8.57}$$

$$A_k = G_k + G_{-k} \qquad 1 \le k \le \frac{N-1}{2} \tag{8.58}$$

$$B_k = i(G_k - G_{-k}) \qquad 1 \le k \le \frac{N-1}{2} \tag{8.59}$$

We leave the proof of the equivalence of Equations 8.55 and 8.56 to 8.59 to the Exercises.

To obtain an expression for G_m, we calculate the sum

$$\begin{aligned} &\sum_{n=-(N-1)/2}^{(N-1)/2} g(n\Delta t)\exp\left(-\frac{2\pi imn}{N}\right) \\ &= \sum_{n=-(N-1)/2}^{(N-1)/2} \exp\left(-\frac{2\pi imn}{N}\right) \sum_{k=-(N-1)/2}^{(N-1)/2} G_k \exp\left(\frac{2\pi ikn}{N}\right) \\ &= \sum_{k=-(N-1)/2}^{(N-1)/2} G_k \sum_{n=-(N-1)/2}^{(N-1)/2} \exp\left[\frac{2\pi in(k-m)}{N}\right] \end{aligned} \tag{8.60}$$

Now let $X = \exp[2\pi i(k - m)/N]$. We can see that the summation over n on the right-hand side of Equation 8.60 has the form QS, where $Q = X^{-(N-1)/2}$ and $S = 1 + X + X^2 + X^3 + \cdots + X^{N-1}$. Hence $SX = X + X^2 + X^3 + \cdots + X^N$. We conclude that $S(1 - X) = 1 - X^N$. Now if $k = m$, $X = 1$, $Q = 1$, and $S = N$. Since the maximum possible value of $|k - m|$ is $N - 1$, $X = 1$ only when $k = m$. It should also be clear that $X^N = 1$. Hence $S = 0$ when $k \neq m$. We conclude that

$$\sum_{n=-(N-1)/2}^{(N-1)/2} \exp\left[\frac{2\pi in(k-m)}{N}\right] = N\delta_{km} \quad \text{for } -\frac{N-1}{2} \leq k, m \leq \frac{N-1}{2} \tag{8.61}$$

It follows from Equations 8.60 and 8.61 that

$$G_m = \frac{1}{N} \sum_{n=-(N-1)/2}^{(N-1)/2} g(n\Delta t) \exp\left(-\frac{2\pi imn}{N}\right) \tag{8.62}$$

The right-hand side of Equation 8.62 is called the *discrete Fourier transform* of g, and the right-hand side of Equation 8.55 is the *inverse discrete Fourier transform* of G.

We can now define for the discrete case the convolution of g and h as follows:

$$g * h(k\Delta t) = \frac{1}{N} \sum_{n=-(N-1)/2}^{(N-1)/2} g(n\Delta t) h[(k-n)\,\Delta t] \tag{8.63}$$

The discrete Fourier transform of $g * h(k\Delta t)$, which we will call $G * H_k$, is then given by

$$G * H_k = \frac{1}{N} \sum_{m=-(N-1)/2}^{(N-1)/2} \exp\left(-\frac{2\pi imk}{N}\right) \times \left(\frac{1}{N} \sum_{n=-(N-1)/2}^{(N-1)/2} g(n\Delta t) h[(m-n)\,\Delta t]\right) \tag{8.64}$$

where it is assumed that $h[(m - n)\,\Delta t]$ is cyclic such that $h[(m - n)\,\Delta t] =$

$h[(m-n\pm N)\,\Delta t]$. We then have

$$
\begin{aligned}
G*H_k &= \frac{1}{N}\sum_{n=-(N-1)/2}^{(N-1)/2}\exp\left(-\frac{2\pi ink}{N}\right)g(n\Delta t)\\
&\quad\times\left(\frac{1}{N}\sum_{m=-(N-1)/2}^{(N-1)/2}\exp\left[-\frac{2\pi i(m-n)k}{N}\right]h[(m-n)\,\Delta t]\right)\\
&= \frac{1}{N}\sum_{n=-(N-1)/2}^{(N-1)/2}\exp\left(-\frac{2\pi ink}{N}\right)g(n\Delta t)\\
&\quad\times\left(\frac{1}{N}\sum_{p=-(N-1)/2-n}^{(N-1)/2-n}\exp\left(-\frac{2\pi ipk}{N}\right)h(p\Delta t)\right)\\
&= G_k H_k
\end{aligned}
\tag{8.65}
$$

The last equality in Equation 8.65 follows from the fact that both functions in the second summation are cyclic or assumed to be cyclic, and therefore the value of the summation is independent of the value assigned to n and, in particular, is equal to the value of the summation when $n = 0$, that is, H_k. We thus conclude that the discrete Fourier transform of the convolution of two discrete functions is the product of the discrete Fourier transforms of the functions, that is, $g*h(k\Delta t)\subset G_k H_k$. It is also true (see Exercise 8.7) that

$$
g(k\Delta t)h(k\Delta t)\subset\sum_{n=-(N-1)/2}^{(N-1)/2}G_n H_{k-n}
\tag{8.66}
$$

Thus, as was the case for sample records taken continuously over an infinite time interval, we conclude in the case of discrete sampling over a finite time interval that:

1. The Fourier transform of the convolution of two functions is the product of their Fourier transforms.
2. The Fourier transform of the product of two functions is the convolution of their Fourier transforms.

The equation defining convolution in the time domain (Eq. 8.63) is consistent with the notion of convolution as a smoothing operation. For example, if $h = 1$, then

$$
g*h(k\Delta t)=\frac{1}{N}\sum_{n=-(N-1)/2}^{(N-1)/2}g(n\Delta t)
$$

and $g * h(k\Delta t)$ becomes the average of the N values of g in the sample record. Furthermore, if $h = 1$ it is straightforward to show using Equations 8.61 and 8.62 that $H_{k-n} = \delta_{kn}$ and hence from Equation 8.66 that $g(k\Delta t)h(k\Delta t) \subset G_k$, as would logically be the case if $h = 1$. Hence Equation 8.66 defining convolution in the frequency domain is a logical analog of Equation 8.63 defining convolution in the time domain.

We have already noted that the variance of a continuous periodic function can be expressed as a sum of components, each of which is associated with a sinusoidal function having a period equal to the fundamental period of the function or an integral multiple thereof (Eq. 8.35). The analog of this theorem in the case of discrete sampling is derived as follows. We first note that $E[g(n\Delta t)] = A_0/2 = G_0$. Hence

$$g(n\Delta t) - E[g(n\Delta t)] = \sum_{k=1}^{(N-1)/2} \left(G_k \exp\left(\frac{2\pi ikn}{N}\right) + G_{-k} \exp\left(-\frac{2\pi ikn}{N}\right)\right) \tag{8.67}$$

However, we can see from Equation 8.62 that if g is a real-valued function $G_{-k} = G_k^*$, where G_k^* is the complex conjugate of G_k. Thus the functions being summed in Equation 8.67 are just twice the real part of

$$G_k \exp\left(\frac{2\pi ikn}{N}\right) = 2\left[G_{kR} \cos\left(\frac{2\pi kn}{N}\right) + G_{kI} \sin\left(\frac{2\pi kn}{N}\right)\right]$$

where G_{kR} and G_{kI} are the real and imaginary parts, respectively, of G_k. Therefore

$$V(g) = \frac{4}{N} \sum_{n=-(N-1)/2}^{(N-1)/2} \sum_{k=1}^{(N-1)} \left(G_{kR} \cos\left(\frac{2\pi kn}{N}\right) + G_{kI} \sin\left(\frac{2\pi kn}{N}\right)\right)$$
$$\times \sum_{j=1}^{(N-1)/2} \left(G_{jR} \cos\left(\frac{2\pi jn}{N}\right) + G_{jI} \sin\left(\frac{2\pi jn}{N}\right)\right) \tag{8.68}$$

The cross product of the sin and cos terms, when summed over n, equals 0. The same is true of

$$\cos(2\pi kn/N)\cos(2\pi jn/N) \quad \text{and} \quad \sin(2\pi kn/N)\sin(2\pi jn/N),$$

unless $k = j$, in which case both sums over n equal $N/2$. This result is the discrete analog of Equations 8.31 and 8.32. We conclude that

$$V(g) = \frac{4}{N} \sum_{k=1}^{(N-1)/2} \frac{N}{2}\left(G_{kR}^2 + G_{kI}^2\right) \tag{8.69}$$

Since $G_{kR}^2 + G_{kI}^2 = |G_k|^2$, it follows that

$$V(g) = 2 \sum_{k=1}^{(N-1)/2} |G_k|^2 \tag{8.70}$$

We see again that $V(g)$ can be partitioned into a sum of components associated with the Fourier transform of g, and the orthogonality relationship, Equation 8.61, ensures that the components are independent of one another. It would seem clear in the discrete case that each G_k is associated with a frequency $k/\tau = k/(N\Delta t)$, where $k < N/2$. In the discrete case, however, the association between Fourier coefficients and frequencies is not so simple. The reason is that our discrete analysis prevents us from distinguishing contributions from a Fourier sinusoidal function with a frequency of k/τ and any other sinusoidal function with a frequency of

$$\frac{m}{\tau} = \frac{n}{\Delta t} \pm \frac{k}{\tau} = \frac{Nn \pm k}{\tau}$$

where n is any positive integer. The explanation is the aliasing problem which we previously discussed with respect to Figure 8.17 and can be traced to the fact that the left-hand side of Equation 8.61 will equal N whenever $|m - k| = jN$, where j is any integer. Clearly, if $m - k = jN$, then $|m - k| = jN$. However, we can see from Equations 8.58 and 8.59 that the contribution of the sinusoidal function with frequency k/τ is determined by both G_k and G_{-k}. Obviously, $m - (N - k) = m + k - N$ will be an integral multiple of N if $m + k$ is an integral multiple of N. Hence if either $m = jN - k$ or $m = jN + k$, the amplitude of the kth Fourier sinusoid will be incremented by the sinusoid with frequency m/τ. Thus, assuming $g(t)$ is a periodic function with period τ, each $|G_k|^2$ in Equation 8.70 is, in fact, equal to

$$|G_k|^2 = |G_k|^2 + \sum_{j=1}^{\infty} |G_{jN \pm k}|^2 \tag{8.71}$$

where the $|G_k|^2$ on the right-hand side of Equation 8.71 is the true contribution of the kth sinusoid to the spectrum and the $|G_k|^2$ on the left-hand side is what we calculate based on sampling at discrete points in time. We have no way to determine how important the higher-frequency sinusoids are without sampling at more closely spaced intervals.

Parseval's Theorem and Error Bounds for the Discrete Power Spectrum

Using an approach similar to that used to derive Equation 8.70, we can show that

$$\sum_{n=-(N-1)/2}^{(N-1)/2} [g(n\Delta t)]^2$$

$$= \sum_{n=-(N-1)/2}^{(N-1)/2} \sum_{k=-(N-1)/2}^{(N-1)/2} G_k \exp\left(\frac{2\pi ikn}{N}\right) \sum_{j=-(N-1)/2}^{(N-1)/2} G_j \exp\left(\frac{2\pi ijn}{N}\right)$$

$$= \sum_{k=-(N-1)/2}^{(N-1)/2} G_k \sum_{j=-(N-1)/2}^{(N-1)/2} G_j \sum_{n=-(N-1)/2}^{(N-1)/2} \exp\left[\frac{2\pi i(k+j)n}{N}\right]$$

$$= N \sum_{k=-(N-1)/2}^{(N-1)/2} G_k G_{-k} = N \sum_{k=-(N-1)/2}^{(N-1)/2} |G_k|^2 \tag{8.72}$$

Equation 8.72 is known as *Parseval's theorem* for Fourier series, and the summation $\Sigma_{k=-(N-1)/2}^{(N-1)/2}|G_k|^2$ in the limit as $N \to \infty$ ($\Delta t \to 0$ for finite τ) is the discrete power spectrum of g. When N is finite, as is always the case in practice, $\Sigma_{k=0}^{N-1}|G_k|^2$ is often referred to as the discrete power spectrum of g, but with the obvious caveat of Equation 8.71.

Although Equation 8.70 provides a very convenient way of partitioning the variance of g, it has little statistical meaning unless we can provide some estimate of the uncertainty associated with each component of the spectrum. The G_k are, in general, complex numbers, and we can therefore write

$$|G_k|^2 = G_{kR}^2 + G_{kI}^2 \tag{8.73}$$

where G_{kR} is the real part of G_k and G_{kI} is the imaginary part. If $\{g(t)\}$ is Gaussian, then it can be shown that G_{kR} and G_{kI} are uncorrelated random variables with zero means and equal variances. It follows from the definition of the χ^2 distribution function that $|\tilde{G}_k|^2/\sigma_k^2$ will follow the χ^2 distribution with 2 degrees of freedom, where $\tilde{G}_k$ is an estimate of G_k and σ_k^2 is the common variance of G_{kR} and G_{kI}. Since G_{kR} and G_{kI} have means of 0, it follows that $|G_k|^2$, the expectation value of $|\tilde{G}_k|^2$, is equal to $2\sigma_k^2$. Hence $2|\tilde{G}_k|^2/|G_k|^2$ will follow the χ^2 distribution with 2 degrees of freedom. Since the variance of a χ^2 variable with n degrees of freedom is $2n$, it follows that

$$\frac{\sigma\left(|\tilde{G}_k|^2\right)}{|G_k|^2} = \frac{\sqrt{2n}}{2} = \frac{\sqrt{4}}{2} = 1 \tag{8.74}$$

In other words, the standard deviation of the estimate is equal to the true value of the parameter. This is not a very acceptable state of affairs. There are basically two approaches to dealing with this problem:

1. Divide the sample record into n ensembles and calculate a discrete power spectrum for each ensemble. Let $\hat{G}_k$ be the average of the n estimates of G_k obtained from the n ensembles. Since each $\tilde{G}_k$ is associated with 2 degrees of freedom, it follows that $\Sigma_{m=1}^{n}|\tilde{G}_k|_m^2/\sigma_k^2$ will follow the χ^2 distribution with $2n$ degrees of freedom. Substituting $\Sigma_{m=1}^{n}|\tilde{G}_k|_m^2 = n|\hat{G}_k|^2$ and $\sigma_k^2 = \frac{1}{2}|G_k|^2$, we conclude that $2n|\hat{G}_k|^2/|G_k|^2$ will follow a χ^2 distribution with $2n$ degrees of freedom. Hence

$$\frac{\sigma\left(|\hat{G}_k|^2\right)}{|G_k|^2} = \frac{\sqrt{2(2n)}}{2n} = \frac{1}{\sqrt{n}} \tag{8.75}$$

The lower and upper confidence limits are then given by

$$\text{Lower confidence limit} = \frac{2n|\hat{G}_k|^2}{\chi^2_{2n;\,\alpha/2}} \tag{8.76}$$

$$\text{Upper confidence limit} = \frac{2n|\hat{G}_k|^2}{\chi^2_{2n;\,1-\alpha/2}} \tag{8.77}$$

where $\chi^2_{2n;\,\alpha/2}$ is the critical value of the χ^2 statistic with $2n$ degrees of freedom at confidence level $\alpha/2$. Note that this is a two-tailed test, in contrast to the one-tailed χ^2 test we used for goodness of fit to the normal distribution function.

2. Compute the discrete power spectrum for the entire sample record and then average $|\tilde{G}_k|^2$ for n consecutive values of k. For example, we might average the first five $|\tilde{G}_k|^2$ values, the second five $|\tilde{G}_k|^2$ values, and so forth. The df associated with each average value will be $2n$, and the confidence limits are calculated using Equations 8.76 and 8.77, where $|\hat{G}_k|^2$ is now the average of the n $|\tilde{G}_k|^2$ values centered at frequency k/τ.

Figure 8.19 illustrates the application of Equations 8.76 and 8.77 to the power spectra of the data in Figures 8.3 and 8.9*B*. In each case the sample records were divided into n ensembles. In the case of the SOI data, n was taken to be 8. Hence each ensemble spanned 14 years. For the Mauna Loa CO_2 data, n was taken to be 7. Hence each ensemble spanned 5 years. Since the ensembles in both cases span an integral number of years, N is even, and the spectral decomposition of $V(g)$ can be expressed using only Fourier coefficients with indices up to and including $N/2$. The frequency associated

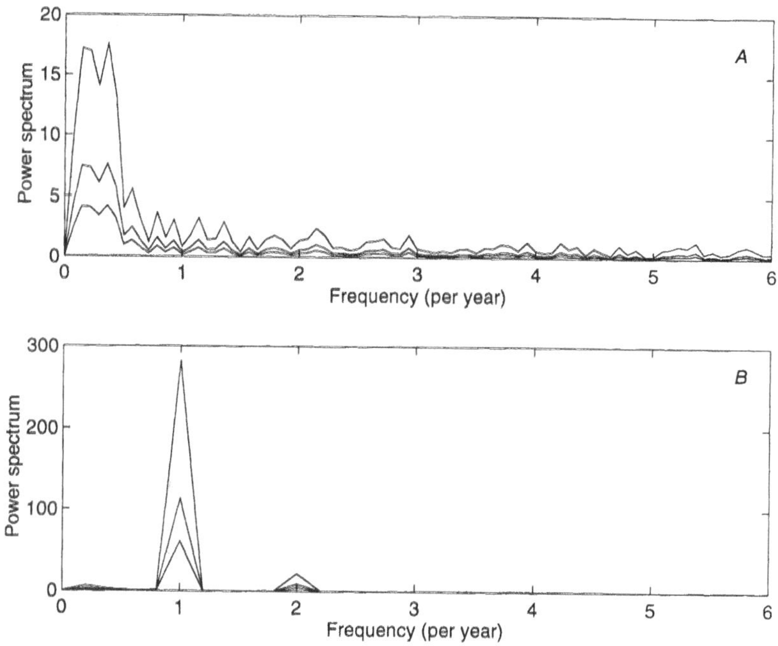

Figure 8.19 Discrete power spectrum of Southern Oscillation Index data (A) and detrended Mauna Loa CO_2 data (B) and error bounds calculated by dividing the sample record into eight and seven ensembles, respectively. The middle curve in each case is the discrete power spectrum. The upper and lower curves are the lower and upper 95% confidence intervals calculated from Equations 8.75 and 8.76, respectively.

with the sinusoid with index $N/2$ is $1/(2\Delta t)$ and is called the *Nyquist frequency*. Since $\Delta t = 1$ month $= 1/12$ year, the Nyquist frequency is $12/2 = 6$ per year. This is the highest frequency of any sinusoid in the discrete power spectra of the two sample records.

In the case of the SOI power spectrum, Figure 8.19A, we see a peak at frequencies between $2/14$ and $6/14$ per year, or at periods of $7/3$ to 7 years. The error bounds for $\alpha = 0.05$ provide an indication of whether the peaks are significant relative to adjacent troughs. We can see in this case that the lower error bounds to the peaks are higher than the upper error bounds to not only the troughs but also the peaks at all frequencies higher than $8/14$ per year. It therefore seems safe to conclude that there is a significant peak in the discrete power spectrum at periods of roughly 2 to 7 years. The error analysis in the case of the detrended Mauna Loa CO_2 data make it clear that the peaks at frequencies of 1 and 2 per year are significant. There is virtually nothing in the power spectrum at any other frequency.

ALIASING

As previously noted, whenever a time series analysis is based on data collected at discrete points in time separated by a time interval Δt, a problem called aliasing can arise. In our analysis of Figure 8.17, we concluded that aliasing would occur whenever Δt was greater than $1/(2f_{\max})$, where $f_{\max}$ is the highest frequency of any sinusoidal component in the Fourier transform of the sample function. More precisely, we discovered in the analysis leading up to Equation 8.71 that a Fourier frequency $f_a = k/\tau = k/N\Delta t)$ can be aliased by a higher frequency f_a' which satisfies the equation

$$f_a' = \frac{n}{\Delta t} \pm f_a \tag{8.78}$$

where n is any positive integer. Once data have been collected, there is no way to eliminate the effects of aliasing. Hence it is important when analyzing data to be aware of the problem and to anticipate the effects of periodic phenomena which might alias into the Fourier spectrum. The most effective time to deal with aliasing, however, is in the design phase of the experiment. There are basically two approaches:

1. If the highest frequency $f_{\max}$ in the spectrum of $g(t)$ is known, then choose the time interval Δt so that the Nyquist frequency is greater than or equal to $f_{\max}$; that is, choose $\Delta t \leq 1/(2f_{\max})$.
2. Prefilter $g(t)$ to remove all frequencies higher than $1/(2\Delta t)$.

The rationale behind these two approaches is straightforward enough. Since the highest frequency of any sinusoid in the discrete Fourier spectrum is the Nyquist frequency $1/(2\Delta t)$ and the smallest aliasing frequency f_a' is $(1/\Delta t) - f_a$, it follows that the smallest aliasing frequency must be greater than the Nyquist frequency, $1/(2\Delta t)$. Put another way, if none of the variance of a stationary process $g(t)$ is associated with frequencies higher than $1/(2\Delta t)$, then a complete characterization of $g(t)$ is possible by sampling at times separated by an interval Δt. Prefiltering is an option available when a process is being monitored continuously or at a very high frequency but analysis is anticipated at a lower frequency. For example, if data are collected electronically in the laboratory at a frequency of 100 per second and if the equipment is not properly shielded, then one can anticipate aliasing from 60 cycles per second (cps) background noise. The reason is that the Nyquist frequency in this example is $100/2 = 50$ cps, which is less than 60 cps. The aliasing would occur at a frequency equal to $(1/\Delta t) - f_a' = 100 - 60 = 40$ cps. The solution would be to filter out the background 60 cps noise by properly shielding the equipment.

Figure 8.20 illustrates how aliasing can occur. In this example the sampling interval Δt is 200, and a sinusoid with a period of 1000 is part of the Fourier

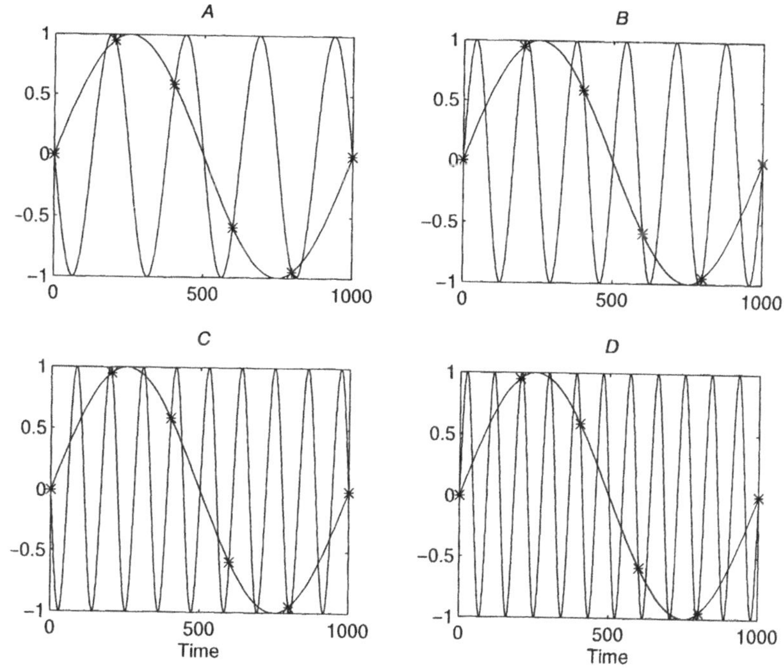

Figure 8.20 Sinusoids with frequencies of 4 (A), 6 (B), 9 (C), and 11 (D) per 1000 superimposed on a sinusoid with a frequency of 1 per 1000. Asterisks correspond to samples collected at time 0 and at subsequent time intervals of 200.

spectrum. This would be the case if the length of the sample record were some integer multiple of 1000. Inserting $\Delta t = 200$ and $f_a = 1/1000$ into Equation 8.67, we conclude that aliasing of the power spectrum at $f_a = 1/1000$ could result from the following frequencies in excess of the Nyquist frequency of $1/400$:

$$n = 1: f_a' = \frac{4}{1000} \text{ or } \frac{6}{1000}$$

$$n = 2: f_a' = \frac{9}{1000} \text{ or } \frac{11}{1000}$$

and, in general,

$$f_a' = \frac{5n \pm 1}{1000}$$

Sinusoids having the four frequencies corresponding to $n = 1$ and 2 are superimposed on the Fourier sinusoid with a frequency of $1/1000$ in Figure 8.20. It is apparent from the asterisks in the figure that these four functions

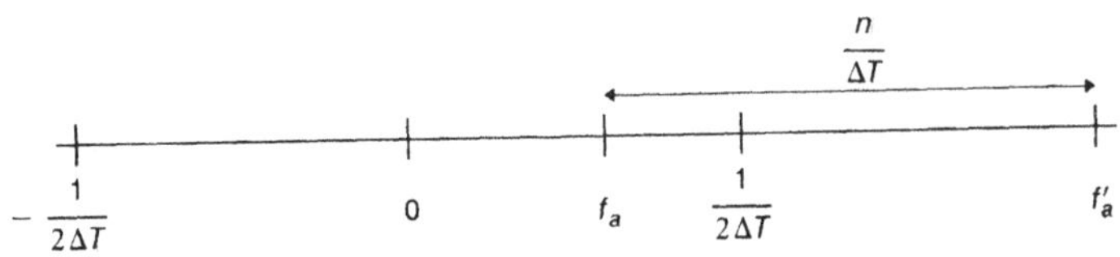

Figure 8.21 Relationship between aliasing frequency f'_a and aliased frequency f_a in the interval $[-1/(2\Delta t), 1/(2\Delta t)]$.

intersect the Fourier sinusoid at exactly the times when the process is being sampled, that is, at time intervals of 200. Hence it would be impossible, with a sampling time interval of 200, to distinguish these four sinusoids from the sinusoid with a frequency of 1/1000.

Another way to think of the problem of aliasing is shown in Figure 8.21. Here we initially consider both negative and positive Fourier frequencies, all of which lie in the window $[-1/(2\Delta t), 1(2\Delta t)]$. The relationship between a Fourier frequency f_a and an aliasing frequency f'_a must be such that $|f_a| = |f'_a - n/\Delta t|$. Given f'_a, we can find the frequency f_a which would be aliased by f'_a by first finding the integer value of n that satisfies the condition

$$\left| f'_a - \frac{n}{\Delta t} \right| \leq \frac{1}{2\Delta t}$$

It is straightforward to show (see Exercise 8.8) that the integer value of n which satisfies this condition is $f'_a\Delta t$ rounded to the nearest integer. The positive frequency aliased by f'_a is then equal to $|f'_a - n/\Delta t|$. In our example of 60 cps electronic noise, $f'_a\Delta t = 60/100$. Hence $n = 1$ and $|f'_a - n/\Delta t| = |60 - 100| = 40$ cps. Voila.

LEAKAGE

Leakage, as previously noted, is a problem associated with the fact that sample records invariably span a finite record length. The nature of the problem can easily be appreciated using an approach similar to the method we used in analyzing the characteristics of the Fourier transform of $\hat{g}(t)$ in Equation 8.52. Assume that $g(t)$ is a continuous function on the interval $[-\infty, \infty]$ and that the Fourier transform of $g(t)$ is $G(f)$. Assume that our sample record of $g(t)$ extends over the time interval $[-\tau/2, \tau/2]$. We will call this sample record $g_\tau(t)$. It follows from our definition of the Box function in Equation 8.45 that

$$g_\tau(t) = \text{Box}\left(\frac{t}{\tau}\right) g(t) \tag{8.79}$$

It follows from our analysis of convolution that the Fourier transform of $g_T(t)$, which we will call $G_T(f)$, will be equal to the convolution of $G(f)$ and the Fourier transform of Box(t/τ). It is straightforward to show (see Exercise 8.10) that the Fourier transform of Box(t/τ) is equal to $\sin(\pi f\tau)/\pi f$. The function $\sin(\pi x)/\pi x$ is defined to be sinc(x). Hence the Fourier transform of Box(t/τ) is equal to τ sinc$(f\tau)$. The behavior of this function is shown in Figure 8.22. The function is characterized by a central peak with a height of τ at $f = 0$ and an infinite number of side peaks whose amplitude decreases rapidly with increasing f. The width of the central peak is $2/\tau$. As τ becomes large τ sinc$(f\tau)$ begins to look more and more like a Dirac δ function, and indeed it can be shown that $\lim_{T\to\infty} \tau \operatorname{sinc}(\tau f) = \delta(f)$. We conclude that

$$G_T(f) = G * \tau \operatorname{sinc}(f\tau) = \int_{-\infty}^{\infty} G(x)\tau \operatorname{sinc}[\tau(f-x)]\, dx \qquad (8.80)$$

and that

$$\lim_{T\to\infty} G_T(f) = \int_{-\infty}^{\infty} G(x)\, \delta(f-x)\, dx = G(f) \qquad (8.81)$$

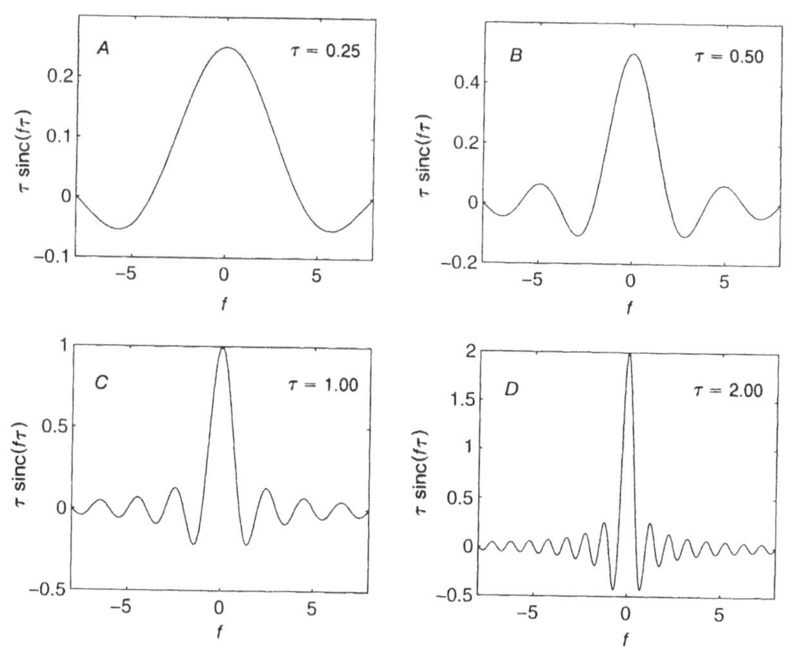

Figure 8.22 Behavior of the function τ sinc(tf) for four values of τ and $-8 \leq f \leq 8$.

For finite τ, $G_\tau(f)$ will obviously be an imperfect representation of $G(f)$, and we here consider a few examples to illustrate some characteristics of the imperfection. First consider the simple case where $g(t) = \cos(2\pi f_0 t)$. It is straightforward to show (see Exercise 8.11) that the Fourier transform of $\cos(2\pi f_0 t)$ is $\frac{1}{2}[\delta(f+f_0) + \delta(f-f_0)]$. It then follows from Equation 8.80 that

$$G_\tau(f) = \frac{\tau}{2}\{\mathrm{sinc}[(f+f_0)\tau] + \mathrm{sinc}[(f-f_0)\tau]\} \tag{8.82}$$

Graphs of the right-hand side of Equation 8.82 are shown in Figure 8.23 for $f_0 = 2$ and several values of τ. For large τ the Fourier transform of $\cos(2\pi f_0 t)$ does begin to approach one-half the sum of two δ functions centered at $f = \pm f_0$. However, for any finite value of τ, the δ function peaks leak out, producing oscillatory waves and reducing the height of the main peaks.

As a second example, consider the simple function $g(t) = \cos(2\pi f_1 t) + \cos(2\pi f_2 t)$. In this case the exact Fourier transform is one-half the sum of four δ functions centered at $f = \pm f_1$ and $\pm f_2$, and $G_\tau(f)$ will be $\tau/2$ times the sum of four sinc functions also centered at $f = \pm f_1$ and $\pm f_2$. Graphs of $G_\tau(f)$ are shown in Figure 8.24 for $f_1 = 2$, $f_2 = 4$, and several values of τ.

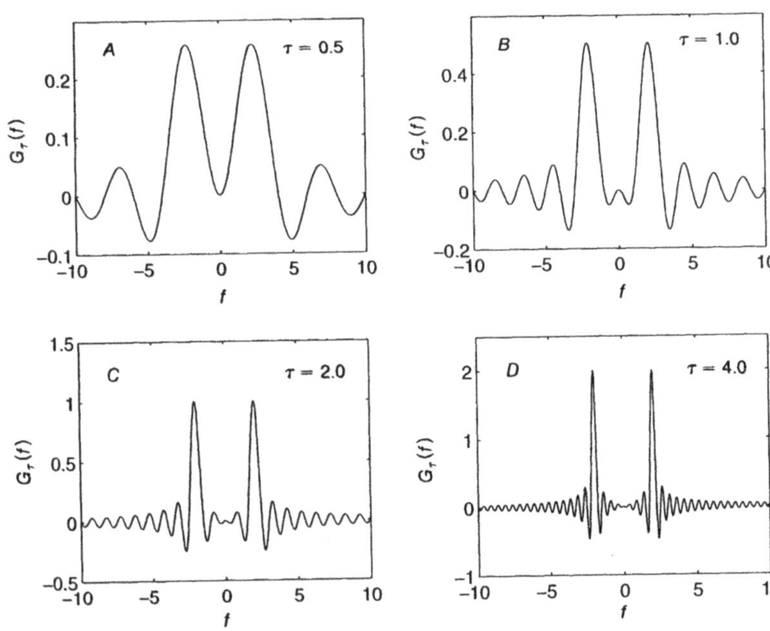

Figure 8.23 The Fourier transform of $\cos(2\pi f_0 t)$ with $f_0 = 2$ based on a sample record of length (A) $\tau = 0.5$, (B) $\tau = 1.0$, (C) $\tau = 2.0$, and (D) $\tau = 4.0$.

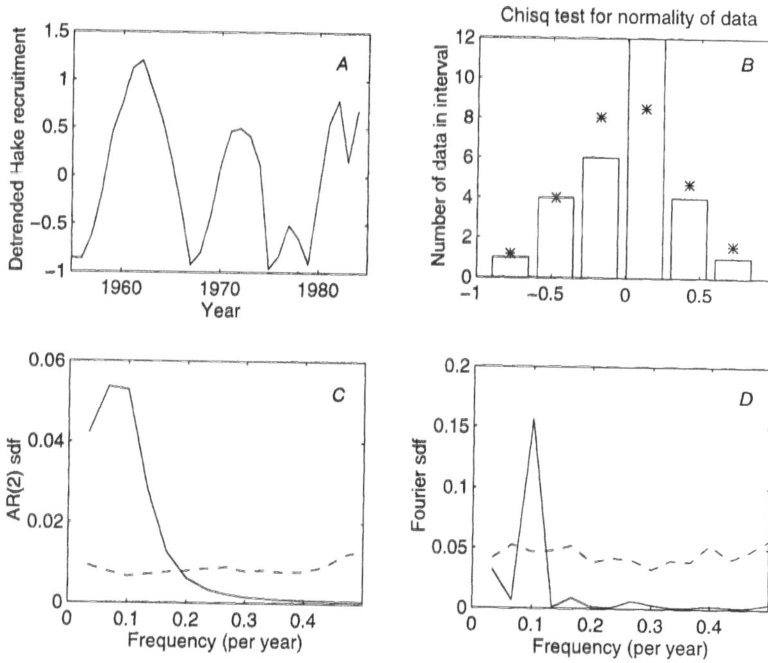

Figure 8.24 The Fourier transform of $\cos(2\pi f_1 t) + \cos(2\pi f_2 t)$ with $f_1 = 2$ and $f_2 = 4$ based on a sample record of length (A) $\tau = 0.5$, (B) $\tau = 1.0$, (C) $\tau = 2.0$, and (D) $\tau = 4.0$.

One particularly interesting result is that the two peaks at $f = 2$ and 4 and at $f = -2$ and -4 are completely unresolved in Figure 8.24*A*. The reason can easily be traced to something we already know about the sinc function, namely, that the width of the central peak is $2/\tau$. In the case of Figure 8.24*A*, $\tau = 0.5$ and hence $2/\tau = 4$. Since $f_2 - f_1 = 2$, there is considerable overlap of the peaks. In order to resolve two adjacent peaks, the gap between the peaks must be more than the half-width of the peaks, that is, $1/\tau$. In Figure 8.24*B*, for example, τ has been increased to 1, and the gap between the peaks is therefore twice the half-width of the peaks. Two peaks are now clearly discernible at both positive and negative frequencies, but the resolution of the peaks still leaves something to be desired. The resolution of the peaks improves as τ is further increased in Figure 8.24*C*–*D*.

Finally, we will consider a sample function dominated by low-frequency variability whose Fourier transform, for the sake of simplicity, we will assume to be triangular in shape with a peak at $f = 0$ declining to 0 at $f = \pm 0.5$ (Fig. 8.25*A*). The convolution of this Fourier transform with $\tau\,\mathrm{sinc}(f\tau)$ is shown in Figure 8.25*B*–*D* for several values of τ. With a record length of 8, we get a reasonably good representation of $G(f)$ (Fig. 8.25*B*), but as the record length is shortened $G_\tau(f)$ becomes increasingly rounded with a lower

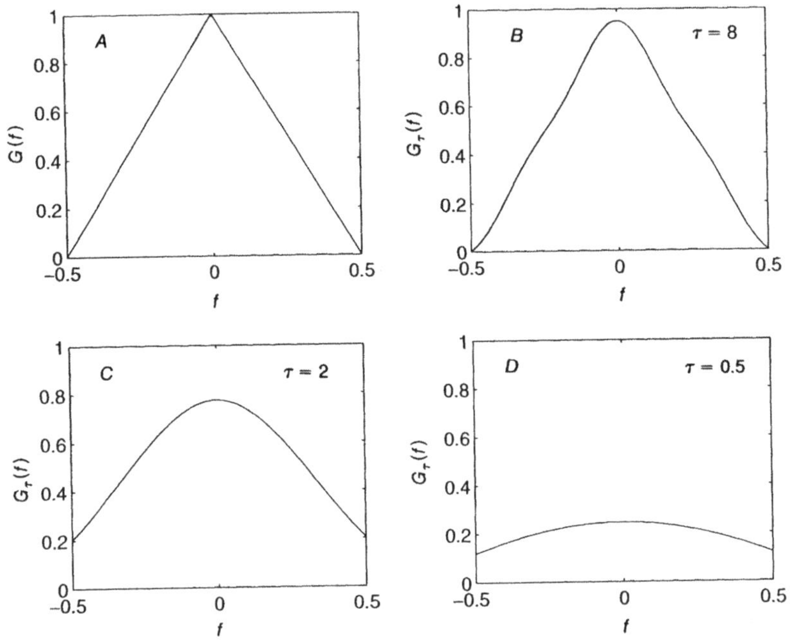

Figure 8.25 The Fourier transform of a sample function dominated by low-frequency variability based on a sample record of length (A) $\tau = \infty$, (B) $\tau = 8$, (C) $\tau = 2$, and (D) $\tau = 0.5$.

height and greater width than $G(f)$. At $\tau = 0.5$, $G_\tau(f)$ bears little resemblance to $G(f)$.

What we learn from this analysis is that leakage changes the shape of peaks in the Fourier transform, making them more rounded, less distinct, with a lower height and greater width. Peaks which are separated by a frequency less than $1/\tau$ will be merged, and good resolution of neighboring peaks will require that the record length be greater than approximately $2/\Delta f$, where Δf is the difference in the frequency of adjacent peaks.

What can we do to minimize artifacts caused by leakage? One obvious solution is to make τ as large as possible, but practical considerations impose a limit on the duration of the sample record. Given the fact that τ will always be finite, the following are some possible ways to minimize artifacts caused by leakage.

1. If sharp, well-defined periodicities are known to exist in the sample record, they may be filtered out before proceeding with the Fourier analysis. In the case of sea level records, for example, we might choose to filter out the tidal cycle. In the cases of the El Niño Southern Oscillation Index and the Mauna Loa CO_2 sample records, we might filter out the annual cycle. In

some cases this can be done by calculating running averages. In other cases we may choose a least-squares technique, that is, fitting the sample record with a function of the form $A \sin(\omega t) + B \cos(\omega t)$, where ω is the frequency we wish to filter out of the spectrum. The Fourier analysis is then carried out on the residuals, that is, $g(t) - A \sin(\omega t) - B \cos(\omega t)$.

2. We can modify the kernel in the convolution in Equation 8.80. The kernel in this case, $\tau \operatorname{sinc}(f\tau)$, is the result of our choosing $\operatorname{Box}(t/\tau)$ to produce $g_\tau(t)$ from $g(t)$. When used in this context, $\operatorname{Box}(t/\tau)$ is referred to as a data window or data taper. There is, in fact, nothing sacred about using $\operatorname{Box}(t/\tau)$ as a data window, and several other possibilities are illustrated in Figure 8.26. The ideal data window would have a Fourier transform that closely approximates a Dirac δ function, that is, a transform with a narrow central peak and very small side lobes. Unfortunately, no such ideal data window exists, and a compromise must invariably be struck between obtaining a Fourier transform with a narrow central peak and obtaining a Fourier transform with small side lobes. This fact is apparent from an examination of Figure 8.27, which shows the absolute values of the Fourier transforms of the data windows in Figure 8.26. The Blackman, Hamming, and Hanning windows are all constructed using cosine functions. The Bartlett window is

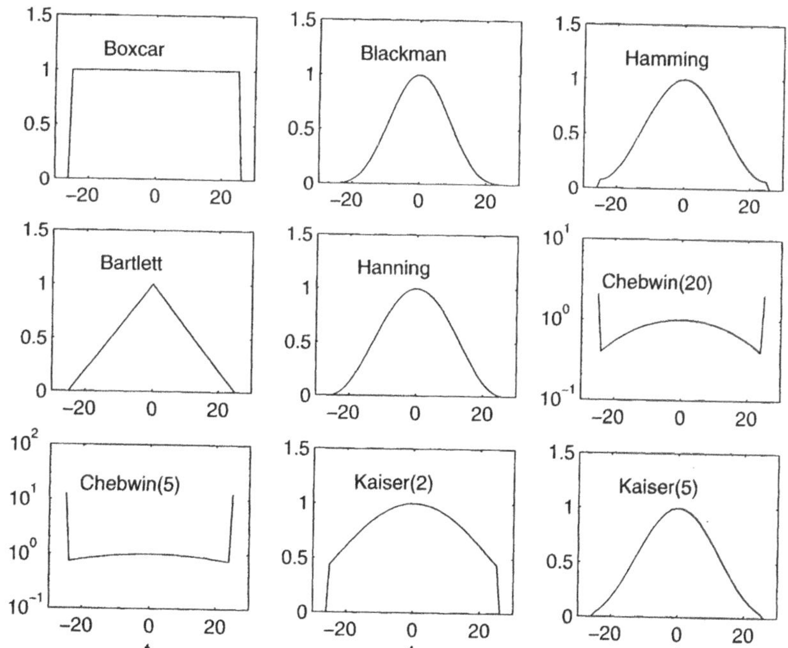

Figure 8.26 Shapes of various data windows or data tapers for a sample record of duration $\tau = 51$.

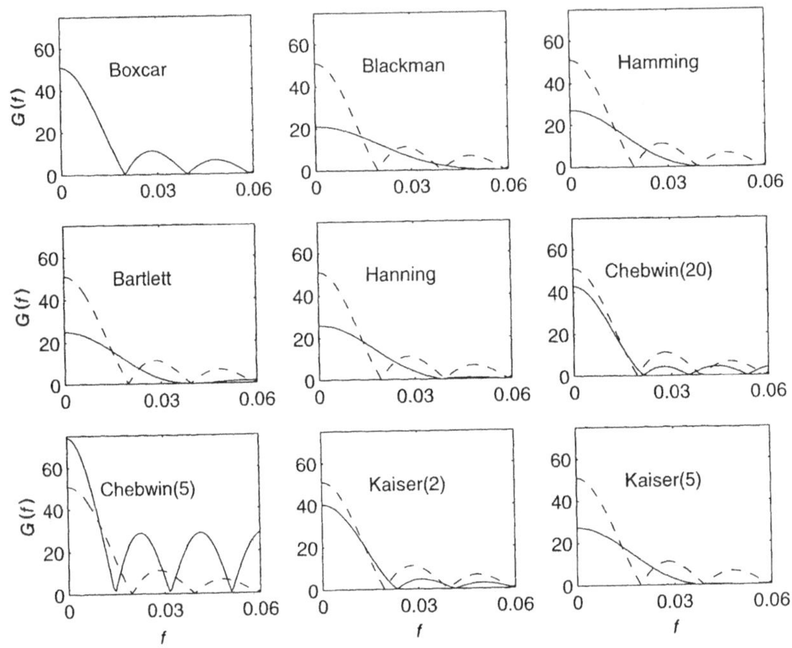

Figure 8.27 The absolute values of the Fourier transforms of the data windows shown in Figure 8.26. For comparison, the dashed curve shows the absolute value of the Fourier transform of the Boxcar data window.

triangular. These four windows produce relatively broad central peaks with very small side lobes. The Chebyshev and Kaiser windows are more complex data tapers which allow some manipulation of the characteristics of the central peak and side lobes. The Chebyshev window minimizes the width of the central peak, given a particular side lobe height. In Figures 8.26 and 8.27, Chebwin(20) and Chebwin(5) are Matlab Chebyshev windows, the magnitudes of whose Fourier transform side lobe heights are 20 and 5 dB,[4] respectively, below the height of the magnitude of the central peak. The Kaiser window is an approximation to a prolate spheroidal window and maximizes the ratio of the central peak energy to the side lobe energy. In Figures 8.26 and 8.27 Kaiser(2) and Kaiser(5) are Matlab Kaiser windows. Increasing the argument of the Kaiser subroutine widens the central peak and decreases the amplitude of the side lobes. Kaiser(0) is a Boxcar.

In order to get some feeling for the effect of using data windows other than the Boxcar, consider the Blackman window. The Blackman window

[4]The decibel as used here is 20 times the common logarithm of the ratio of the numbers. Thus 20 dB corresponds to a factor of 10, 40 dB to a factor of 100, and so forth.

Black(t) is defined by

$$\text{Black}(t) = \begin{cases} 0.42 + 0.5\cos\left(\dfrac{2\pi t}{\tau}\right) + 0.08\cos\left(\dfrac{4\pi t}{\tau}\right) & \text{for } -\dfrac{\tau}{2} \le t \le \dfrac{\tau}{2} \\ 0 & \text{otherwise} \end{cases} \tag{8.83}$$

It is straightforward to show (see Exercise 8.12) that the Fourier transform of Black(t) is

$$0.42\tau\,\text{sinc}(f\tau) + 0.5\frac{\tau}{2}[\text{sinc}(1 - f\tau) + \text{sinc}(1 + f\tau)]$$
$$\times 0.08\frac{\tau}{2}[\text{sinc}(2 - f\tau) + \text{sinc}(2 + f\tau)]$$

Using this fact and the convolution theorem, we can easily recalculate the analog of Figure 8.24 for the case in which our data window is the Blackman window rather than Box(t/τ). The result is shown in Figure 8.28. The side lobes are now gone from the Fourier transform, but the central peaks are

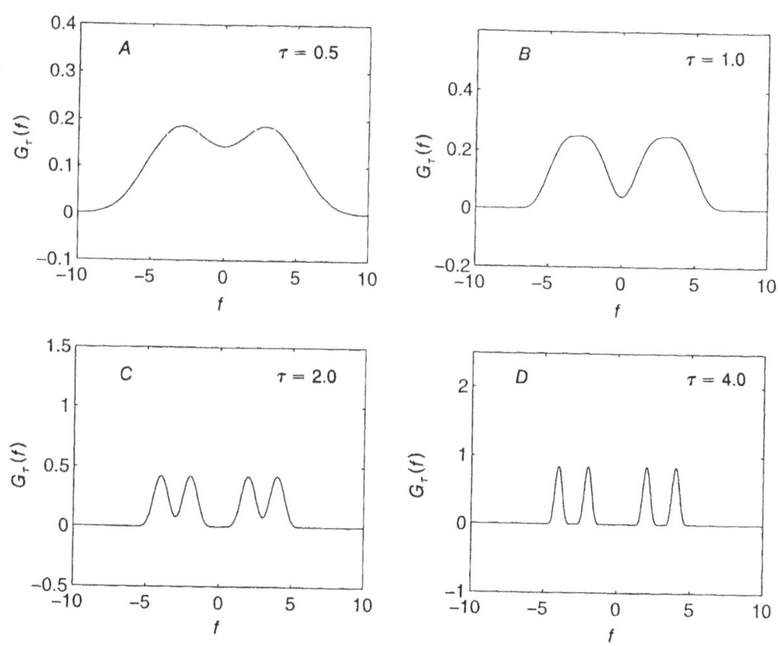

Figure 8.28 The analog of Figure 8.24 using a Blackman data window rather than Box(t/τ).

broader and not as high as Figure 8.24. A longer sample record is now needed to resolve the peaks in the spectrum. Alternatively, we could have used a data window with a narrower central peak in its transform, but the price would have been higher side lobes. There is no perfect data window.

3. Another strategy for dealing with leakage is a procedure called prewhitening. To understand how prewhitening works, we first need to derive a few facts about Fourier transforms. It is straightforward to show from Equations 8.39 and 8.40 that if $g(t) \subset G(f)$ and $h(t) \subset H(f)$, then $c_1 g(t) + c_2 h(t) \subset c_1 G(f) + c_2 H(f)$, where c_1 and c_2 are constants, independent of t. It is also easy to show that $g(t + T) \subset G(f)\exp(2\pi i T f)$. This latter fact is known as the *shift theorem*. It follows from these facts that a function of the form $\Sigma c_k g(t + k\Delta x)$ will have the Fourier transform $G(f)\Sigma c_k \exp(2\pi i k \Delta x f)$, where k is a running index with potentially both positive and negative values. Now define $x_k \equiv k\Delta x$ and let $c_k = \rho(x_k)\,\Delta x$. We then have

$$\sum \rho(x_k) g(t + x_k)\,\Delta x \subset G(f) \sum \rho(x_k)\exp(2\pi i x_k f)\,\Delta x$$

When we now take the limit as $\Delta x \to 0$, it follows from what we learned about integration in Chapter 6 that

$$\int \rho(x) g(t + x)\,dx \subset G(f) \int \rho(x)\exp(2\pi i f x)\,dx \tag{8.84}$$

where, in practice, the range of integration is assumed to be some relatively small interval in the vicinity of $x = 0$. Now what does this have to do with prewhitening?

First, we can define a function $h(t) \equiv \int \rho(x) g(t + x)\,dx$. It follows from Equation 8.84 that $H(f)$ will equal $G(f)\int \rho(x)\exp(2\pi i f x)\,dx$. In the case of prewhitening we are presumably dealing (or think we are dealing) with a function $g(t)$ which has certain undesirable characteristics from the standpoint of Fourier analysis. In particular, $G(f)$ may have a wide dynamic range, where the *dynamic range* is simply a measure of the ratio of the contributions to the power spectrum of the maximum and minimum frequencies in the spectrum. Because of leakage a sample record from a process with a wide dynamic range may give a misleading picture of the importance of certain frequencies to the spectrum. By selecting the right $\rho(x)$, however, it may be possible to define a new function $h(t)$ with a smaller dynamic range. We then carry out our Fourier analysis on $h(t)$, obtaining $H(f)$, and then back-transform using the fact that

$$G(f) = \frac{H(f)}{\int \rho(x)\exp(2\pi i f x)\,dx} \tag{8.85}$$

The term prewhitening comes from the fact that so-called *white noise* has a power spectrum that is constant, that is, the same for all frequencies, the analogy being to white light, which contains roughly equal contributions from all frequencies of light in the visible spectrum. Ideally, the process of prewhitening, that is, $g(t) \to h(t)$, would convert $g(t)$ into white noise, but if we knew enough about $g(t)$ to be able to transform it into white noise via $\int \rho(x) g(t+x)\,dx$, we would not have to do the Fourier analysis in the first place. Thus prewhitening is, in practice, just a step in the direction of whitening the time series, but it can be an important step if it effectively deals with a leakage problem. The final transformation using Equation 8.85 is referred to as *postcoloring*, that is, recovering the transform of $g(t)$ after carrying out the Fourier analysis on the partially whitened function $h(t)$.

We will consider prewhitening in more detail when we discuss parametric spectral analysis, but for now let us consider a simple example. Suppose we have a sample record of a sample function which we have reason to believe is dominated by low-frequency variability, but we are much more interested in the variability at higher frequencies. Is there a simple transformation which could be used to prewhiten the function and effectively reduce leakage from the low-frequency portion of the spectrum?

One simple approach to this problem is to carry out the Fourier analysis on the derivative of $g(t)$, realizing (from Eqs. 8.39 and 8.40) that $dg/dt \subset 2\pi i f G(f)$. Thus the Fourier transform of dg/dt will deemphasize contributions to the spectrum of $G(f)$ from low frequencies and emphasize contributions from high frequencies. Within the context of the formalism we have worked out, taking the derivative of $g(t)$ means setting $\rho(x)$ equal to the derivative of the Dirac delta function, that is, $\rho(x) = \delta'(x)$, where $\delta'(x-a)$ is defined by the equation

$$\int_{-\infty}^{\infty} \delta'(x-a) f(x)\,dx = f'(a) \tag{8.86}$$

As an example of a function dominated by low-frequency variability, consider the function $g(t) = 10\cos(2\pi f_1 t) + \cos(2\pi f_2 t)$, where $f_1 = 0.01$ and $f_2 = 1$. The Fourier transform of this function consists of four Dirac delta functions centered at $f = \pm 0.01$ and $f = \pm 1$, as indicated in Figure 8.29*A*. If our sample record has a length of 8, then our estimate of the Fourier transform of $g(t)$ obtained from this sample record is as shown in Figure 8.29*B*. It is the sum of four sinc functions centered at $f = \pm 0.01$ and $f = \pm 1$. The side lobes from the sinc functions centered at $f = \pm 0.01$ have almost completely obscured the central peaks from the smaller sinc functions centered at $f = \pm 1$. The Fourier transform of the derivative of $g(t)$ is shown in Figure 8.29*C*. Now the central peaks of the sinc functions centered at $f = \pm 1$ are the dominant peaks in the spectrum. The postcolored transformation is shown in Figure 8.29*D*. The unresolved peak associated with the central peaks of the sinc functions centered at $f = \pm 0.01$ is now again the dominant

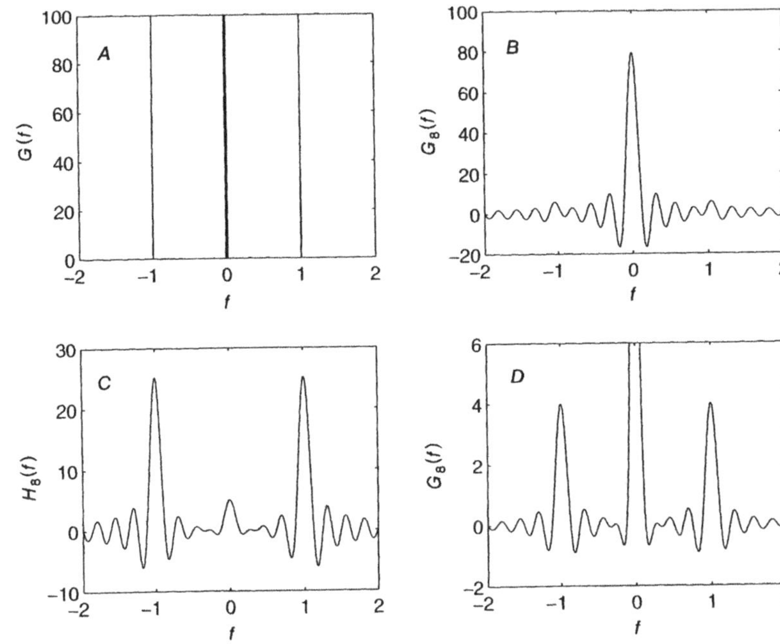

Figure 8.29 (A) The exact Fourier transform of $g(t) = 10\cos(2\pi f_1 t) + \cos(2\pi f_2 t)$, where $f_1 = 0.01$ and $f_2 = 1$, (B) the Fourier transform of the same function estimated from a sample record of length 8, (C) the Fourier transform of the derivative of $g(t)$ estimated from a sample record of length 8, and (D) the Fourier transform of $g(t)$ estimated by postcoloring the Fourier transform in (C) using Equation 8.85.

peak, as should be the case. However, the peaks associated with the sinc functions centered at $f = \pm 1$ now clearly rise above the side lobes of the sinc functions centered at $f = \pm 0.01$. We have thus effectively reduced the leakage associated with the low-frequency variability in the spectrum.

In practice, the technique of prewhitening and postcoloring certainly has its limitations. First, there is something of a chicken-and-egg problem. In order to make an intelligent choice of $\rho(x)$, we must know something about the spectrum, which is, of course, what we are trying to find out by carrying out the Fourier analysis. In fact, we sometimes have reason to believe that we do know something about the spectrum, either, for example, from previous experience or from theoretical considerations. In the absence of such knowledge, we may attempt to estimate $\rho(x)$ from the time series itself using an autoregressive model, an approach which we will explore in the following section on parametric spectral analysis. Even if we do succeed in identifying an effective prewhitening filter $\rho(x)$, the range of integration of the integral $\int \rho(x) g(t + x)\, dx$ may substantially reduce the record length of $h(t)$ compared to $g(t)$. For example, if the range of integration is $[x_1, x_2]$ and if the record length of $g(t)$ is τ, then the record length of $h(t)$ will be $\tau - (x_2 - x_1)$. Since leakage is a problem associated with the finite duration of the record length, substantially reducing the record length in order to prewhiten the time series tends to work against overcoming leakage problems. In summary, selection of an effective prewhitening filter $\rho(x)$ is by no means trivial.

PARAMETRIC SPECTRAL ANALYSIS

Spectral analysis is a technique for estimating the spectrum of a stationary, stochastic process. Fourier analysis, as described in this chapter, has traditionally been the basic tool of spectral characterization. It is essentially a nonparametric technique in that no model is assumed a priori. Admittedly, expressing a function as a sum of sines and cosines having a certain fundamental frequency or integral multiples thereof might seem like a model, but, in fact, it is a completely general representation of a function and is no more a model than, for example, the Taylor series expansion of a function about a point. Parametric spectral analysis is an alternative approach to spectral analysis based on the assumption that the process being studied can be described by a model whose general form is assumed a priori. The model is a function of certain parameters, and hence the name *parametric spectral analysis*.

As an example of a process whose spectral characteristics might be logically investigated using parametric spectral analysis, we will consider the time series of the number of adult fish in the stock, for example, of North Atlantic cod. For the purposes of this example, we will consider the state of affairs before the fishing activities of humans decimated the stock, that is, we will consider the population dynamics of the virgin stock. We assume in our model that each year a certain percentage of the fish disappear from the stock due to natural mortality and that each year fish are "recruited" to the adult stock based on their having reached a certain size, become sexually mature, or whatever. Now it is a fact that year-to-year variations in recruitment to an *r*-selected species such as cod tend to be rather large due to the presumed degree of spatial and/or temporal match/mismatch of resources at critical stages in the life cycle of the juvenile cod (Cushing, 1982). Natural mortality of the adult cod, however, tends to remove a relatively constant percentage of the adult stock, since the adults are less sensitive to the vagaries of the environment. Although it should be obvious that recruitment must depend to some extent on the size of the adult stock, as a first approximation we can ignore this dependence in the virgin population, since the adult stock is relatively stable, and year-to-year differences in recruitment are primarily controlled by the random match/mismatch of resources during the early life of the juvenile cod. A model of the population dynamics of the adult cod might therefore be

$$g_t = \alpha g_{t-1} + z_t \tag{8.87}$$

where g_t is the number of adult cod in year t, α is the fraction of adult cod that survive from one year to the next, and z_t is a normally distributed random variable with a specified mean and variance. For the purposes of this example, we will assume that $\alpha = 0.8$ and that z_t has a mean of 2 and a variance of 1. A simulated 500-year sample record obtained using Equation 8.87 is shown in Figure 8.30. The mean population size is 10, and the variance (see the following discussion) is $1/0.36 = 2.8$. When we plot the autocorrelation coefficient function $\rho_{xx}(\Delta t)$ against lag time, we find that it decays in a more-or-less exponential manner to values that are statistically insignificant at time lags of 5 to 8 years (Fig. 8.31).

In order to determine the Fourier transform of this process, we first consider the function $h_t = g_t - \alpha g_{t-1}$. The function h_t can be obtained from the function g_t by defining the function $\rho(x)$ in Equation 8.85 to be $\delta(x-t) - \alpha\delta(x-t+1)$. Since in the discrete case the only frequencies in the Fourier spectrum are integral multiples of $1/N$, it follows that the

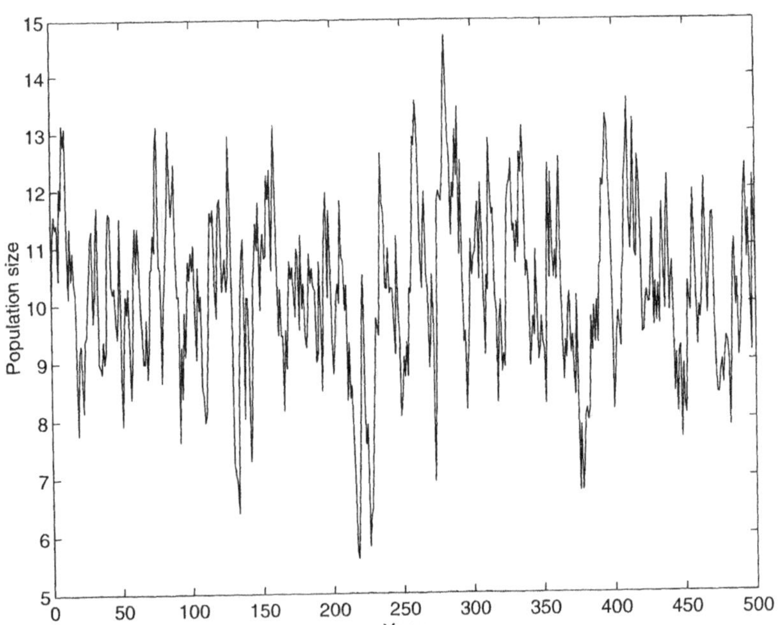

Figure 8.30 A 500-year sample record of the number of North Atlantic cod (arbitrary units) obtained using the model in Equation 8.87.

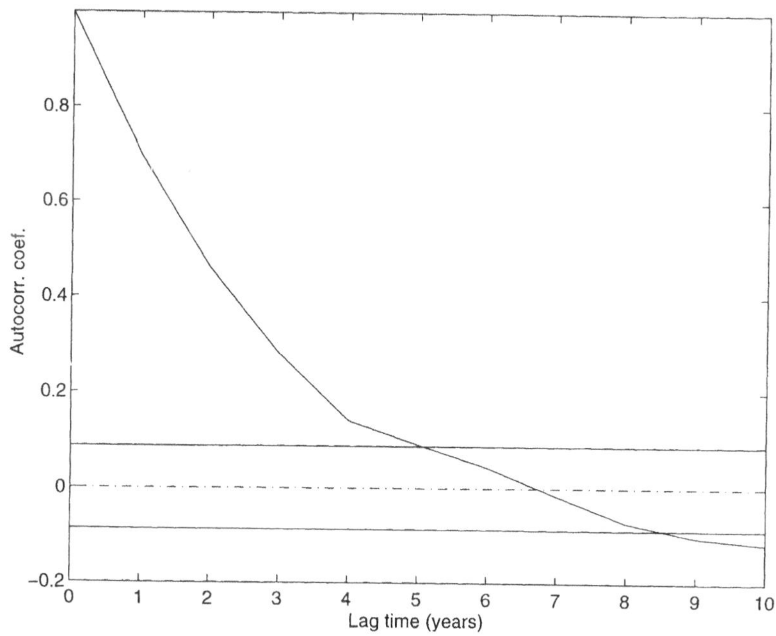

Figure 8.31 Autocorrelation coefficient function for the time series in Figure 8.30. The solid lines at approximately ± 0.09 are the 95% confidence limits to $\rho_{xx}(\Delta t)$ calculated using Equation 8.25.

denominator of Equation 8.84 becomes

$$\int \rho(x)\exp\left(\frac{2\pi ifx}{N}\right) dx = \exp\left(\frac{2\pi ift}{N}\right) - \alpha\exp\left(\frac{2\pi if(t-1)}{N}\right)$$
$$= \exp\left(\frac{2\pi ift}{N}\right)\left[1 - \alpha\exp\left(-\frac{2\pi if}{N}\right)\right]$$

where f is now understood to take on only integral values from $-\frac{N-1}{2}$ to $\frac{N-1}{2}$. We conclude that in this case

$$\left|\int \rho(x)\exp(2\pi ifx)\, dx\right| \to \left|1 - \alpha\exp\left(-\frac{2\pi if}{N}\right)\right|$$

It follows from Equation 8.85 that

$$|G_f| = \frac{|H_f|}{\left|1 - \alpha\exp\left(-\frac{2\pi if}{N}\right)\right|}$$

where G_f and H_f are the discrete Fourier transforms of g and h, respectively. Now in this case it is obvious from Equation 8.87 that $h_t = z_t$, which is a normally distributed random variable. It is straightforward to show (see Exercise 8.15) that if $h(t)$ has the Fourier transform $H(f)$, then the Fourier transform of the autocorrelation function of $h(t)$ is $|H(f)|^2$. Similarly, if the discrete Fourier transform of h_t is H_f, then the discrete Fourier transform of the autocorrelation function of h_t is $|H_f|^2$. It follows that in the discrete case

$$|H_f|^2 = \frac{1}{N} \sum_{k=-(N-1)/2}^{(N-1)/2} R_{hh}(k) \exp\left(-\frac{2\pi ifk}{N}\right) \tag{8.88}$$

where

$$R_{hh}(k) = \frac{1}{N} \sum_{t=1}^{N} h_t h_{t+k} \tag{8.89}$$

Now if h is a set of normally distributed random numbers with a mean of μ_z and a variance of σ^2, it follows that, for $k \neq 0$, $E[R_{hh}(k)] = \mu_z^2$, and for $k = 0$, $E[R_{hh}(0)] = \mu_z^2 + \sigma^2$. From Equation 8.88 it follows that, for $f \neq 0$,

$$E\left[|H_f|^2\right] = \frac{1}{N}\left(\sigma^2 + \sum_{k=-(N-1)/2}^{(N-1)/2} \mu_z^2 \exp\left(-\frac{2\pi ifk}{N}\right)\right) = \frac{\sigma^2}{N}$$

For $f = 0$,

$$E\left[|H_0|^2\right] = \frac{1}{N}(\sigma^2 + N\mu_z^2) = \mu_z^2 + \frac{\sigma^2}{N}$$

It now follows that, for $f \neq 0$,

$$E\left[|G_f|^2\right] = \frac{\sigma^2}{N\left|1 - \alpha \exp\left(-\frac{2\pi if}{N}\right)\right|^2} \tag{8.90}$$

and, for $f = 0$,

$$E\left[|G_0|^2\right] = \frac{\sigma^2 + N\mu_z^2}{N(1-\alpha)^2} \tag{8.91}$$

Now if g_t is a stationary process, it follows from Equation 8.87 that $E[g_t] = \mu_z/(1 - \alpha)$, and it is clear from Equation 8.91 that

$$\lim_{N \to \infty} E\left[|G_0|^2\right] = \frac{\mu_z^2}{(1 - \alpha)^2} = (E[g_t])^2$$

This result is to be expected, since it follows from Equation 8.62 that G_0 will always be the average value of the sample record.

Figure 8.32 shows the spectrum of the process defined by Equation 8.87 for sample records of length 500 years. The smooth curve in each case is the expectation value of the spectrum calculated from Equation 8.90. The dots represent the discrete Fourier spectrum estimated from 1, 5, 10, or 20 sample records. The figure illustrates two important points. First, if a large enough number of sample records is included in the analysis, the spectrum does approach the theoretical curve. Second, if the analysis involves only one or a few sample records, the calculated spectrum may deviate greatly from the expectation value of the spectrum. This latter observation underscores the

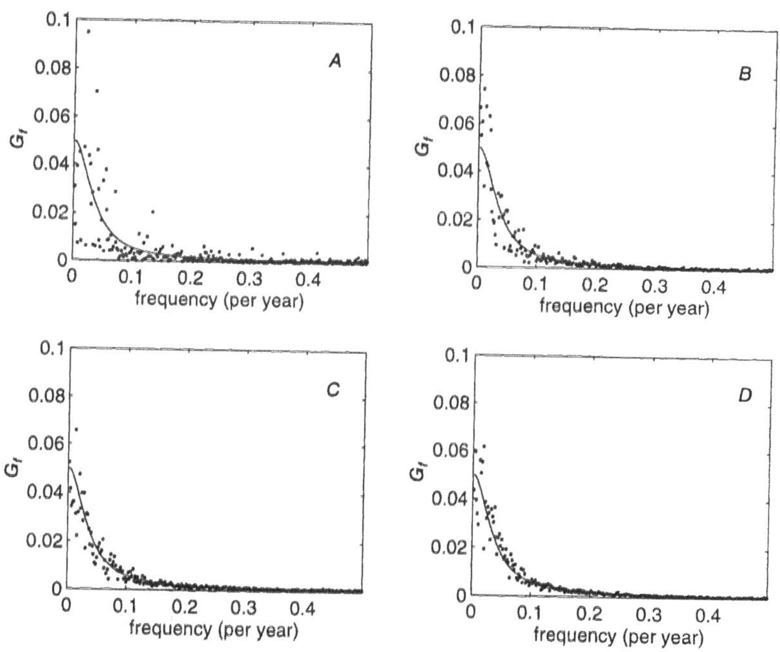

Figure 8.32 Discrete Fourier power spectrum for frequencies greater than 0 of 500-year sample records of the model in Equation 8.87. The solid line is the expectation value of the spectrum based on Equation 8.90. The dots are average spectra based on (A) 1 sample record, (B) 5 sample records, (C) 10 sample records, and (D) 20 sample records.

point made previously in this chapter that some sort of averaging process is often needed to reduce the noise level in a Fourier spectrum. Otherwise it may be impossible to characterize the spectrum in a meaningful way.

Moving Average and Autoregressive Models

The most commonly used models in parametric spectral analysis are moving average (MA) and autoregressive (AR) models. The model of North Atlantic cod which we just examined is a simple example of an AR process. In fact, both MA and AR processes are straightforward to understand and can be shown to be related to each other in a rather interesting way. For the purposes of our discussion, we consider the MA model first.

Moving Average Processes. In an MA model the value of a variable X at time t, X_t, is assumed to be described by an equation of the form

$$X_t = \mu + \beta_0 Z_t + \beta_1 Z_{t-1} + \beta_2 Z_{t-2} + \cdots + \beta_q Z_{t-q} \tag{8.92}$$

where μ is a constant, Z_k are independent random numbers, each with a mean of 0 and a variance of σ^2, and β_k are a set of constant coefficients. If a process is described by such a model, it is said to be a moving average process of order q, typically abbreviated MA(q). It should be obvious that $E(X_t) = \mu$ and that $V(X_t) = \sigma^2 \Sigma_{k=0}^{q} \beta_k^2$. It is also straightforward to show that for such a process

$$R_{xx}(k) = \begin{cases} \mu^2 & |k| > q \\ \mu^2 + \sigma^2 \sum_{m=0}^{q-k} \beta_m \beta_{m+k} & |k| = 0, 1, \ldots, q \end{cases} \tag{8.93}$$

where $R_{xx}(k) = E(X_t X_{t+k})$. It follows from Equation 8.93 that for an MA(q) process

$$\rho_{xx}(k) = \begin{cases} 1 & k = 0 \\ \sum_{m=0}^{q-k} \beta_m \beta_{m+k} \Big/ \sum_{n=0}^{q} \beta_n^2 & |k| = 1, \ldots, q \\ 0 & |k| > q \end{cases} \tag{8.94}$$

An important characteristic of MA(q) processes is that the autocorrelation coefficient function is 0 for $|k| > q$. Thus if one has reason to believe that a time series is a sample record from a moving average process, examination of the autocorrelation coefficient function may be useful in determining the order of the MA process.

There are no restrictions on the coefficients β_m in the moving average model to ensure that the process is stationary, but a restriction is customarily imposed to ensure that a given ρ_{xx} can be associated with one and only one MA(q) process. To understand why the association is not unique, consider the simple MA(1) process $X_t = \mu + \beta_0 Z_t + \beta_1 Z_{t-1}$. In this case we have

$$\rho_{xx}(k) = \begin{cases} 1 & k = 0 \\ \beta_0 \beta_1 / (\beta_0^2 + \beta_1^2) & |k| = 1 \\ 0 & |k| > 1 \end{cases} \tag{8.95}$$

Fair enough, but it is straightforward to show (see Exercise 8.17) that exactly the same $\rho_{xx}(k)$ is obtained for the MA(1) process $X_t = \mu + \beta_1 Z_t + \beta_0 Z_{t-1}$. In fact, for any set of autocorrelation coefficients $\rho_{xx}(k)$, $k = 0, 1, \dots, q$, there is always more than one set of coefficients β_k, $k = 0, 1, \dots, q$, that produce the given set of autocorrelation coefficients in an MA(q) model. In the case of an MA(1) model, for example, if β_0 and β_1 are one set of coefficients, then an alternative set of coefficients α_0 and α_1 must satisfy the constraint equations $\alpha_0^2 + \alpha_1^2 = \beta_0^2 + \beta_1^2$ and $\alpha_0 \alpha_1 = \beta_0 \beta_1$. The solutions to these two constraint equations are

$$\alpha_0^2 = \frac{\beta_0^2 + \beta_1^2 \pm (\beta_0^2 - \beta_1^2)}{2} \quad \text{and} \quad \alpha_1^2 = \left(\frac{\beta_0 \beta_1}{\alpha_0} \right)^2$$

Taking the plus sign before the parentheses gives us back the original coefficients, that is, $\alpha_0 = \beta_0$ and $\alpha_1 = \beta_1$. Taking the minus sign gives the alternative solution $\alpha_0 = \beta_1$ and $\alpha_1 = \beta_0$. The ambiguity regarding which set of coefficients to choose can be eliminated if we impose a condition called *invertibility*. To understand what is meant by invertibility, we rewrite Equation 8.92 in the form

$$X_t - \mu = (\beta_0 B^0 + \beta_1 B^1 + \beta_2 B^2 + \cdots + \beta_q B^q) Z_t \tag{8.96}$$

where B is a so-called backward shift operator defined by $B^k Z_t = Z_{t-k}$. The term invertibility comes from the idea of inverting Equation 8.96 to solve for Z_t, the solution taking the form of a polynomial in B operating on $X_t - \mu$. For example, in the case of the MA(1) process $X_t = \mu + \beta_0 Z_t + \beta_1 Z_{t-1}$, it is straightforward to show by successive substitutions that

$$Z_t = \frac{1}{\beta_0} \sum_{k=0}^{\infty} \left(\frac{-\beta_1}{\beta_0} \right)^k (X_{t-k} - \mu) = \frac{1}{\beta_0} \sum_{k=0}^{\infty} \left(\frac{-\beta_1}{\beta_0} B \right)^k (X_t - \mu) \tag{8.97}$$

If we regard B as a complex variable rather than an operator, Equation 8.96

can be shown to be invertible if the roots of the polynomial $\beta_0 B^0 + \beta_1 B^1 + \beta_2 B^2 + \cdots + \beta_q B^q$ all lie outside the unit circle. In the case of an MA(1) process, the polynomial is $\beta_0 + \beta_1 B$, and the root of the polynomial is $B = -\beta_0/\beta_1$. The requirement that Equation 8.96 be invertible then implies that $|-\beta_0|/|\beta_1| > 1$ or $|\beta_1| < |\beta_0|$. It should be obvious that if this condition is satisfied, then the alternative model $X_t = \mu + \beta_1 Z_t + \beta_0 Z_{t-1}$ is not invertible.

Autoregressive Processes. A typical autoregressive model takes the form

$$X_t - \mu = \alpha_1(X_{t-1} - \mu) + \alpha_2(X_{t-2} - \mu) + \cdots + \alpha_p(X_{t-p} - \mu) + Z_t \tag{8.98}$$

where μ is the expectation value of the process if the process is stationary, α_k are constant coefficients, and Z_t is a random variable with a mean of 0 and a variance of σ^2. If there are p coefficients α_k, $k = 1, \ldots, p$, in the model, the process is referred to as an autoregressive process of order p and is designated AR(p). Equation 8.98 may be recast in the form

$$X_t = \alpha_1 X_{t-1} + \alpha_2 X_{t-2} + \cdots + \alpha_p X_{t-p} + Z'_t \tag{8.99}$$

where Z'_t is a random variable with a variance of σ^2 and a mean of $\mu(1 - \sum_{k=1}^{p} \alpha_k)$. Using the backward shift operator, it is straightforward to show from Equation 8.98 that

$$Z_t = \left(1 - \sum_{k=1}^{p} \alpha_k B^k\right)(X_t - \mu) \tag{8.100}$$

Assuming that Equation 8.100 can be inverted, we conclude that

$$X_t - \mu = \left(1 + \sum_{k=1}^{\infty} \beta_k B^k\right) Z_t \tag{8.101}$$

Equation 8.101 describes an infinite-order MA process. Since all MA processes are stationary, it follows that an AR process is stationary if Equation 8.100 can be inverted.

The simplest form of an autoregressive process is an AR(1) process, also commonly referred to as a Markov process. In this case we have

$$X_t - \mu = \alpha_1(X_{t-1} - \mu) + Z_t \tag{8.102}$$

The model we adopted to describe the population dynamics of North Atlantic cod (Eq. 8.87) is one example of an AR(1) process. By successive substitutions it is straightforward to show from Equation 8.102 that

$$X_t - \mu = \sum_{k=0}^{\infty} \alpha_1^k Z_{t-k} \tag{8.103}$$

Equation 8.103 is the analog of Equation 8.101 in the specific case of an AR(1) process. From Equation 8.103 it follows that $V(X_t) = \sigma^2 \sum_{k=0}^{\infty} \alpha_1^{2k}$. The variance is therefore finite as long as $|\alpha_1| < 1$. In this case $V(X_t) = \sigma^2/(1 - \alpha_1^2)$. In the case of the North Atlantic cod, for example, we had $\sigma^2 = 1$ and $\alpha_1 = 0.8$. Hence $V(X_t) = 1/(1 - 0.64) = 2.8$.

Using Equation 8.103, it is straightforward to show that for the AR(1) process

$$R_{xx}(k) = \mu^2 + \sigma^2 \sum_{m=0}^{\infty} \alpha_1^{2m+k} = R_{xx}(-k) \tag{8.104}$$

If $|\alpha_1| < 1$ Equation 8.104 simplifies to

$$R_{xx}(k) = \mu^2 + \frac{\sigma^2 \alpha_1^{|k|}}{1 - \alpha_1^2} \tag{8.105}$$

From Equation 8.88 we now have, for $f \neq 0$,

$$\begin{aligned} |X_f|^2 &= \frac{1}{N} \sum_{k=-(N-1)/2}^{(N-1)/2} \left(\mu^2 + \frac{\sigma^2}{1 - \alpha_1^2} \alpha_1^{|k|} \right) \exp\left(-\frac{2\pi i f k}{N} \right) \\ &= \frac{1}{N} \frac{\sigma^2}{1 - \alpha_1^2} \left(1 + \sum_{k=1}^{(N-1)/2} \alpha_1^k \left[\exp\left(-\frac{2\pi i f k}{N} \right) + \exp\left(\frac{2\pi i f k}{N} \right) \right] \right) \end{aligned} \tag{8.106}$$

Taking the limit of large N and assuming $|\alpha_1| < 1$, it is straightforward to show that the right-hand side of Equation 8.106 becomes identical to the right-hand side of Equation 8.90. Furthermore, for $f = 0$,

$$\begin{aligned} |X_0|^2 &= \frac{1}{N} \sum_{k=-(N-1)/2}^{(N-1)/2} \left(\mu^2 + \frac{\sigma^2}{1 - \alpha_1^2} \alpha_1^{|k|} \right) \\ &= \mu^2 + \frac{1}{N} \frac{\sigma^2}{1 - \alpha_1^2} \left(1 + 2 \sum_{k=1}^{(N-1)/2} \alpha_1^k \right) \end{aligned} \tag{8.107}$$

Again taking the limit of large N, it is straightforward to show that the right-hand side of Equation 8.107 becomes identical to the right-hand side of

Equation 8.91 if allowance is made for the fact that the μ^2 in Equation 8.107 is to be compared to $\mu_z^2/(1-\alpha)^2 = (E[g_t])^2$ in Equation 8.91. The right-hand sides of both Equations 8.91 and 8.107 $\rightarrow (E[g_t])^2$ as $N \rightarrow \infty$.

In the general case of an AR(p) model, it is easy to show from Equation 8.101 that

$$R_{xx}(k) = \sigma^2 \sum_{m=0}^{\infty} \beta_m \beta_{m+|k|} \tag{8.108}$$

with the understanding that $\beta_0 = 1$. The problem with Equation 8.108 is that it is often algebraically difficult to solve for the β_k in Equation 8.101. The most common method of circumventing this problem is to assume the process is stationary, multiply both sides of Equation 8.98 by $X_{t-k} - \mu$, and take the expectation value of both sides. Taking advantage of the fact that $R_{xx}(k) = R_{xx}(-k)$, we conclude that

$$R_{xx}(k) - \mu^2 = \sum_{m=1}^{p} \alpha_m \left(R_{xx}(k-m) - \mu^2\right) \quad \text{for all } k > 0 \tag{8.109}$$

and

$$R_{xx}(0) - \mu^2 = \sigma^2 + \sum_{m=1}^{p} \alpha_m \left(R_{xx}(m) - \mu^2\right) \tag{8.110}$$

For example, consider the simple case where $p = 1$. In this case Equation 8.109 would tell us that

$$R_{xx}(k) - \mu^2 = \alpha_1 \left(R_{xx}(k-1) - \mu^2\right) \tag{8.111}$$

and Equation 8.110 tells us that

$$R_{xx}(0) - \mu^2 = \alpha_1 \left(R_{xx}(1) - \mu^2\right) + \sigma^2 \tag{8.112}$$

Substituting $k = 1$ in Equation 8.111 gives

$$R_{xx}(1) - \mu^2 = \alpha_1 \left(R_{xx}(0) - \mu^2\right)$$

Substituting the right-hand side of this equation for $R_{xx}(1) - \mu^2$ in the right-hand side of Equation 8.112 and solving for $R_{xx}(0) - \mu^2$ gives

$$R_{xx}(0) - \mu^2 = \frac{\sigma^2}{1 - \alpha_1^2}$$

which is consistent with Equation 8.105. One can then solve for all values of $k > 0$ by repeatedly substituting into Equation 8.111.

Equations 8.109 and 8.110 or modifications thereof are known as the Yule–Walker equations. They amount to a bootstrap technique for determining the autocorrelation function for an AR(p) process. One begins with $p + 1$ equations which can be solved to determine $R_{xx}(0)$ through $R_{xx}(p)$. Then $R_{xx}(k)$ for $k > p$ can be determined using the appropriate recursion relationship. For $p = 2$, for example, the relevant equations are

$$\begin{aligned}
R_{xx}(1) - \mu^2 &= \alpha_1\big(R_{xx}(0) - \mu^2\big) + \alpha_2\big(R_{xx}(1) - \mu^2\big)\\
R_{xx}(2) - \mu^2 &= \alpha_1\big(R_{xx}(1) - \mu^2\big) + \alpha_2\big(R_{xx}(0) - \mu^2\big)\\
R_{xx}(0) - \mu^2 &= \alpha_1\big(R_{xx}(1) - \mu^2\big) + \alpha_2\big(R_{xx}(2) - \mu^2\big) + \sigma^2
\end{aligned} \tag{8.113}$$

and, in general,

$$R_{xx}(k) - \mu^2 = \alpha_1\big(R_{xx}(k-1) - \mu^2\big) + \alpha_2\big(R_{xx}(k-2) - \mu^2\big) \tag{8.114}$$

Working with autoregressive processes in the frequency domain is simpler. In this case we can set up the problem with the use of the equation

$$Z_t = \int \rho(x)\big(X(t+x) - \mu\big)\, dx \tag{8.115}$$

where Z_t is a random variable with a mean of 0 and a variance of σ^2. We now define $\rho(x)$ to be the sum of $p + 1$ δ functions as follows:

$$\rho(x) = \delta(x) - \sum_{k=1}^{p} \alpha_k\, \delta(x + k) \tag{8.116}$$

With this definition for $\rho(x)$, Equation 8.115 becomes identical to Equation 8.98. It then follows from Equations 8.85 and 8.90 and our previous discussion that, for $f \neq 0$,

$$E\big[|G_f|^2\big] = \frac{\sigma^2}{N\left|1 - \sum_{k=1}^{p} \alpha_k \exp\left(-\frac{2\pi i f k}{N}\right)\right|^2} \tag{8.117}$$

and, for $f = 0$,

$$E\big[|G_0|^2\big] = \mu^2 + \frac{\sigma^2}{N\left(1 - \sum_{k=1}^{p} \alpha_k\right)^2} \tag{8.118}$$

We have previously noted (Eq. 8.101) that any stationary AR process can be described as an infinite-order MA process. It is also true that any invertible MA process can be described as an infinite-order AR process. This follows from the fact that if Equation 8.96 can be inverted, then

$$Z_t = \frac{1}{\beta_0}\left(X_t - \mu - \sum_{k=1}^{\infty} \alpha_k B^k (X_t - \mu)\right)$$

$$= \frac{1}{\beta_0}\left(X_t - \mu - \sum_{k=1}^{\infty} \alpha_k (X_{t-k} - \mu)\right) \tag{8.119}$$

It follows immediately from Equation 8.119 that

$$X_t - \mu = \sum_{k=1}^{\infty} \alpha_k (X_{t-k} - \mu) + \beta_0 Z_t \tag{8.120}$$

which, of course, describes an infinite-order AR process, the random variable in this case having a variance $\beta_0^2 \sigma^2$.

Given the fact that a finite-order MA process is equivalent to an infinite-order AR process and vice versa, one may wonder in practice how one distinguishes a time series from an MA process from a time series from an AR process. The answer is that this is not always easy, but there are certainly differences. First, by either applying the shift theorem or taking the discrete Fourier transform of the right-hand side of Equation 8.93 (Exercise 8.22), it is easy to show that for an MA(q) process

$$E\left[|X_f|^2\right] = \frac{\sigma^2}{N}\left|\sum_{k=0}^{q} \beta_k \exp\left(-\frac{2\pi ikf}{N}\right)\right|^2 \quad \text{for } f \neq 0 \tag{8.121}$$

This is somewhat different from the spectrum for an AR(p) process, as is apparent from Equation 8.117. Second, for an MA(q) process, the expectation value of $\rho_{xx}(k)$ is 0 for $k > q$. This sort of behavior is not expected for an AR(p) process, since, in general, for such a process the expectation value of $R_{xx}(k) - \mu^2 \neq 0$. One should bear in mind, however, that estimates of the power spectrum or autocorrelation coefficient function derived from a single sample record will be influenced by noise (Fig. 8.32), and this fact can seriously confound our ability to determine with what type of process we are dealing.

ARMA Models. If we have no a priori reason to believe that we are dealing with a purely AR or MA process, we may, for the purposes of identifying peaks in the spectrum, choose to model the process as a combination AR and

MA process, or ARMA. In this case the underlying model is

$$X_t - \mu = \sum_{k=1}^{p} \alpha_k (X_{t-k} - \mu) + \sum_{m=0}^{q} \beta_m Z_{t-m} \qquad (8.122)$$

The model in this case is referred to as an ARMA(p, q) model, the p and q indicating the order of the AR and MA components, respectively. We can rewrite Equation 8.122 in the form

$$\sum_{m=0}^{q} \beta_m Z_{t-m} = \int (X(t+y) - \mu) \rho(y) \, dy \qquad (8.123)$$

with the understanding that $\rho(y) = \delta(y) - \sum_{k=1}^{p} \alpha_k \, \delta(y+k)$. It then follows from our previous discussion that

$$E\left[|X_f|^2\right] = \frac{\sigma^2}{N} \frac{\left| \sum_{m=0}^{q} \beta_m \exp\left(-\frac{2\pi imf}{N}\right)\right|^2}{\left|1 - \sum_{k=1}^{p} \alpha_k \exp\left(-\frac{2\pi ikf}{N}\right)\right|^2} \qquad (8.124)$$

Model Order and Parameter Estimation. Two fundamental problems in ARMA modeling are choosing the order of the model and estimating the parameters in the model. The first problem is greatly simplified if we know or think we know the mechanism governing the behavior of X. The second problem is relatively straightforward in the case of an AR process, but less so in the case of an MA or mixed ARMA model.

We will first consider the problem of estimating the parameters in an AR model. Thinking of Z_t in Equation 8.98 as a random error term, it would seem reasonable to choose the parameters in the model so as to minimize

$$\sum_{t=p+1}^{N} Z_t^2 = \sum_{t=p+1}^{N} \left(X_t - \mu - \sum_{k=1}^{p} \alpha_k (X_{t-k} - \mu) \right)^2 \qquad (8.125)$$

Note that the summation extends from $t = p + 1$ to N, since there is no way to estimate Z_t for $t \leq p$. While it is straightforward to set up the least-squares equations in this way, the solution is not as straightforward, since the model involves the product of μ and α_k. In the case of a simple AR(1) model, differentiating Equation 8.125 with respect to μ and α_1 gives

$$\hat{\mu} = \frac{\sum_{t=2}^{N} (X_t - \hat{\alpha}_1 X_{t-1})}{(1 - \hat{\alpha}_1)(N - 1)} \tag{8.126}$$

$$\hat{\alpha}_1 = \frac{\sum_{t=2}^{N} (X_t - \hat{\mu})(X_{t-1} - \hat{\mu})}{\sum_{t=2}^{N} (X_{t-1} - \hat{\mu})(X_{t-1} - \hat{\mu})} \tag{8.127}$$

where $\hat{\mu}$ and $\hat{\alpha}_1$ are the least-squares estimates of μ and α_1, respectively. These equations can be manipulated to obtain expressions for $\hat{\mu}$ and $\hat{\alpha}_1$ in terms of the X_k (Exercise 8.23), but a simpler, albeit approximate, solution can be obtained by noting that $[1/(N-1)]\Sigma_{t=2}^{N} X_t$ is the average of the last $N - 1$ values of X and that $[1/(N-1)]\Sigma_{t=2}^{N} X_{t-1}$ is the average of the first $N - 1$ values of X. Both are likely to be very close to $\bar{X} = (1/N)\Sigma_{t=1}^{N} X_t$. Assuming this to be the case, the approximate least-squares solutions become

$$\hat{\mu} \cong \bar{X}$$

$$\hat{\alpha}_1 \cong \frac{\sum_{t=2}^{N} (X_t - \bar{X})(X_{t-1} - \bar{X})}{\sum_{t=2}^{N} (X_{t-1} - \bar{X})(X_{t-1} - \bar{X})} \tag{8.128}$$

In the more general case of an AR(p) model, we find that

$$\hat{\mu} = \frac{\sum_{t=p+1}^{N} X_t - \sum_{k=1}^{p} \alpha_k \sum_{t=p+1}^{N} X_{t-k}}{(N-p)\left(1 - \sum_{k=1}^{p} \alpha_k\right)} \tag{8.129}$$

and that the $\hat{\alpha}_k$ satisfy the matrix equation $\mathbf{M}\hat{\boldsymbol{\alpha}} = \mathbf{N}$, where $\hat{\boldsymbol{\alpha}}$ is a column vector containing the p values of $\hat{\alpha}$, $\mathbf{N}$ is a column vector whose nth element is $\Sigma_{t=p+1}^{N}(X_t - \hat{\mu})(X_{t-n} - \hat{\mu})$, and $\mathbf{M}$ is a square matrix with p rows and columns having $M_{jm} = \Sigma_{t=p+1}^{N}(X_{t-m} - \hat{\mu})(X_{t-j} - \hat{\mu})$. It should be obvious that the summations involving X in the numerator of Equation 8.129, when divided by $N - p$, are the averages of the first to the last $N - p$ values of X in the sample record. If p is small compared to N, as should always be the case in practice, these averages should be very similar and comparable to $\bar{X}$. Assuming this to be the case, it follows that $\hat{\mu} \cong \bar{X}$, and an approximate $\boldsymbol{\alpha}$

vector may be obtained by solving the matrix equation $\mathbf{M}\hat{\boldsymbol{\alpha}} = \mathbf{N}$ with $\overline{X}$ replacing $\hat{\mu}$ in the **M** and **N** matrices. An exact solution to the least-squares equations can be obtained by iterating this process, using $\overline{X}$ as the initial guess of $\hat{\mu}$, solving the matrix equation for $\boldsymbol{\alpha}$, and then using the estimated $\boldsymbol{\alpha}$ vector in Equation 8.129 to make an improved estimate of $\hat{\mu}$, which in turn is used to make an improved estimate of $\boldsymbol{\alpha}$, and so forth. In practice, this process converges very rapidly, since $\overline{X}$ is typically very close to $\hat{\mu}$.

Once we have the $\boldsymbol{\alpha}$ vector, we can estimate the spectral density function (sdf) using Equations 8.117 and 8.118. If any of the α_k are nonzero, we will see some peaks in the spectrum, and the question then arises as to whether the peaks are significant or not in a statistical sense. A theoretical treatment of the problem of calculating error bounds for AR spectral estimators is given by Percival and Walden (1993, Section 9.12). The error bounds are exact only in the limit as $N \to \infty$, and the equations for the error bounds, not to mention the derivation of them, are very complicated. In short, the theoretical treatment is beyond an introductory course on mathematical methods. A reasonable alternative to estimating the statistical significance of the peaks is as follows.[5]

1. Randomly select values of X from your data to create a set of random sample records.
2. Calculate the coefficients of the appropriate AR(p) model that gives the best description of each random sample record and calculate the corresponding sdf, in each case taking p to be identical to the order of the AR model fit to the original sample record.
3. For each frequency f rank the sdf from smallest to largest and pick the value above which 5% or less of the values lie. If the sdf of the original sample record lies above this value, then it is reasonable to say that there is a significant ($p \leq 0.05$) periodicity at this frequency in the sample record.

The validity of this alternative procedure rests in part on the assumption that the distribution of X values in your sample record is representative of the distribution of X values in the ensemble of X. The procedure is straightforward to implement and on modern high-speed digital computers is a practical alternative to the more cerebral theoretical approach.

A final issue with AR models is, of course, to determine the order of the model. The choice of the order of the model is to some extent an art, but one can cite at least a few reasonable guidelines. First, given that Z_t is a random error term, it certainly makes no sense to extend the model to the point that we are fitting our model to noise. Thus our desire is to choose the order of the model no greater than is necessary to describe the patterns in the data.

[5] I am indebted to Dr. Peter Howd of the Duke University Marine Laboratory for pointing out this method.

At each step of the fitting process, we should therefore examine the residuals for evidence of randomness. One reasonable criterion might be that the distribution of residuals not be significantly different from that expected for a normal distribution of errors and that $\rho_{zz}(\Delta t) \cong 0$ for $\Delta t \neq 0$. The first condition can be tested by a χ^2 test. The second condition can be tested by calculating $\rho_{zz}(\Delta t)$, for which the approximate 95% confidence limits are $\pm 2/\sqrt{N}$ (Chatfield, 1989), meaning that $\rho_{zz}(\Delta t)$ is significant if its magnitude exceeds $2/\sqrt{N}$. Once we are satisfied that the residuals are behaving like random numbers, it is probably time to stop adding coefficients to the model. At that point we have, to the best of our ability, prewhitened the data. If the residuals do not behave like random numbers, we must decide whether to extend the model by increasing the order of the AR(p) model or fit an MA(q) model to the residuals.

As an example of the use of an AR model, consider the time series of hake recruitment in Figure 8.2. A casual examination of the time series reveals two patterns: First, there are peaks in recruitment separated by about 11 years; second, the height of the peaks becomes progressively smaller with time. We know from our previous discussion that our sample record should span at least twice the period of a sinusoid if we are going to have any chance at all of identifying the corresponding frequency in our sdf. In this case the sample record spans 30 years, so we do have a chance of picking out the 11-year periodicity. The decline in the amplitude of the peaks, however, must obviously be treated as a trend, which we should try to remove from the sample record before going ahead with the analysis. Even if the pattern is part of a long-term cycle, our sample record is certainly not long enough to allow use to isolate the associated frequency in our sdf. The trend will therefore bias our results at low frequency unless we remove it.

We remove the trend in two steps. First, we postulate that the amplitude of the peaks, expressed as a percentage of the average recruitment over each roughly 11-year cycle, is approximately constant. If this is the case, then log-transforming the data should remove much of the trend in the amplitudes. Having log-transformed the data, we find that there is still some trend apparent. We fit a low-order polynomial to the log-transformed data and find that a second-order polynomial removes virtually all the remaining trend. The resultant detrended time series is shown in Figure 8.33*A*. We then fit the detrended time series using an AR model and find that an AR(2) model gives a set of residuals that are very similar to the pattern expected for a normal distribution of errors (Fig. 8.33*B*). An examination of $\rho_{xx}(\Delta t)$ reveals that for time lags up to 8 years, none of the $\rho_{xx}(\Delta t)$ exceeds in magnitude the critical value of $2/\sqrt{28} = 0.378$. Hence there is no reason to reject the null hypothesis that the errors are random and normally distributed. The sdf for the AR(2) model, calculated using Equation 8.117, is shown in Figure 8.33*C*. The dashed curve is the value of the sdf that would be exceeded no more than 5% of the time by sample records drawn randomly from a set of X values with the same probability density function as the sample record in

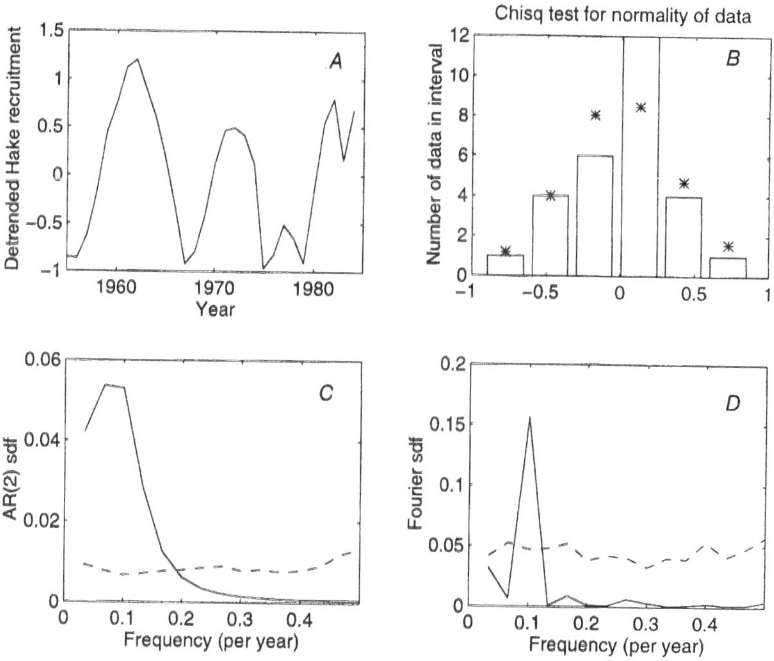

Figure 8.33 (A) Detrended hake recruitment data, (B) χ^2 test for normality of residuals after fitting the data in (A) to an AR(2) model, (C) spectral density function for the AR(2) model (continuous line) and bootstrap estimation of values above which the sdf is significant at 95% confidence, and (D) analysis similar to (C) using Fourier analysis.

Figure 8.2. Based on this criterion, our sample sdf is judged to be significant at frequencies of 0.033 to 0.165 yr^{-1}, that is, periods of 6 to 30 years. Had we simply carried out a Fourier analysis on the detrended data (Fig. 8.33A), we would have obtained the results in Figure 8.33D. In this case only one frequency in the sdf is judged to be significant, namely, 0.1 yr^{-1}.

An important point to note about Figure 8.33C and D is that the signal-to-noise ratio is about twice as high in Figure 8.33C. In other words, we can detect the peak in the AR(2) spectrum more easily than is the case if we rely on standard Fourier analysis. This is one of the appealing features of AR models and explains why they are often used to analyze short sample records. With a short sample record, it is difficult with Fourier analysis to detect peaks in the sdf unless the periodicity is obvious, as happens to be the case here.

The task of estimating the parameters in a moving average model is more complex for several reasons. First, we have no way to calculate Z_k for $1 \le k \le q$ in terms of the X's and the parameters in the model. Second, even if we make the arbitrary assumption that $Z_k = 0$ for $1 \le k \le q$, the expressions for the remaining Z_k are complex functions of the β's and not well suited to a least-squares treatment. For example, in the case of an MA(1) model we find that, for $n > 1$,

$$Z_n = Z_1\left(\frac{-\beta_1}{\beta_0}\right)^{n-1} + \sum_{k=2}^{n}\frac{(-\beta_1)^n}{\beta_0^{n-k+1}}(X_k - \mu_k) \tag{8.130}$$

An alternative approach is to choose the β's so that the values of $\rho_{xx}(k)$ estimated from the right-hand side of Equation 8.94 equal the values of $\rho_{xx}(k)$ estimated from the sample record. For the purposes of implementing this approach, we can assume without loss of generality that $\sigma^2 = 1$. With $\sigma^2 = 1$ it follows that $V(X_t) = \Sigma_{k=0}^{q} \beta_k^2$ and that

$$\left(\sum_{k=0}^{q}\beta_k\right)^2 = V(X_t)\left(1 + 2\sum_{k=1}^{q}\rho_{xx}(k)\right)$$

Given this information, it is straightforward to write down the equations that must be satisfied by the β's. In the case of an MA(1) model, for example, we have

$$\beta_0\beta_1 = V(X_t)\rho_{xx}(1) \tag{8.131}$$

and

$$\beta_0 + \beta_1 = \pm\sqrt{V(X_t)[1 + 2\rho_{xx}(1)]} \tag{8.132}$$

In general, there are four possible solutions to these two equations, depending first on whether we take the + or − sign on the right-hand side of Equation 8.132 and second on which of the two roots of the resultant quadratic equation we assign to β_0. Specifically, if β_0 and β_1 are acceptable parameters, then so are β_1 and β_0, $-\beta_0$ and $-\beta_1$, and $-\beta_1$ and $-\beta_0$. This ambiguity need not concern us, since all four combinations are consistent with the observed $V(X_t)$ and $\rho_{xx}(1)$. If we arbitrarily take the + sign on the right-hand side of Equation 8.132, then only one of the remaining two combinations of β_0 and β_1 is invertible, for reasons discussed previously.

Although it is easy to manipulate Equations 8.131 and 8.132 to solve for β_0 and β_1, solving the corresponding equations for higher values of q by algebraic means becomes forbiddingly complicated. For example, for $q = 3$ we have

$$\begin{aligned}\beta_0\beta_1 + \beta_1\beta_2 + \beta_2\beta_3 &= V(X_t)\rho_{xx}(1)\\ \beta_0\beta_2 + \beta_1\beta_3 &= V(X_t)\rho_{xx}(2)\\ \beta_0\beta_3 &= V(X_t)\rho_{xx}(3)\\ \beta_0 + \beta_1 + \beta_2 + \beta_3 &= \pm\sqrt{V(X_t)[1 + 2(\rho_{xx}(1) + \rho_{xx}(2) + \rho_{xx}(3))]}\end{aligned} \tag{8.133}$$

Fortunately, this set of equations, and similar equations corresponding to other values of q, can be easily solved by an iterative procedure such as the Gauss–Newton method described in Chapter 5. An effective procedure for making the initial guess of the β's is to assign β_1 through β_q a small initial value such as 0.01 and then set β_0 equal to $V(X_t)\rho_{xx}(q)/\beta_q$. All the β's are then multiplied by a constant scaling factor to make their sum identical to the right-hand side of the last line of Equation 8.133. One then iterates using the Gauss–Newton procedure. Since differentiating the left-hand sides of Equations 8.133 with respect to each β is trivial, the method is easy to implement and converges rapidly.

An ARMA(p, q) model may be used in the case where the residuals from an AR(p) fit do not appear to be random and extending the order of the AR(p) component of the model does not seem justified. In this case the initial AR(p) fit can be accomplished by the least-squares procedure described previously. An MA(q) model of appropriate order is then fit to the residuals based on an examination of the autocorrelation coefficient functions of the residuals.

As an example of an ARMA analysis, we consider the results shown in Figure 8.34. In this case a sample record of length 256 was generated using sinusoids with frequencies of 0.1, 0.2, 0.3, and 0.4. The data were noise corrupted using an MA(2) process. The sampling interval was assumed to be

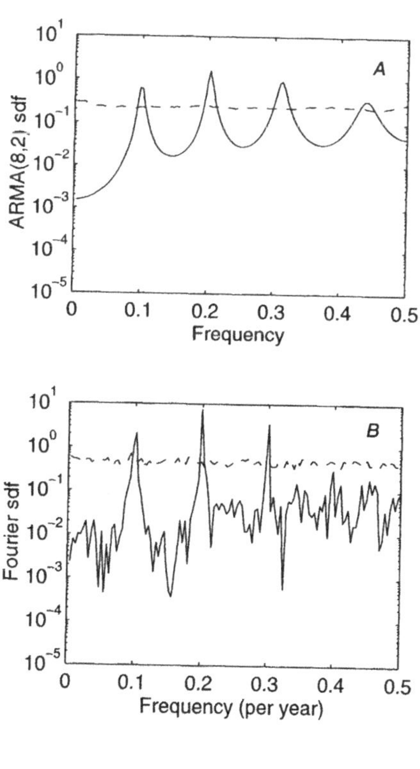

Figure 8.34 (A) ARMA(8, 2) analysis of noise-corrupted sample record including sinusoidal components having frequencies of 0.1, 0.2, 0.3, and 0.4. The solid line is the sdf. (B) Fourier analysis of the same data set. The solid line is the Fourier sdf. In both (A) and (B) the dashed line is the bootstrap estimation of values above which the sdf is significant at 95% confidence.

1. An ARMA(8) model fit to the data clearly shows four peaks, but an examination of the autocorrelation coefficient functions reveals that the first few values exceed the critical value of $2/\sqrt{256} = 0.125$ and are therefore significant. The model was therefore extended to an ARMA(8, 2). The sdf of the ARMA(8, 2) model is shown in Figure 8.34*A*. All four peaks lie above the 95% confidence level based on the previously discussed bootstrap calculation. The two highest-frequency peaks are shifted to frequencies greater than 0.3 and 0.4 in part because of the spectral characteristics of the MA(2) process used to noise corrupt the data and in part because our AR model is marginal for detecting four peaks. A traditional Fourier analysis of the same sample record is shown in Figure 8.34*B*. The Fourier analysis does detect the three lowest frequency peaks, all of which are judged to be significant at 95% confidence. However, the highest-frequency peak is lost in the Fourier analysis. This example again illustrates the better signal-to-noise characteristics of ARMA analysis compared to Fourier analysis.

There remains, however, the somewhat thorny issue of choosing the order of the ARMA model. A useful rule of thumb is to choose the order of the AR model to be equal to or somewhat greater than twice the expected number of peaks in the spectrum. In the case of Figure 8.34, for example, we

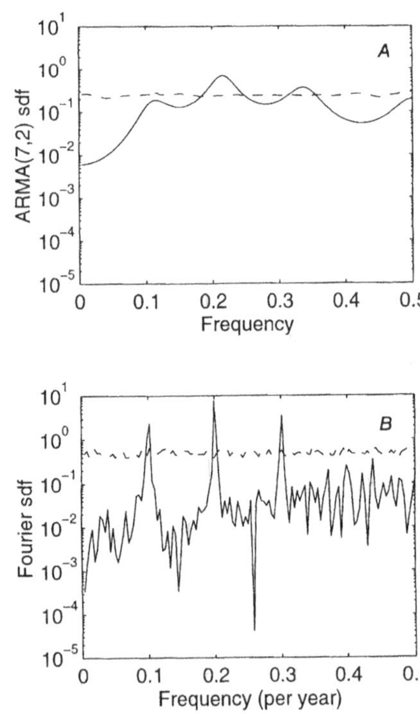

Figure 8.35 Analysis similar to Figure 8.34, but in (*A*) using an ARMA(7, 2) model rather than an ARMA(8, 2) model.

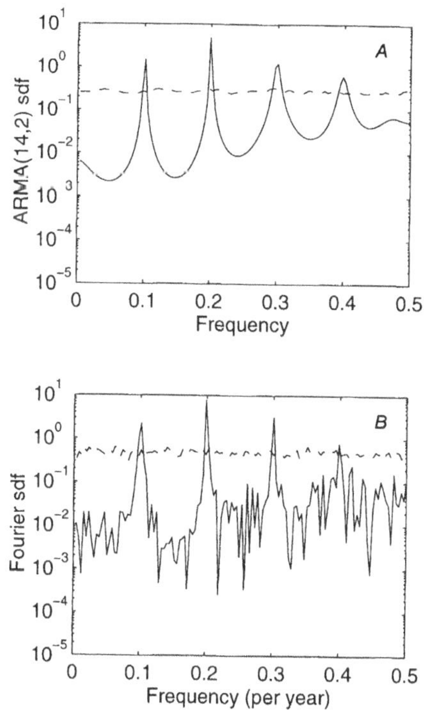

Figure 8.36 Analysis similar to Figure 8.34, but in (A) using an ARMA(14, 2) model rather than an ARMA(8, 2) model.

might expect from an examination of the Fourier spectrum that there were three significant periodicities in the time series. However, had we chosen to fit the data with an ARMA(7, 2) model, for example, we would have failed to detect the fourth significant frequency, as illustrated in Figure 8.35. It is necessary to use at least an AR(8) model to detect four peaks. In this case, extending the order helps to better define the location of the peaks (Figure 8.36). Extending the order of the model too far, however, can create spurious peaks if we are not careful to check for their statistical significance.

In summary, parametric analysis is an appealing alternative to the more traditional Fourier analysis when one is not dealing with large data sets. There is, however, nothing analogous to the fast Fourier transform (Percival and Walden, 1993) for rapidly analyzing large data sets using parametric analysis. If for no other reason, Fourier analysis will undoubtedly continue to be the method of choice for carrying out spectral analyses of large data sets.

EXERCISES

8.1 What are $\mu_x(t)$ and $R_{xx}(t, \Delta t)$ for the function defined by Equation 8.19?

8.2 Consider the following variations of the random process illustrated in Figure 8.6:

a. constant amplitude and phase angle, random period

b. constant amplitude and period, random phase angle

c. constant period and phase angle, random amplitude

d. constant amplitude, random phase angle and period

e. constant phase angle, random amplitude and period

Which of these would be stationary processes? Of the stationary processes, which ones would be ergodic?

8.3 Using the definition of the Fourier transform in Equation 8.40 and the Dirac delta function in Equation 8.48, show that the Fourier transform of $\delta(t)$ is 1.

8.4 Use the result of Exercise 8.3 and the definition of the inverse Fourier transform in Equation 8.39 to show that $\int_{-\infty}^{\infty} \exp(2\pi ift)\,df = \delta(t)$.

8.5 Use the result of Exercise 8.4 to show that the convolution of the Fourier transforms of two functions is equal to the Fourier transform of the product of the two functions.

8.6 Prove that Equations 8.56 to 8.59 are equivalent to Equation 8.55.

8.7 Prove that if g and h are discretely sampled functions on a time interval $N\Delta t$, then the Fourier transform of the product of g and h is equal to the convolution of their Fourier transforms. *Hint*: The proof follows directly from Equations 8.55 and 8.62 along with Equation 8.61.

8.8 Show that the integer value of n which satisfies the inequality

$$\left| f_a' - \frac{n}{\Delta t} \right| \leq \frac{1}{2\Delta t}$$

is equal to $f_a'\Delta t$ rounded to the nearest integer.

8.9 A student collects data every 17 days. What frequencies less than the Nyquist frequency would be aliased by the following tidal frequencies:

a. lunar diurnal with a period of 25.8 hours

b. lunar semidiurnal with a period of 12.4 hours

c. solar diurnal with a period of 24.1 hours

d. solar semidiurnal with a period of 12.05 hours

8.10 Show that the Fourier transform of $\mathrm{Box}(t/\tau)$ is equal to $\sin(\pi f\tau)/\pi f$.

8.11 Show that the Fourier transform of $\cos(2\pi f_0 t)$ is $\frac{1}{2}[\delta(f+f_0)+\delta(f-f_0)]$. *Hint*: Use Equation 8.40, the fact that

$$\cos(2\pi f_0 t) = \frac{1}{2}(\exp(2\pi i f_0 t) + \exp(-2\pi i f_0 t))$$

and the result of Exercise 8.4.

8.12 Show that the Fourier transform of Black(t) defined in Equation 8.83 is equal to

$$0.42\tau \sin c(f\tau) + 0.5\frac{\tau}{2}[\sin c(1-f\tau) + \sin c(1+f\tau)]$$

$$\times 0.08\frac{\tau}{2}[\sin c(2-f\tau) + \sin c(2+f\tau)]$$

8.13 Show, using Equations 8.39 and 8.40, that if $g(t) \subset G(f)$ and $h(t) \subset H(f)$, then $c_1 g(t) + c_2 h(t) \subset c_1 G(f) + c_2 H(f)$.

8.14 Show, using Equation 8.40, that if $g(t) \subset G(f)$, then $g(t+T) \subset G(f)\exp(2\pi i T f)$.

8.15 Show that if $g(t)$ has the Fourier transform $G(f)$, then the Fourier transform of the autocorrelation function of $g(t)$ is $|G(f)|^2$. *Hint*: The autocorrelation function $R_{gg}(x)$ of $g(t)$ in the continuous case is defined by

$$R_{gg}(x) = \int_{-\infty}^{\infty} g^*(t) g(t+x)\, dt$$

where $g^*(t)$ is the complex conjugate of $g(t)$. Your job is to show that

$$R_{gg}(x) = \int_{-\infty}^{\infty} |G(f)|^2 \exp(2\pi i f x)\, df$$

Solving the problem requires you to use Equation 8.39 and the result of Exercise 8.4.

8.16 Show that for any real-valued function $f(t)$ whose Fourier transform is $F(x)$, $F^*(s) = F(-s)$.

8.17 Given $R_{xx}(k) = E(X_t X_{t+k})$, show that the MA(1) process $X_t = \mu + \beta_1 Z_t + \beta_0 Z_{t-1}$ has the $\rho_{xx}(k)$ given by Equation 8.95.

8.18 Let $X_t = \mu + \beta_0 Z_t + \beta_1 Z_{t-1} + \beta_2 Z_{t-2}$ represent an MA(2) process. Show that the alternative MA(2) model $X_t = \mu + \beta_2 Z_t + \beta_2 Z_t + \beta_1 Z_{t-1} + \beta_0 Z_{t-2}$ has the same autocorrelation coefficient function.

8.19 Prove that Equation 8.104 is true.

8.20 Show that the right-hand side of Equation 8.106 becomes identical to the right-hand side of Equation 8.90 in the limit of large N if $|\alpha_1| < 1$.

8.21 Solve Equations 8.113 for $R_{xx}(0)$.

8.22 By taking the discrete Fourier transform of the right-hand side of Equation 8.93, show that for an MA(q) processes

$$E\left[|X_f|^2\right] = \frac{\sigma^2}{N}\left|\sum_{k=0}^{q} \beta_k \exp\left(-\frac{2\pi ikf}{N}\right)\right|^2 \quad \text{for } f \neq 0$$

Hint: First show that $R_{xx}(-k) = R_{xx}(k)$. Using this fact, you can write

$$\frac{1}{N}\sum_{k=-(N-1)/2}^{(N-1)/2} R_{xx}(k)\exp\left(-\frac{2\pi ikf}{N}\right)$$

$$= \frac{1}{N}\left(R_{xx}(0) + 2\sum_{k=1}^{(N-1)/2} R_{xx}(k)\cos\left(\frac{2\pi kf}{N}\right)\right)$$

8.23 Substitute the right-hand side of Equation 8.127 for $\hat{\alpha}_1$ in Equation 8.126 and solve the resultant equation for $\hat{\mu}$.

8.24 Show that for an MA(q) process

$$\left(\sum_{k=0}^{q} \beta_k\right)^2 = V(X_t)\left(1 + 2\sum_{k=1}^{q} \rho_{xx}(k)\right) \quad \text{if } \sigma^2 = 1$$

REFERENCES

Bendat, J. S., and A. G. Piersol. 1986. *Random Data Analysis and Measurement Procedures*, 2nd ed. Wiley–Interscience, New York.

Chatfield, C. 1989. *The Analysis of Time Series: An Introduction*, 4th ed. Chapman & Hall. London.

Cushing, D. H. 1982. *Climate and Fisheries*. Academic, New York.

Davis, J. C. 1986. *Statistics and Data Analysis in Gedogy*, 2nd ed. Wiley, New York.

Percival, D. B., and A. T. Walden. 1993. *Spectral Analysis for Physical Applications*. Cambridge University Press, Cambridge.

Sissenwine, M. P., M. J. Fogarty, and W. J. Overholtz. 1988. Some fisheries management implications of recruitment variability. In J. A. Gulland (ed.), *Fish Population Dynamics*, 2nd ed. pp.129–152. Wiley–Interscience, New York.

APPENDIX A

Probabilities associated with the normal distribution function. Each number in the table is the probability that Z will exceed Z_α if Z is a standard normal variate. The probability that $|Z| > Z_\alpha$ is twice the value of the entry in the table, since the normal distribution function is symmetric about $Z = 0$.

Z_α	0.00	0.01	0.02	0.03	0.04	0.05	0.06	0.07	0.08	0.09
0.0	0.50000	0.49601	0.49202	0.48803	0.48405	0.48006	0.47608	0.47210	0.46812	0.46414
0.1	0.46017	0.45620	0.45224	0.44828	0.44433	0.44038	0.43644	0.43251	0.42858	0.42465
0.2	0.42074	0.41683	0.41294	0.40905	0.40517	0.40129	0.39743	0.39358	0.38974	0.38591
0.3	0.38209	0.37828	0.37448	0.37070	0.36693	0.36317	0.35942	0.35569	0.35197	0.34827
0.4	0.34458	0.34090	0.33724	0.33360	0.32997	0.32636	0.32276	0.31918	0.31561	0.31207
0.5	0.30854	0.30503	0.30153	0.29806	0.29460	0.29116	0.28774	0.28434	0.28096	0.27760
0.6	0.27425	0.27093	0.26763	0.26435	0.26109	0.25785	0.25463	0.25143	0.24825	0.24510
0.7	0.24196	0.23885	0.23576	0.23270	0.22965	0.22663	0.22363	0.22065	0.21770	0.21476
0.8	0.21186	0.20897	0.20611	0.20327	0.20045	0.19766	0.19489	0.19215	0.18943	0.18673
0.9	0.18406	0.18141	0.17879	0.17619	0.17361	0.17106	0.16853	0.16602	0.16354	0.16109
1.0	0.15866	0.15625	0.15386	0.15151	0.14917	0.14686	0.14457	0.14231	0.14007	0.13786
1.1	0.13567	0.13350	0.13136	0.12924	0.12714	0.12507	0.12302	0.12100	0.11900	0.11702
1.2	0.11507	0.11314	0.11123	0.10935	0.10749	0.10565	0.10383	0.10204	0.10027	0.09853
1.3	0.09680	0.09510	0.09342	0.09176	0.09012	0.08851	0.08691	0.08534	0.08379	0.08226
1.4	0.08076	0.07927	0.07780	0.07636	0.07493	0.07353	0.07215	0.07078	0.06944	0.06811
1.5	0.06681	0.06552	0.06426	0.06301	0.06178	0.06057	0.05938	0.05821	0.05705	0.05592
1.6	0.05480	0.05370	0.05262	0.05155	0.05050	0.04947	0.04846	0.04746	0.04648	0.04551
1.7	0.04457	0.04363	0.04272	0.04182	0.04093	0.04006	0.03920	0.03836	0.03754	0.03673
1.8	0.03593	0.03515	0.03438	0.03362	0.03288	0.03216	0.03144	0.03074	0.03005	0.02938
1.9	0.02872	0.02807	0.02743	0.02680	0.02619	0.02559	0.02500	0.02442	0.02385	0.02330
2.0	0.02275	0.02222	0.02169	0.02118	0.02068	0.02018	0.01970	0.01923	0.01876	0.01831
2.1	0.01786	0.01743	0.01700	0.01659	0.01618	0.01578	0.01539	0.01500	0.01463	0.01426
2.2	0.01390	0.01355	0.01321	0.01287	0.01255	0.01222	0.01191	0.01160	0.01130	0.01101
2.3	0.01072	0.01044	0.01017	0.00990	0.00964	0.00939	0.00914	0.00889	0.00866	0.00842
2.4	0.00820	0.00798	0.00776	0.00755	0.00734	0.00714	0.00695	0.00676	0.00657	0.00639
2.5	0.00621	0.00604	0.00587	0.00570	0.00554	0.00539	0.00523	0.00508	0.00494	0.00480
2.6	0.00466	0.00453	0.00440	0.00427	0.00415	0.00402	0.00391	0.00379	0.00368	0.00357
2.7	0.00347	0.00336	0.00326	0.00317	0.00307	0.00298	0.00289	0.00280	0.00272	0.00264
2.8	0.00256	0.00248	0.00240	0.00233	0.00226	0.00219	0.00212	0.00205	0.00199	0.00193
2.9	0.00187	0.00181	0.00175	0.00169	0.00164	0.00159	0.00154	0.00149	0.00144	0.00139

APPENDIX B

Critical values Z_α of the argument of the t distribution function. The heading above each column gives the probability α that $Z > Z_\alpha$ assuming that the probability density function for Z is $t(Z, n)$. The probability that $|Z| > Z_\alpha$ is 2α, since the normal distribution function is symmetric about $Z = 0$.

	α				
df = n	0.10	0.05	0.025	0.010	0.005
1	3.078	6.3141	2.706	31.821	63.657
2	1.886	2.920	4.303	6.965	9.925
3	1.638	2.353	3.182	4.541	5.841
4	1.533	2.132	2.776	3.747	4.604
5	1.476	2.015	2.571	3.365	4.032
6	1.440	1.943	2.447	3.143	3.707
7	1.415	1.895	2.365	2.998	3.499
8	1.397	1.860	2.306	2.896	3.355
9	1.383	1.833	2.262	2.821	3.250
10	1.372	1.812	2.228	2.764	3.169
11	1.363	1.796	2.201	2.718	3.106
12	1.356	1.782	2.179	2.681	3.055
13	1.350	1.771	2.160	2.650	3.012
14	1.345	1.761	2.145	2.624	2.977
15	1.341	1.753	2.131	2.602	2.947
16	1.337	1.746	2.120	2.583	2.921
17	1.333	1.740	2.110	2.567	2.898
18	1.330	1.734	2.101	2.552	2.878
19	1.328	1.729	2.093	2.539	2.861
20	1.325	1.725	2.086	2.528	2.845
21	1.323	1.721	2.080	2.518	2.831
22	1.321	1.717	2.074	2.508	2.819
23	1.319	1.714	2.069	2.500	2.807
24	1.318	1.711	2.064	2.492	2.797
25	1.316	1.708	2.060	2.485	2.787
26	1.315	1.706	2.056	2.479	2.779
27	1.314	1.703	2.052	2.473	2.771
28	1.313	1.701	2.048	2.467	2.763
29	1.311	1.699	2.045	2.462	2.756
30	1.310	1.697	2.042	2.457	2.750
40	1.303	1.684	2.021	2.423	2.704
60	1.296	1.671	2.000	2.390	2.660
120	1.289	1.658	1.980	2.358	2.617
∞	1.282	1.645	1.960	2.326	2.576

APPENDIX C

Critical values r_α of the product–moment correlation coefficient. The heading above each column is the probability α that r will equal or exceed the value in the table if $\rho = 0$ when r is calculated from a random sampling of $N\,[X, Y]$ pairs. The number of degrees of freedom df is $N - 2$. It is assumed that either X or Y is normally distributed. The probability that $|r| \geq r_\alpha$ when $\rho = 0$ is 2α, since the probability density function for r is symmetric about 0 when $\rho = 0$.

	α					α			
df = $N - 2$	0.05	0.025	0.01	0.005	df = $N - 2$	0.05	0.025	0.01	0.005
1	0.988	0.997	1.000	1.000	24	0.330	0.388	0.453	0.496
2	0.900	0.950	0.980	0.990	25	0.323	0.381	0.445	0.487
3	0.805	0.878	0.934	0.959	26	0.317	0.374	0.437	0.478
4	0.729	0.811	0.882	0.917	27	0.311	0.367	0.430	0.470
5	0.669	0.754	0.833	0.874	28	0.306	0.361	0.423	0.463
6	0.621	0.707	0.789	0.834	29	0.301	0.355	0.416	0.456
7	0.582	0.666	0.750	0.798	30	0.296	0.349	0.409	0.449
8	0.549	0.632	0.715	0.765	35	0.275	0.325	0.381	0.418
9	0.521	0.602	0.685	0.735	40	0.257	0.304	0.358	0.393
10	0.497	0.576	0.658	0.708	45	0.243	0.288	0.338	0.372
11	0.476	0.553	0.634	0.684	50	0.231	0.273	0.322	0.354
12	0.458	0.532	0.612	0.661	60	0.211	0.250	0.295	0.325
13	0.441	0.514	0.592	0.641	70	0.195	0.232	0.274	0.302
14	0.426	0.497	0.574	0.623	80	0.183	0.217	0.257	0.283
15	0.412	0.482	0.558	0.606	90	0.173	0.205	0.242	0.267
16	0.400	0.468	0.543	0.590	100	0.164	0.195	0.230	0.254
17	0.389	0.456	0.529	0.575	125	0.147	0.174	0.206	0.228
18	0.378	0.444	0.516	0.561	150	0.134	0.159	0.189	0.208
19	0.369	0.433	0.503	0.549	200	0.116	0.138	0.164	0.181
20	0.360	0.423	0.492	0.537	300	0.095	0.113	0.134	0.148
21	0.352	0.413	0.482	0.526	400	0.082	0.098	0.116	0.128
22	0.344	0.404	0.472	0.515	500	0.073	0.088	0.104	0.115
23	0.337	0.396	0.462	0.505	1000	0.052	0.062	0.073	0.081

APPENDIX D

Critical values r_α of the Spearman rank correlation coefficient, r_s. If $\rho = 0$ the entries in the table are the smallest values of r_s that will be equaled or exceeded with a probability less than α, except as noted, when r_s is calculated from a random sampling of N $[X, Y]$ pairs. For a two-tailed test, the entries in the table are the smallest values of $|r_s|$ that will be equaled or exceeded with a probability less than 2α when $\rho = 0$, since the probability distribution function for r_s is symmetric about 0 when $\rho = 0$. Critical values were determined from all permutations of the first N integers for df ≤ 9. For $10 \leq$ df ≤ 60 the critical values were estimated from 200,000 random permutations of the first N integers. For larger df the Spearman rank and product–moment correlation coefficients are identical to three significant figures.

	α					α			
df = $N - 2$	0.05	0.025	0.01	0.005	df = $N - 2$	0.05	0.025	0.01	0.005
1	—	—	—	—	19	0.369	0.436	0.509	0.556
2	—	—	—	—	20	0.361	0.425	0.498	0.543
3	0.900	1.000	1.000	—	21	0.353	0.415	0.486	0.532
4	0.771[a]	0.886	0.943	1.000	22	0.344	0.406	0.477	0.521
5	0.714	0.786	0.893	0.929	23	0.338	0.399	0.467	0.511
6	0.643	0.783	0.810[b]	0.881	24	0.331	0.389	0.458	0.501
7	0.600	0.683[c]	0.767[d]	0.833	25	0.324	0.382	0.449	0.491
8	0.564	0.648	0.733	0.794	26	0.319	0.376	0.441	0.483
9	0.527	0.618	0.700	0.755	27	0.313	0.368	0.433	0.475
10	0.503	0.587	0.671	0.727	28	0.307	0.362	0.426	0.467
11	0.478	0.560	0.643	0.698	29	0.302	0.356	0.419	0.459
12	0.459	0.539	0.622	0.675	30	0.297	0.350	0.412	0.451
13	0.443	0.521	0.600	0.654	35	0.275	0.325	0.382	0.421
14	0.426	0.503	0.582	0.632	40	0.257	0.305	0.358	0.396
15	0.412	0.488	0.564	0.615	45	0.243	0.288	0.338	0.374
16	0.399	0.474	0.548	0.600	50	0.231	0.274	0.322	0.356
17	0.389	0.460	0.535	0.582	60	0.211	0.251	0.295	0.326
18	0.379	0.447	0.520	0.568	70	0.195	0.232	0.274	0.302

[a] $\alpha = 0.051$.
[b] $\alpha = 0.011$.
[c] $\alpha = 0.0252$.
[d] $\alpha = 0.011$.

APPENDIX E

Critical values τ_α of the Kendall rank correlation coefficient, τ. If $\rho = 0$ the entries in the table are the smallest values of τ that will be equaled or exceeded with a probability less than α when τ is calculated from a random sampling of N $[X, Y]$ pairs. For a two-tailed test, the entries in the table are the smallest values of $|\tau|$ that will be equaled or exceeded with a probability less than 2α when $\rho = 0$. For df ≤ 150 the τ_a were calculated from the cumulative probability distribution function for τ, which is based on all possible permutations of the first N integers. For df ≥ 200 the τ_a were calculated using the normal approximation given by R. R. Sokal and F. J. Rohlf, *Biometry*, 2nd ed., W. H. Freeman, New York (1981), page 606.

	α					α			
df = $N - 2$	0.05	0.025	0.01	0.005	df = $N - 2$	0.05	0.025	0.01	0.005
1	—	—	—	—	24	0.237	0.280	0.329	0.360
2	1.000	—	—	—	25	0.231	0.271	0.322	0.356
3	0.800	1.000	1.000	—	26	0.228	0.265	0.312	0.344
4	0.733	0.867	0.867	1.000	27	0.222	0.261	0.310	0.340
5	0.619	0.714	0.810	0.905	28	0.218	0.255	0.301	0.333
6	0.571	0.643	0.714	0.786	29	0.213	0.252	0.295	0.325
7	0.500	0.556	0.667	0.722	30	0.210	0.246	0.290	0.323
8	0.467	0.511	0.600	0.644	35	0.192	0.228	0.267	0.297
9	0.418	0.491	0.564	0.600	40	0.181	0.211	0.249	0.276
10	0.394	0.455	0.545	0.576	45	0.166	0.198	0.235	0.260
11	0.359	0.436	0.513	0.564	50	0.157	0.187	0.222	0.246
12	0.363	0.407	0.473	0.516	60	0.143	0.171	0.203	0.224
13	0.333	0.390	0.467	0.505	70	0.132	0.158	0.187	0.207
14	0.317	0.383	0.433	0.483	80	0.124	0.147	0.175	0.194
15	0.309	0.368	0.426	0.471	90	0.117	0.139	0.165	0.182
16	0.294	0.346	0.412	0.451	100	0.110	0.132	0.156	0.173
17	0.287	0.333	0.392	0.439	125	0.099	0.118	0.139	0.154
18	0.274	0.326	0.379	0.421	150	0.090	0.107	0.127	0.141
19	0.267	0.314	0.371	0.410	200	0.078	0.093	0.110	0.122
20	0.264	0.307	0.359	0.394	300	0.063	0.076	0.090	0.099
21	0.257	0.296	0.352	0.391	400	0.055	0.065	0.078	0.086
22	0.246	0.290	0.341	0.377	500	0.049	0.059	0.069	0.077
23	0.240	0.287	0.333	0.367	1000	0.035	0.041	0.049	0.054

APPENDIX F

Critical values of the number of runs above or below the median in a sequence of $2N$ numbers. The probability that the number of runs will be equal to or greater than the high value is less than α. The probability that the number of runs will be equal to or less than the low value is likewise less than α. For a two-tailed test, the probability that the number of runs will lie between the low and high values is greater than $1 - 2\alpha$.

	α							
	0.05		0.025		0.01		0.005	
N	low	high	low	high	low	high	low	high
4	2	8						
5	3	9	2	10	2	10		
6	3	11	3	11	2	12	2	12
7	4	12	3	13	3	13	3	13
8	5	13	4	14	4	14	3	15
9	6	14	5	15	4	16	4	16
10	6	16	6	16	5	17	5	17
11	7	17	7	17	6	18	5	19
12	8	18	7	19	6	20	6	20
13	9	19	8	20	7	21	7	21
14	10	20	9	21	8	22	7	23
15	11	21	10	22	9	23	8	24
16	11	23	11	23	10	24	9	25
17	12	24	11	25	10	26	10	26
18	13	25	12	26	11	27	10	28
19	14	26	13	27	12	28	11	29
20	15	27	14	28	13	29	12	30
21	16	28	15	29	14	30	13	31
22	17	29	16	30	14	32	14	32
23	17	31	16	32	15	33	14	34
24	18	32	17	33	16	34	15	35
25	19	33	18	34	17	35	16	36

APPENDIX G

Critical values of the χ^2 distribution function. The probability that χ^2 with n degrees of freedom will equal or exceed the value in the table is α.

	α					
n = df	0.05	0.025	0.975	0.01	0.005	0.995
1	3.8415	5.0239	0.0010	6.6349	7.8794	0.0000
2	5.9915	7.3778	0.0506	9.2103	10.5966	0.0100
3	7.8147	9.3484	0.2158	11.3449	12.8382	0.0717
4	9.4877	11.1433	0.4844	13.2767	14.8603	0.2070
5	11.0705	12.8325	0.8312	15.0863	16.7496	0.4117
6	12.5916	14.4494	1.2373	16.8119	18.5476	0.6757
7	14.0671	16.0128	1.6899	18.4753	20.2777	0.9893
8	15.5073	17.5345	2.1797	20.0902	21.9550	1.3444
9	16.9190	19.0228	2.7004	21.6660	23.5894	1.7349
10	18.3070	20.4832	3.2470	23.2093	25.1882	2.1559
11	19.6751	21.9200	3.8157	24.7250	26.7568	2.6032
12	21.0261	23.3367	4.4038	26.2170	28.2995	3.0738
13	22.3620	24.7356	5.0088	27.6882	29.8195	3.5650
14	23.6848	26.1189	5.6287	29.1412	31.3193	4.0747
15	24.9958	27.4884	6.2621	30.5779	32.8013	4.6009
16	26.2962	28.8454	6.9077	31.9999	34.2672	5.1422
17	27.5871	30.1910	7.5642	33.4087	35.7185	5.6972
18	28.8693	31.5264	8.2307	34.8053	37.1565	6.2648
19	30.1435	32.8523	8.9065	36.1909	38.5823	6.8440
20	31.4104	34.1696	9.5908	37.5662	39.9968	7.4338
21	32.6706	35.4789	10.2829	38.9322	41.4011	8.0337
22	33.9244	36.7807	10.9823	40.2894	42.7957	8.6427
23	35.1725	38.0756	11.6886	41.6384	44.1813	9.2604
24	36.4150	39.3641	12.4012	42.9798	45.5585	9.8862
25	37.6525	40.6465	13.1197	44.3141	46.9279	10.5197

ANSWERS TO EXERCISES

CHAPTER 1

1.1

$$\frac{dF}{dX} = \lim_{\Delta X\to 0} \frac{F(X+\Delta X) - F(X)}{\Delta X}$$

$$= \lim_{\Delta X\to 0} \frac{\dfrac{X+\Delta X}{K+X+\Delta X} - \dfrac{X}{K+X}}{\Delta X}$$

$$= \lim_{\Delta X\to 0} \frac{(X+\Delta X)(K+X) - X(K+X+\Delta X)}{\Delta X(K+X+\Delta X)(K+X)}$$

$$= \lim_{\Delta X\to 0} \frac{K\,\Delta X}{\Delta X(K+X+\Delta X)(K+X)}$$

$$= \lim_{\Delta X\to 0} \frac{K}{(K+X+\Delta X)(K+X)} = \frac{K}{(K+X)^2}$$

1.2 As in the usual proof by induction, we first show that the theorem is true for $n = 1$. We then show that if the theorem is true for $n = N$, it must be true for $n = N + 1$. To do this, we must first establish that

$$\frac{d}{dX}\left(\frac{1}{K+X}\right) = \frac{-1}{(K+X)^2}$$

Thus

$$\frac{dF}{dX} = \lim_{\Delta X\to 0} \frac{F(X+\Delta X) - F(X)}{\Delta X}$$

$$= \lim_{\Delta X\to 0} \frac{\dfrac{1}{K+X+\Delta X} - \dfrac{1}{K+X}}{\Delta X}$$

$$= \lim_{\Delta X\to 0} \frac{(K+X) - (K+X+\Delta X)}{\Delta X(K+X+\Delta X)(K+X)}$$

$$= \lim_{\Delta X \to 0} \frac{-\Delta X}{\Delta X(K + X + \Delta X)(K + X)}$$

$$= \lim_{\Delta X \to 0} \frac{-1}{(K + X + \Delta X)(K + X)} = \frac{-1}{(K + X)^2}$$

The fact that the theorem is true for $n = 1$ follows directly from Exercise 1.1.

Proof by induction. If the theorem is true for $n = N$, then

$$\left(\frac{d}{dX}\right)^N \left(\frac{X}{K + X}\right) = \frac{KN!}{(-K - X)^{N+1}}$$

For $n = N + 1$, we can write

$$\left(\frac{d}{dX}\right)^{N+1} \left(\frac{X}{K + X}\right) = \frac{d}{dX} \frac{KN!}{(-K - X)^{N+1}}$$

Now let $Y = 1/(K + X)$. Then

$$\frac{d}{dX} \frac{KN!}{(-K - X)^{N+1}} = \frac{KN!}{(-1)^{N+1}} \frac{d}{dX} Y^{N+1}$$

By the chain rule this equals

$$\frac{KN!}{(-1)^{N+1}} \frac{d}{dY} Y^{N+1} \frac{dY}{dX}$$

However,

$$\frac{d}{dY} Y^{N+1} = (N + 1)Y^N = \frac{N + 1}{(K + X)^N}$$

and

$$\frac{dY}{dX} = \frac{d}{dX}\left(\frac{1}{K + X}\right) = \frac{-1}{(K + X)^2}$$

Hence

$$\frac{KN!}{(-1)^{N+1}} \frac{d}{dY} Y^{N+1} \frac{dY}{dX} = \frac{KN!}{(-1)^{N+1}} \frac{N + 1}{(K + X)^N} \frac{-1}{(K + X)^2}$$

$$= \frac{K(N + 1)!}{(-K - X)^{N+2}}$$

1.3 **(a)** (0, 0) is a minimum ($R = 0,\ P = Q = 2;\ R^2 - PQ = -4$).
(b) (1, 1) is a saddle point ($R = -1,\ P = 4,\ Q = -6;\ R^2 - PQ = 25$).
(c) (0, 0) is a saddle point ($R = -5,\ P = 2,\ Q = -2;\ R^2 - PQ = 29$).

1.4

$$\left(\frac{\Delta}{\Delta X}\right)\left(\frac{\Delta G}{\Delta Y}\right) = \left(\frac{\Delta}{\Delta X}\right)\frac{G(X,Y+\Delta Y) - G(X,Y)}{\Delta Y}$$

$$= \frac{1}{\Delta X}\left(\frac{G(X+\Delta X, Y+\Delta Y) - G(X+\Delta X, Y)}{\Delta Y} - \frac{G(X,Y+\Delta Y) - G(X,Y)}{\Delta Y}\right)$$

$$= \frac{G(X+\Delta X, Y+\Delta Y) - G(X+\Delta X, Y) - G(X, Y+\Delta Y) + G(X,Y)}{\Delta X \Delta Y}$$

$$\left(\frac{\Delta}{\Delta Y}\right)\left(\frac{\Delta G}{\Delta X}\right) = \left(\frac{\Delta}{\Delta Y}\right)\frac{G(X+\Delta X,Y) - G(X,Y)}{\Delta X}$$

$$= \frac{1}{\Delta Y}\left(\frac{G(X+\Delta X, Y+\Delta Y) - G(X, Y+\Delta Y)}{\Delta X} - \frac{G(X+\Delta X,Y) - G(X,Y)}{\Delta X}\right)$$

$$= \frac{G(X+\Delta X, Y+\Delta Y) - G(X, Y+\Delta Y) - G(X+\Delta X, Y) + G(X,Y)}{\Delta X \Delta Y}$$

$$= \left(\frac{\Delta}{\Delta X}\right)\left(\frac{\Delta G}{\Delta Y}\right)$$

1.5

$$P = P_m(1 - e^{-\alpha I})$$

$$I = I_0 e^{-\beta Z}$$

$$\frac{dP}{dZ} = \frac{dP}{dI}\frac{dI}{dZ} = (\alpha P_m e^{-\alpha I})(-\beta I_0 e^{-\beta Z}) = -\alpha\beta(P_m - P)I$$

1.6 For $N = 1$ we have $G_1(X) = F_1(X)$ and

$$\frac{dG_1(X)}{dX} = \frac{dF_1(X)}{dX} = \frac{dF_1(X)}{dX}\frac{F_1(X)}{F_1(X)}$$

Hence the theorem is true for $N = 1$. Now assume that the theorem is true for N and ask whether the fact that it is true for 1 and N imply that it is true for $N + 1$.

$$G_{N+1}(X) = \prod_{k=1}^{N+1} F_k(X) = F_{N+1} \prod_{k=1}^{N} F_k(X)$$

Hence

$$\begin{aligned}
\frac{dG_{N+1}(X)}{dX} &= \left(\prod_{k=1}^{N} F_k(X)\right)\left(\frac{dF_{N+1}}{dX}\right) + F_{N+1}\frac{d}{dX}\left(\prod_{k=1}^{N} F_k(X)\right) \\
&= F_{N+1}\left(\frac{dF_{N+1}}{dX}\right)\frac{1}{F_{N+1}(X)}\left(\prod_{k=1}^{N} F_k(X)\right) \\
&\quad + F_{N+1}\sum_{k=1}^{N}\frac{dF_k(X)}{dX}\frac{1}{F_k(X)}\prod_{m=1}^{N} F_m(X) \\
&= \left(\frac{dF_{N+1}}{dX}\frac{1}{F_{N+1}(X)} + \sum_{k=1}^{N}\frac{dF_k(X)}{dX}\frac{1}{F_k(X)}\right)\left(\prod_{m=1}^{N+1} F_m(X)\right) \\
&= \left(\sum_{k=1}^{N+1}\frac{dF_k(X)}{dX}\frac{1}{F_k(X)}\right)\left(\prod_{m=1}^{N+1} F_m(X)\right)
\end{aligned}$$

This completes the proof by induction.

1.7 The figures are shown on page 288. The local minimum and maximum of P occur at depths of approximately 50 and 84.5 m, respectively. The expression for dP/dZ is as follows:

$$\begin{aligned}
\frac{dP}{dZ} &= -\mathrm{chl}\frac{\exp(-Z/20)}{(0.1 + I)^2} \\
&\quad - \mathrm{PI}\left(\frac{\exp(-Z/20)}{200} + \frac{0.4}{700}(Z - 100)\exp\left[\frac{-(Z - 100)^2}{700}\right]\right)
\end{aligned}$$

where chl is the chlorophyll concentration at depth Z and PI is the productivity index at light intensity $I(Z)$.

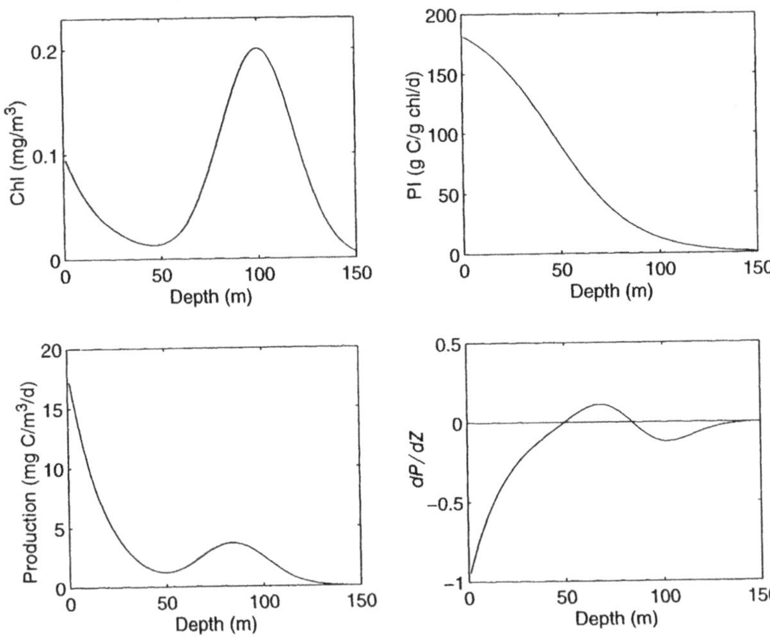

1.8 The values of X, Y, and $F(X, Y)$ are as follows:

X	1.1	1.1	1.1	1.0	1.0	1.0	0.9	0.9	0.9
Y	3.1	3.0	2.9	3.1	3.0	2.9	3.1	3.0	2.9
$F(X,Y)$	8.97	9.0	9.05	9.01	9.0	9.01	9.05	9.0	8.97

X	2.1	2.1	2.1	2.0	2.0	2.0	1.9	1.9	1.9
Y	4.1	4.0	3.9	4.1	4.0	3.9	4.1	4.0	3.9
$F(X,Y)$	8.03	8.02	8.03	8.01	8.0	8.01	8.03	8.02	8.03

X	3.1	3.1	3.1	3.0	3.0	3.0	2.9	2.9	2.9
Y	3.1	3.0	2.9	3.1	3.0	2.9	3.1	3.0	2.9
$F(X,Y)$	9.05	9.0	8.97	9.01	9.0	9.01	8.97	9.0	9.05

The results are indeed consistent with the conclusion that $(1, 3)$ and $(3, 3)$ are saddle points and that $(2, 4)$ is a local minimum.

CHAPTER 2

2.1 **(a)** Since $\overline{X} = \Sigma_{k=1}^{N} X_k$, it follows immediately that $\Sigma_{k=1}^{N} x_k = 0$ and hence that $\Sigma_{k=1}^{N} c_k = 0$.

(b)

$$\sum_{k=1}^{N} c_k^2 = \frac{\sum x_k^2}{\left(\sum x_m^2\right)^2} = \frac{1}{\sum x_m^2}$$

(c)

$$\sum_{k=1}^{N} c_k(\alpha + \beta X_k) = \alpha \sum_{k=1}^{N} c_k + \beta \sum_{k=1}^{N} c_k X_k$$

$$= 0 + \frac{\beta}{\sum x_m^2} \sum_{k=1}^{N} x_k\left(x_k + \overline{X}\right)$$

$$= 0 + \frac{\beta}{\sum x_m^2} \sum x_k^2 + 0 = \beta$$

(d)

$$N \sum_{k=1}^{N} d_k^2 = N \sum_{k=1}^{N} \left(\frac{1}{N} - c_k \overline{X}\right)^2$$

$$= N\left(\frac{N}{N^2} + \overline{X}^2 \sum c_k^2\right)$$

$$= 1 + N\overline{X}^2 \frac{1}{\sum x_m^2} = 1 + \frac{\sum X_k^2 - \sum x_k^2}{\sum x_m^2}$$

$$= 1 + \frac{\sum X_k^2}{\sum x_m^2} - 1 = \frac{\sum X_k^2}{\sum x_m^2}$$

(e)

$$\sum_{k=1}^{N} d_k = \sum_{k=1}^{N} \left(\frac{1}{N} - c_k \overline{X}\right) = \frac{N}{N} = 1$$

2.2 From Exercise 2.1(c), it follows that

$$\sum_{m=1}^{N} c_m(\alpha + \beta X_m)\overline{X} = \beta\overline{X} = \frac{1}{N}\sum_{k=1}^{N} \beta X_k$$

The theorem then follows immediately.

2.3

$$\frac{dV(\hat{Y}_0)}{dX_0} = \frac{2V(Y|X)(X_0 - \overline{X})}{\sum x_k^2}$$

$$\frac{d^2V(\hat{Y}_0)}{dX_0^2} = \frac{2V(Y|X)}{\sum x_k^2}$$

The first derivative is clearly 0 if $X_0 = \overline{X}$. The second derivative is clearly positive as long as $V(Y|X)$ is positive. Hence $V(\hat{Y}_0)$ is a minimum when $X_0 = \overline{X}$ as long as $V(Y|X)$ is positive. Obviously, $V(Y|X)$ is positive or 0. The latter situation occurs only when the regression line gives a perfect fit to the data, in which case $V(\hat{Y}_0)$ is 0 for all values of X_0.

An example of the graph asked for in this exercise is shown below. The Y values and 95% confidence intervals will vary from one simulated repetition of the experiment to the next, depending on the random

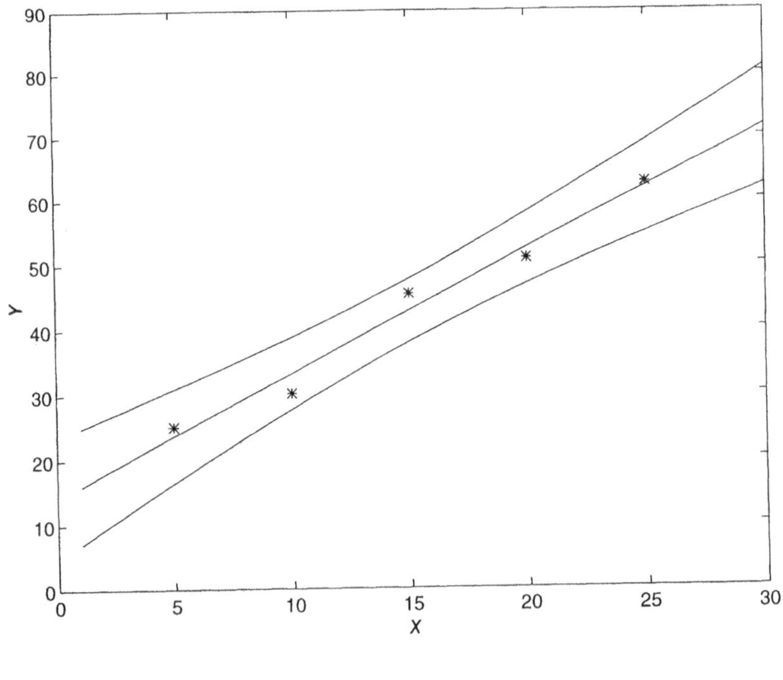

numbers chosen for the errors. In all cases, however, the minimum uncertainty should occur at $X = 15$.

2.4 The choice of sampling time will have no effect on $V(Y|X)$, and the strategy is therefore to maximize Σx_k^2. This will be maximized when five of the samples are taken at the beginning of the experiment (time 0) and five are taken at the end of the experiment (time 24 hours).

2.5 (a)

$$\frac{1}{10}\frac{\sum X_k^2}{\sum x_n^2} = \frac{\sum x_k^2 + 10\overline{X}^2}{10\sum x_n^2} = \frac{1}{10} + \frac{\overline{X}^2}{\sum x_n^2}$$

It follows that $(1/10)\Sigma X_k^2/\Sigma x_n^2$ is a minimum when $\overline{X} = 0$. The minimum value is $1/10$. If the first priority is to minimize $V(A)$, then the values of X should be chosen so that $\overline{X} = 0$. If our second priority is to minimize $V(B)$, we select the values of X so that Σx_k^2 is a maximum. To accomplish this, we pick three values of X at -20, six values of X at 10, and one value of X at 0.

(b) $\Sigma X_k^2 = (N - M)X^2$ and $\overline{X} = [(N - M)/N]X$. Hence

$$\sum x_k^2 = \sum X_k^2 - N\overline{X}^2 = (N - M)X^2 - \frac{(N - M)^2}{N}X^2$$

Hence

$$\frac{1}{N}\frac{\sum X_k^2}{\sum x_n^2} = \frac{1}{N}\frac{N - M}{N - M - \dfrac{(N - M)^2}{N}}$$

$$= \frac{1}{N}\frac{1}{1 - \dfrac{N - M}{N}} = \frac{1}{N - N + M} = \frac{1}{M}$$

The value of M that minimizes $V(A)$ is $N - 1$. In other words, all but one of the measurements should be made at $X = 0$. The final measurement should be made at $X = 20$ in order to maximize Σx_k^2 and hence minimize $V(B)$.

2.6

$$\sum_{k=1}^{N} (Y_k - A - BX_k)^2 = \sum_{k=1}^{N} \left[Y_k - \bar{Y} - B(X_k - \bar{X})\right]^2$$

$$= \sum_{k=1}^{N} \left[y_k - \frac{\sum x_m y_m}{\sum x_n^2} x_k \right]^2$$

$$= \sum y_k^2 - 2\frac{\left(\sum x_k y_k\right)^2}{\sum x_n^2} + \frac{\left(\sum x_k y_k\right)^2}{\sum x_n^2}$$

$$= \sum y_k^2 - \frac{\left(\sum x_k y_k\right)^2}{\sum x_n^2}$$

2.7 The Monte Carlo procedure should produce estimates that are close to the following values: $A = 10$; $B = 2$; $V(A) = 9.9$; $V(B) = 0.036$; $V(Y|X) = 9$. Note that the student is to estimate $V(A)$ and $V(B)$ by two different procedures, first by simply averaging the estimates of $V(A)$ and $V(B)$ calculated using Equations 2.64 and 2.65 and second by using the 1000 estimates of A and B. Both methods should give unbiased estimates of $V(A)$ and $V(B)$. The student should find that the true values of A and B lie outside the calculated 95% confidence limits for A and B in about 50 cases (5%) out of the 1000 simulations.

2.8 In this case the student should again get good unbiased estimates of A, B, and $V(Y|X)$ from the 1000 simulations. The two estimates of $V(A)$ should be about 32, and the two estimates of $V(B)$ should be about 0.13. The student should again find that the true values of A and B lie outside the calculated 95% confidence limits for A and B in about 50 cases (5%) out of the 1000 simulations. The estimates of $V(A)$ and $V(B)$ calculated by plugging the expectation values of $V(Y|X)$, ΣX^2, and Σx^2 into Equations 2.64 and 2.65 are 17.4 and 0.0675. These, as noted in the question, are quite different from the true expectation values and reflect the fact that the expectation value of a quotient is seldom the quotient of the expectation values.

2.9 The results are similar to the results from Exercise 2.8, but in this case the two estimates of $V(A)$ should be about 28, and the two estimates of $V(B)$ should be about 0.11.

2.10 We consider Equation 2.30 first. In the limit as $N \to \infty$, the term $(1/N)\Sigma_{k=1}^{N} \varepsilon_k$ just becomes the expectation value of ε, which is of course 0. For the term $\bar{X}\Sigma_{m=1}^{N} c_m \varepsilon_m$, we assume that the limit of the product is the product of the limits. As $N \to \infty$, $\bar{X} \to \mu$, where μ is the expectation value of X. As noted in the hint for the problem, $\sum_{m=1}^{N} c_m \varepsilon_m$ can be written as

$$\left(\frac{N}{\sum x_n^2}\right)\left(\frac{1}{N}\sum x_k \varepsilon_k\right)$$

Again assuming the limit of the product to be the product of the limits, we see that the limit of the first term in parentheses as $N \to \infty$ is $1/V(X)$ and the limit of the second term as $N \to \infty$ is the expectation value of $x\varepsilon$. Since the errors are independent of X, it follows that $E(x\varepsilon) = E(x)E(\varepsilon)$. Since $E(x)$ and $E(\varepsilon)$ both equal 0, it follows that $\lim_{N \to \infty} \overline{X} \sum_{m=1}^{N} c_m \varepsilon_m = 0$. The analogous proof in the case of Equation 2.28 follows similar reasoning.

CHAPTER 3

3.1 For $N = 2$ we have

$$\sum_{k=1}^{1} k = 1 = \frac{2(2-1)}{2} = 1$$

Hence the theorem is true for $N = 2$. Now assume the theorem is true for $N = M$. Then

$$\sum_{k=1}^{M+1-1} k = \sum_{k=1}^{M-1} k + M = \frac{M(M-1)}{2} + M = \frac{M^2 - M + 2M}{2}$$

$$= \frac{M^2 + M}{2} = \frac{(M+1)M}{2}$$

3.2 The maximum value of Q occurs when the Y ranks occur in the opposite order from the X ranks. In that case we have the following arrangement of ranks:

X	1	2	3	4	5	6	7	$\cdots$	$N-1$	N
Y	N	$N-1$	$N-2$	$N-3$	$N-4$	$N-5$	$N-6$	$\cdots$	2	1

When we count the number of ranks to the right of each rank in the Y row that are less than the given rank, we find that the sum of these counts equals $(N-1) + (N-2) + (N-3) + \cdots + 1$. Clearly this is the sum of the integers from 1 to $N-1$, and from Exercise 3.1 this sum equals $N(N-1)/2$. Thus the maximum value of Q is $N(N-1)/2$. Plugging this value of Q into Equation 3.16 gives a τ value of -1.

3.3 Clearly the ranks in this case will be the integers from 1 to N. From Exercise 3.1 it follows that the sum of the ranks equals $(N + 1)N/2$ and hence the average value of the ranks equals

$$\frac{1}{N}\frac{(N+1)N}{2} = \frac{N+1}{2}$$

For the sum of the squares of the ranks, we have, for $N = 1$,

$$\sum_{k=1}^{1} L_k^2 = 1 = \frac{1(1+1)(2+1)}{6} = \frac{6}{6} = 1$$

Hence the theorem is true for $N = 1$. If the theorem is true for $N = M$, then

$$\begin{aligned}
\sum_{k=1}^{M+1} L_k^2 &= \sum_{k=1}^{M} L_k^2 + (M+1)^2 \\
&= \frac{M(M+1)(2M+1)}{6} + (M+1)^2 \\
&= \frac{[M+1][M(2M+1) + 6(M+1)]}{6} \\
&= \frac{(M+1)(2M^2 + 7M + 6)}{6} = \frac{(M+1)(M+2)(2M+3)}{6} \\
&= \frac{[M+1][M+2][2(M+1)+1]}{6}
\end{aligned}$$

It follows that

$$\begin{aligned}
\sum_{k=1}^{N} L_k^2 - N\overline{L}^2 &= \frac{N(N+1)(2N+1)}{6} - N\frac{(N+1)^2}{4} \\
&= \frac{2N(N+1)(2N+1) - 3N(N+1)(N+1)}{12} \\
&= \frac{N(N+1)(4N+2-3N-3)}{12} \\
&= \frac{N(N+1)(N-1)}{12} = \frac{N(N-1)^2}{12}
\end{aligned}$$

Since $\overline{L}$ and $\overline{M}$ are identical, it follows that

$$\begin{aligned}
\sum_{k=1}^{N} (L_k - \overline{L})(M_k - \overline{M}) &= \sum_{k=1}^{N} L_k M_k - \overline{L}N\overline{M} - \overline{M}N\overline{L} + N\overline{L}\overline{M} \\
&= \sum_{k=1}^{N} L_k M_k - N\overline{L}^2
\end{aligned}$$

It also follows that

$$\sum_{k=1}^{N} (L_k - \overline{L})^2 = \sum_{k=1}^{N} (M_k - \overline{M})^2$$

Furthermore,

$$\sum_{k=1}^{N} (L_k - \overline{L})^2 = \sum_{k=1}^{N} L_k^2 - N\overline{L}^2$$

Hence

$$r_s = \frac{\sum L_k M_k - N\overline{L}^2}{\sum L_m^2 - N\overline{L}^2} = \frac{12}{N} \frac{\sum L_k M_k - N\overline{L}^2}{N^2 - 1}$$

We now note that

$$\begin{aligned}\sum_{k=1}^{N} (L_k - M_k)^2 &= \sum_{k=1}^{N} L_k^2 + \sum_{k=1}^{N} M_k^2 - 2 \sum_{k=1}^{N} M_k L_k \\ &= \frac{2N(N+1)(2N+1)}{6} - 2 \sum_{k=1}^{N} M_k L_k\end{aligned}$$

Thus

$$\sum_{k=1}^{N} M_k L_k = \frac{N(N+1)(2N+1)}{6} - \frac{1}{2} \sum_{k=1}^{N} (L_k - M_k)^2$$

We therefore have

$$\begin{aligned}r_s &= \frac{12}{N} \frac{\sum L_k M_k - N\overline{L}^2}{N^2 - 1} \\ &= \frac{12}{N} \frac{N(N+1)(2N+1)}{6(N+1)(N-1)} - \frac{12}{N} \frac{N(N+1)(N+1)}{4(N+1)(N-1)} \\ &\quad - \frac{6}{N} \frac{\sum (L_k - M_k)^2}{N^2 - 1} \\ &= \frac{2(2N+1) - 3(N+1)}{N-1} - \frac{6}{N} \frac{\sum (L_k - M_k)^2}{N^2 - 1} \\ &= 1 - \frac{6}{N(N^2 - 1)} \sum (L_k - M_k)^2\end{aligned}$$

3.4 The smallest value of Q and hence the largest value of τ occurs when lake ELA 222 is assigned C rank 16, lake ELA 304 is assigned C rank 17, and the three ties in the N ranks are assigned ranks so that the lake lower in the N rank column is assigned the higher rank. In that case $Q = 30$ and $\tau = 0.815$. The highest value of Q and hence the smallest value of τ occurs when lake ELA 222 is assigned C rank 17, lake ELA 304 is assigned C rank 16, and the three ties in the N ranks are assigned ranks so that the lake lower in the N rank column is assigned the lower rank. In that case $Q = 34$ and $\tau = 0.791$. Reference to Appendix E shows that both values of τ are significant at $p < 0.01$.

3.5 The results should indeed be consistent with Figure 3.1. The ratio $\bar{r}/\rho$ should be about 0.92, and $s(r)$ should be about 0.42. The results are independent of the values assigned to μ_x, μ_y, σ_x, and σ_y.

3.6 There is not much to add here. The percentage of the confidence intervals that do not include 0.5 is indeed close to 5%.

CHAPTER 4

4.1 The quadratic equation has the form $AZ^2 + BZ - A = 0$, where $A = \Sigma x_k y_k$ and $B = \Sigma x_m^2 - \Sigma y_m^2$. The general solution of the quadratic equation is therefore

$$Z = \frac{-B \pm \sqrt{B^2 + 4A^2}}{2A}$$

Since $A = \Sigma x_k y_k$, it follows that Z will have the same sign as $\Sigma x_k y_k$ as long as $-B \pm \sqrt{B^2 + 4A^2} > 0$. Since $\sqrt{B^2 + 4A^2}$ is clearly greater in magnitude than B, this condition can be satisfied only if we take the + sign on the square root.

4.2 For the purposes of this problem, it is convenient to write Equation 4.16 in the form

$$B_{\text{ma}} = \frac{P - Q + \sqrt{(Q - P)^2 + 4R^2}}{2R}$$

where $P = \Sigma y_n^2$, $Q = \Sigma x_n^2$, and $R = \Sigma x_n y_n$. If all values of y are then multiplied by a factor k, $P \to k^2 P$ and $R \to kR$. We then have

$$B_{\text{ma}} = \frac{k^2P - Q + \sqrt{(Q - k^2P)^2 + 4k^2R^2}}{2kR}$$

$$= \frac{k^2P - Q + \sqrt{Q^2 - 2Qk^2P + k^4P^2 + 4k^2R^2}}{2kR}$$

$$= \frac{1}{2kR}\left(k^2P - Q + k^2P\sqrt{1 + \frac{4R^2 - 2QP}{k^2P^2} + \frac{Q^2}{k^4P^2}}\right)$$

In the limit as $k \to \infty$, the expression in parentheses becomes

$$k^2P - Q + k^2P\left(1 + \frac{2R^2 - QP}{k^2P^2}\right) \to 2k^2P$$

Hence, as $k \to \infty$,

$$B_{\text{ma}} \to \frac{2k^2P}{2kR} = \frac{\sum (ky_n)^2}{\sum (ky_m)x_m}$$

4.3 The approach is similar to Exercise 4.2. In this case

$$B_{\text{ma}} = \frac{P - k^2Q + \sqrt{(k^2Q - P)^2 + 4k^2R^2}}{2kR}$$

$$= \frac{P - k^2Q + \sqrt{k^4Q^2 - 2Qk^2P + P^2 + 4k^2R^2}}{2kR}$$

$$= \frac{1}{2kR}\left(P - k^2Q + k^2Q\sqrt{1 + \frac{4R^2 - 2QP}{k^2Q^2} + \frac{P^2}{k^4Q^2}}\right)$$

In the limit as $k \to \infty$, the expression in parentheses becomes

$$P - k^2Q + k^2Q\left(1 + \frac{2R^2 - QP}{k^2Q^2}\right) \to P + \frac{2R^2}{Q} - P = \frac{2R^2}{Q}$$

Hence, as $k \to \infty$,

$$B_{\text{ma}} \to \frac{2R^2}{2kRQ} = \frac{R}{kQ} = \frac{\sum (kx_m)y_m}{\sum (kx_n)^2}$$

4.4 We first note that

$$\bar{\ddot{X}} = \frac{\bar{X} - \bar{X}}{S_x} = 0 \quad \text{and} \quad \bar{\ddot{Y}} = \frac{\bar{Y} - \bar{Y}}{S_y} = 0$$

Hence $\ddot{X} = \ddot{x}$ and $\ddot{Y} = \ddot{y}$. Hence

$$\sum_{k=1}^{N} \ddot{X}_k\ddot{Y}_k = \sum_{k=1}^{N} \ddot{y}_k\ddot{x}_k$$

Now

$$\sum_{k=1}^{N} \ddot{X}_k \ddot{Y}_k = \frac{1}{S_x S_y} \sum_{k=1}^{N} (X_k - \bar{X})(Y_k - \bar{Y})$$

$$= (N-1) \frac{\sum x_k y_k}{\sqrt{\sum x_m^2 \sum y_n^2}} = (N-1) r$$

4.5 The equation has the form

$$F(B) = -\frac{1}{B} \sum (Y_k - A - BX_k)^2 = -\frac{1}{B} \sum \left[Y_k - \bar{Y} - B(X_k - \bar{X}) \right]^2$$

Differentiating this equation with respect to B and setting the derivative equal to 0 gives

$$\frac{1}{B^2} \sum \left[Y_k - \bar{Y} - B(X_k - \bar{X}) \right]^2$$

$$+ \frac{2}{B} \sum \left[Y_k - \bar{Y} - B(X_k - \bar{X}) \right] (X_k - \bar{X}) = 0$$

$$\Rightarrow \sum (y_k - Bx_k)^2 + 2B \sum (y_k x_k - Bx_k^2) = 0$$

$$\Rightarrow \sum y_k^2 - 2B \sum x_n y_n + B^2 \sum x_m^2 + 2B \sum x_n y_n - 2B^2 \sum x_m^2 = 0$$

$$\Rightarrow \sum y_k^2 - B^2 \sum x_m^2 = 0 \Rightarrow B^2 = \frac{\sum y_k^2}{\sum x_m^2}$$

4.6 Equation 4.16 has the form

$$B_1 = \frac{P - Q + \sqrt{(Q-P)^2 + 4R^2}}{2R}$$

where, as in the answer to Exercise 4.2, $P = \Sigma y_n^2$, $Q = \Sigma x_n^2$, and $R = \Sigma x_n y_n$. If we exchange x and y, R remains unchanged, $P \to Q$ and $Q \to P$. The solution with x and y exchanged is therefore

$$B_2 = \frac{Q - P + \sqrt{(P-Q)^2 + 4R^2}}{2R}$$

We conclude that

$$B_1 B_2 = \frac{(P-Q)^2 - (P-Q)^2 + 4R^2}{4R^2} = 1$$

Hence $B_1 = 1/B_2$.

4.7 **(a)** In this case the major axis method gives a consistent estimate of the slope because $\sigma_\delta = \sigma_\varepsilon$. The reduced major axis method gives an inconsistent estimate, and the estimate tends to underestimate the magnitude of β because $\sigma_\varepsilon/\sigma_\delta < |\beta|$. Bartlett's method also underestimates the magnitude of β.

(b) In this case the reduced major axis gives the consistent estimate because $\sigma_\varepsilon/\sigma_\delta = |\beta|$. The major axis method gives an inconsistent estimate, and the estimate tends to overestimate the magnitude of β because $\sigma_\varepsilon > \sigma_\delta$. Bartlett's method tends to underestimate the magnitude of β.

(c) In this case both the major axis and the reduced major axis methods give consistent estimates because $\sigma_\varepsilon/\sigma_\delta = |\beta| = 1$. Bartlett's method again tends to underestimate the magnitude of β.

CHAPTER 5

5.1 If $\mathbf{HX} = \mathbf{X}$, then $\sum_{k=1}^{N} H_{jk} X_{kl} = X_{jl}$ for all j and l. However, if the first column of $\mathbf{X}$ contains all 1's, then $X_{k1} = X_{j1} = 1$. Hence $\sum_{k=1}^{N} H_{jk} = 1$ for all j; that is, the sum of all the elements in any row of $\mathbf{H}$ equals 1. Similarly, if $\mathbf{X}'\mathbf{H} = \mathbf{X}'$, then $\sum_{j=1}^{N} X_{lj}^{t} H_{jk} = X_{lk}^{t}$ for all k and l. However, $X_{1j}^{t} = X_{j1} = 1$ and $X_{1k}^{t} = X_{k1} = 1$. Hence $\sum_{j=1}^{N} H_{jk} = 1$ for all k; that is, the sum of all the elements in any column of $\mathbf{H}$ equals 1. Now if all the elements of $\mathbf{J}$ are 1, then $\sum_{n=1}^{N} J_{mn} H_{nk} = \sum_{n=1}^{N} H_{nk} = 1$ for all m and k. Hence each element of $\mathbf{JH}$ equals 1, that is, $\mathbf{JH} = \mathbf{J}$. Similarly, $\sum_{n=1}^{N} H_{mn} J_{nk} = \sum_{n=1}^{N} H_{mn} = 1$ for all m and k. Hence each element of $\mathbf{HJ}$ equals 1, that is, $\mathbf{HJ} = \mathbf{J}$.

5.2 If $\mathbf{H} = \mathbf{X}(\mathbf{X}'\mathbf{X})^{-1}\mathbf{X}'$, then

$$\operatorname{tr}(\mathbf{H}) = \sum_{n=1}^{N} \sum_{j=1}^{M+1} \sum_{m=1}^{M+1} X_{nj} a_{jm} X_{mn}^{t}$$

$$= \sum_{j=1}^{M+1} \sum_{m=1}^{M+1} a_{jm} \left(\sum_{n=1}^{N} X_{mn}^{t} X_{nj} \right) = \sum_{j=1}^{M+1} \sum_{m=1}^{M+1} a_{jm} c_{mj}$$

where a_{jm} is the element in row j and column m of $(\mathbf{X}'\mathbf{X})^{-1}$ and c_{mj} is the element in row m and column j of $\mathbf{X}'\mathbf{X}$. Since $\sum_{m=1}^{M+1} a_{jm} c_{mj} = 1$, it follows immediately that $\sum_{j=1}^{M+1} \sum_{m=1}^{M+1} a_{jm} c_{mj} = \sum_{j=1}^{M+1} 1 = M + 1$.

5.3 In model I regression we have $\hat{Y}_j = A + BX_j = \bar{Y} - B\bar{X} + BX_j$. Hence $\hat{Y}_j - \bar{Y} = B(X_j - \bar{X})$. Since the observations are controlled, both X_j and $\bar{X}$ are fixed, and hence $E[B(X_j - \bar{X})] = (X_j - \bar{X})E(B) = (X_j - \bar{X})\beta$. Therefore $\Sigma_{j=1}^{N}[E(\hat{Y}_j - \bar{Y})]^2 = \beta^2\Sigma_{j=1}^{N}(X_j - \bar{X})^2$. Since in linear regression $M = 1$, it then follows immediately from Equation 5.44 that $E(\text{SSR}) = \sigma^2 + \beta^2\Sigma_{j=1}^{N}(X_j - \bar{X})^2$.

5.4 The result is

$$(\mathbf{X}'\mathbf{X})^{-1} = \begin{pmatrix} \dfrac{\sum X^2}{N\sum x^2} & -\dfrac{\bar{X}}{\sum x^2} \\ -\dfrac{\bar{X}}{\sum x^2} & \dfrac{1}{\sum x^2} \end{pmatrix}$$

This result is indeed consistent with Equation 5.25 and Equations 2.64 and 2.65. The diagonal element in the first row, when multiplied by $V(Y|X)$, gives $V(A)$, and the diagonal element in the second row, when multiplied by $V(Y|X)$, gives $V(B)$.

5.5 Y_0 has the form $\mathbf{AY}$, where $\mathbf{A}$ is a constant matrix and $\mathbf{Y}$ is a variable vector. Therefore

$$\begin{aligned} \boldsymbol{\sigma}^2(Y_0) &= \mathbf{A}[\boldsymbol{\sigma}^2(Y)]\mathbf{A}' \\ &= \sigma^2(Y)\mathbf{AA}' = \sigma^2(Y)\mathbf{X}_0'(\mathbf{X}'\mathbf{X})^{-1}\mathbf{X}'\mathbf{X}(\mathbf{X}'\mathbf{X})^{-1}\mathbf{X}_0 \\ &= \sigma^2(Y)\mathbf{X}_0'(\mathbf{X}'\mathbf{X})^{-1}\mathbf{X}_0 \end{aligned}$$

In the case of linear regression ($M = 1$), this equation becomes

$$\boldsymbol{\sigma}^2(Y_0) = \sigma^2(Y)(1 \quad X_0)\begin{pmatrix} \dfrac{\sum X^2}{N\sum x^2} & -\dfrac{\bar{X}}{\sum x^2} \\ -\dfrac{\bar{X}}{\sum x^2} & \dfrac{1}{\sum x^2} \end{pmatrix}\begin{pmatrix} 1 \\ X_0 \end{pmatrix}$$

$$= \sigma^2(Y)(1 \quad X_0)\begin{pmatrix} \dfrac{\sum X^2}{N\sum x^2} - \dfrac{\bar{X}X_0}{\sum x^2} \\ \dfrac{X_0}{\sum x^2} - \dfrac{\bar{X}}{\sum x^2} \end{pmatrix}$$

$$= \sigma^2(Y)\frac{1}{\sum x^2}\left(\frac{\sum X^2}{N} - 2\bar{X}X_0 + X_0^2\right)$$

$$= \sigma^2(Y) \frac{1}{\sum x^2} \left(\frac{\sum X^2}{N} - \overline{X}^2 + \overline{X}^2 - 2\overline{X}X_0 + X_0^2 \right)$$

$$= \sigma^2(Y) \left(\frac{1}{N} + \frac{(X_0 - \overline{X})^2}{\sum x^2} \right)$$

The result is therefore consistent with Equation 2.66 in the case $M = 1$.

5.6 A simple way to answer this question is to note, as in Chapter 2, that the second sample of M measurements is taken after the initial sample of N measurements, and hence the errors in $\hat{Y}_0$ and $\overline{Y}_0$ are independent. Hence $V(\hat{Y}_0 - \overline{Y}_0) = V(\hat{Y}_0) + V(\overline{Y}_0)$. Clearly $V(\overline{Y}_0) = (1/M)\sigma^2(Y)$. Hence

$$V\left(\hat{Y}_0 - \overline{Y}_0\right) = \sigma^2(Y)\left[\frac{1}{M} + \mathbf{X}_0'(\mathbf{X}'\mathbf{X})^{-1}\mathbf{X}_0\right]$$

5.7 The R^2 for 19′-but is 0.91. It is principally correlated with 19′-hex ($r = 0.843$) and chl b ($r = 0.846$). The three pigments needed to describe the chl a data are 19′-hex, zeaxanthin, and chl b. The multiple regression model is chl a = 1.56(19′-hex) + 2.74(zeaxanthin) + 0.61(chl b). The observed chl a concentrations and the concentrations calculated from the model are shown in the following figure. Following reasoning

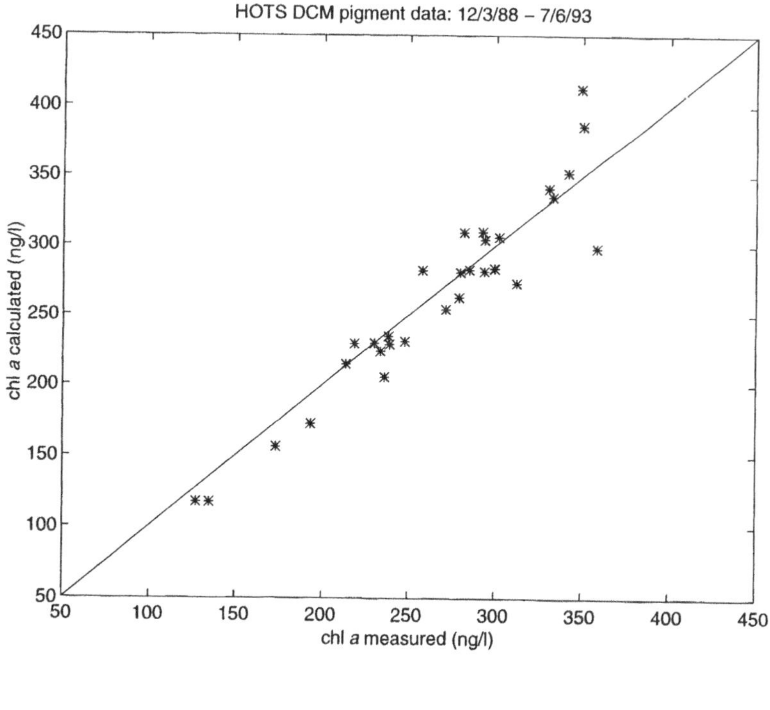

similar to that used in the analysis of the data in Table 5.7, we conclude that the data are consistent with a phytoplankton community composed primarily of *Prochlorococcus* spp., cyanobacteria, and chrysophytes. The correlation between 19′-but and 19′-hex follows from the fact that both pigments are found in chrysophytes (and prymnesiophytes). The correlation between 19′-but and chl *b* suggests that the chrysophytes and *Prochlorococcus* spp. covary in this community.

5.8 The results are $a = 9.57$ with a standard deviation of 1.63, and $k = 0.048$ with a standard deviation of 0.013. A graph of the data and the fitted function is shown below.

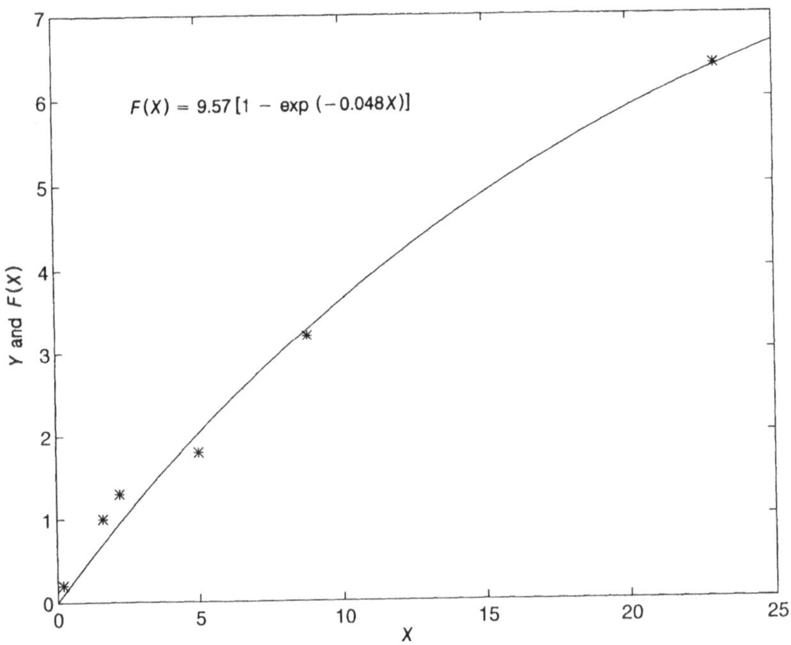

5.9 Equation 5.22 may be written in the form $\mathbf{C} = \mathbf{AY}$, where $\mathbf{A} = (\mathbf{X}^t\mathbf{X})^{-1}\mathbf{X}^t$. The vector $\mathbf{Y}$ can be expressed as $\mathbf{Y} = \mathbf{Y}_t + \boldsymbol{\varepsilon}$, where $\mathbf{Y}_t$ is the vector of true values of Y and $\boldsymbol{\varepsilon}$ is a vector of random errors. It follows that $\mathbf{C} = \mathbf{AY}_t + \mathbf{A}\boldsymbol{\varepsilon}$. Clearly $\mathbf{AY}_t$ is the true value of the vector $\mathbf{C}$, since there are no errors associated with the X's. Thus $\mathbf{C} = \mathbf{C}_t + \mathbf{A}\boldsymbol{\varepsilon}$. Now $E(\mathbf{C}) = E(\mathbf{C}_t) + E(\mathbf{A}\boldsymbol{\varepsilon})$. Since the errors in Y are independent of the X's, $E(\mathbf{A}\boldsymbol{\varepsilon}) = E(\mathbf{A})E(\boldsymbol{\varepsilon})$. However, $E(\boldsymbol{\varepsilon}) = 0$. Hence $E(\mathbf{C}) = E(\mathbf{C}_t) = \mathbf{C}_t$. Hence the expectation value of $\mathbf{C}$ is indeed the true value of $\mathbf{C}$.

5.10 The results will obviously vary a bit from one run to the next, depending on the random numbers that are generated. A typical result is shown in the following graph.

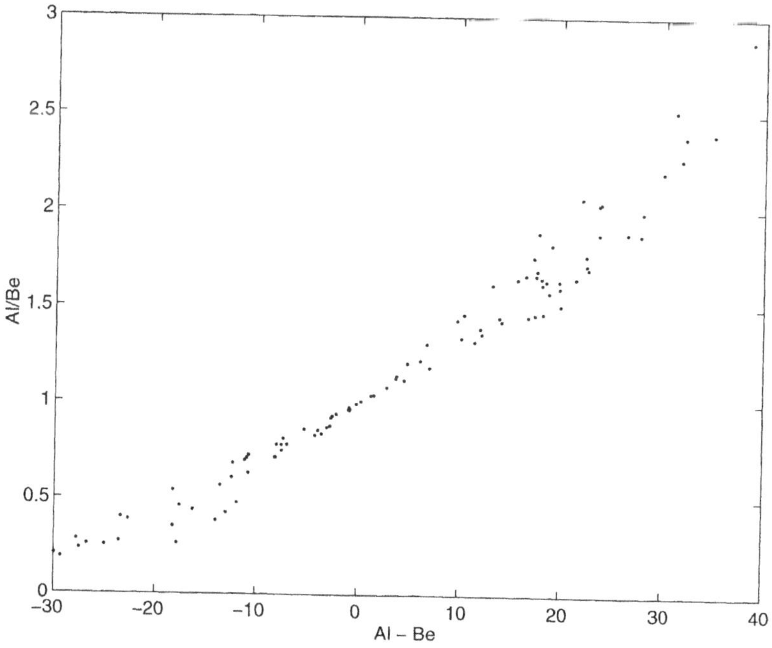

CHAPTER 6

6.1 Assume

$$W(B) = W(A) + \sum_{n=1}^{N} \frac{W^{(n)}(A)(B-A)^n}{n!} + \frac{1}{N!}\int_A^B (B-X)^N \left(\frac{d}{dX}\right)^{N+1} W(X)\, dX$$

Now let

$$\frac{dF}{dX} = (B-X)^N$$

and

$$G(X) = \left(\frac{d}{dX}\right)^{N+1} W(X)$$

Then

$$F(X) = -\frac{1}{N+1}(B-X)^{N+1}$$

and

$$\frac{dG}{dX} = \left(\frac{d}{dX}\right)^{N+2} W(X)$$

Integrating by parts, we then have

$$\frac{1}{N!}\int_A^B (B-X)^N \left(\frac{d}{dX}\right)^{N+1} W(X)\, dX$$

$$= \frac{1}{N!}\int_A^B G(X)\frac{dF}{dX}\, dX$$

$$= \frac{1}{N!}(G(B)F(B) - G(A)F(A)) - \int_A^B F(X)\frac{dG}{dX}\, dX$$

$$= \frac{1}{(N+1)!}(B-A)^{N+1} W^{(N+1)}(A)$$

$$+ \frac{1}{(N+1)!}\int_A^B (B-X)^{N+1}\left(\frac{d}{dX}\right)^{N+2} W(X)\, dX$$

Hence

$$W(B) = W(A) + \sum_{n=1}^{N+1} \frac{W^{(n)}(A)(B-A)^n}{n!}$$

$$+ \frac{1}{(N+1)!}\int_A^B (B-X)^{N+1}\left(\frac{d}{dX}\right)^{N+2} W(X)\, dX$$

6.2

$$\int_0^4 \frac{X}{1+X}\, dX = \int_0^4 \frac{X+1-1}{1+X}\, dX = \int_0^4 \left(1 - \frac{1}{1+X}\right) dX$$

$$= (X - \ln(1+X))|_0^4 = 4 - \ln(5)$$

If we approximate the integral using the expansion in Equation 6.10, we need only consider the contribution from even powers of $(X-2)$. Odd powers of $(X-2)$ are obviously odd functions of $X-2$, and the integral of such odd functions is 0 over any range of integration symmetric about $X = 2$. The approximation therefore becomes

$$\int_0^4 \frac{X}{1+X}\,dX \cong \int_0^4 \left(\frac{2}{3} - \frac{(X-2)^2}{27} - \frac{(X-2)^4}{243} \right) dX$$

$$= \left(\frac{2}{3}X - \frac{(X-2)^3}{81} - \frac{(X-2)^5}{1215} \right) \Bigg|_0^4$$

$$= \frac{8}{3} - \frac{16}{81} - \frac{64}{1215} = \frac{2936}{1215} = 2.4165$$

The exact answer to five significant figures is 2.3906. The difference between the approximate and exact answer is 1%.

6.3 Given the equation $dF/dt = F' = F$, it follows that $F''' = F'' = F$. Using the Taylor series through F''' to project the function, we have

$$F(X) = 1 + X + \frac{X^2}{2} + \frac{X^3}{6}$$

If this process is repeated n times, the predicted value of $F(nX)$ becomes

$$F(nX) = \left(1 + X + \frac{X^2}{2} + \frac{X^3}{6} \right)^n$$

The ratio of the predicted value to the true value will be

$$\frac{\left(1 + X + \frac{X^2}{2} + \frac{X^3}{6} \right)^n}{e^{nX}} = \left(\frac{1 + X + \frac{X^2}{2} + \frac{X^3}{6}}{e^X} \right)^n$$

For $X = 0.3$ and $n = 100$, the value of this ratio is 0.9738. Since in this simple case $F'' = F' = F$, the condition $\Delta t \ll |2F'/F''|$ becomes $\Delta t \ll 2$. This condition is certainly satisfied by $\Delta t = X = 0.1$. If we project with the Euler method 300 times using $X = 0.1$, the approximate solution will be $F(30) \cong (1 + 0.1)^{300}$. The ratio of the approximate to the exact solution in this case is 0.6256. In order to achieve the same accuracy as the third-order Taylor series projection for $F(30)$, we would need to take $X = 0.023$, which is 13 times smaller than 0.3.

6.4 The explanation is simple. When we say that the midpoint and trapezoid rule can exactly integrate a linear function and that Simpson's rule can exactly integrate a cubic polynomial function, the function in question is the integrand. When we say that the modified and improved Euler methods are second-order Runge–Kutta methods and that the classical Runge–Kutta method is a fourth-order method, the order in question refers to the order of the function f, where $F = df/dt$, and F rather than f is the function to be integrated. In other words, when we project a function using Runge–Kutta, the task amounts to estimating $f(T + \Delta T)$, where

$$f(T + \Delta T) = f(T) + \int_{T}^{T+\Delta T} \frac{df}{dt}\, dt$$

Obviously, if $f(t)$ is a fourth-order polynomial in t, then df/dt is a third-order polynomial in t, and if $f(t)$ is a second-order polynomial in t, then df/dt is a linear function of t. Thus the Runge–Kutta methods for making temporal projections are in fact comparable in accuracy to their counterparts for carrying out spatial integrations.

6.5 If $N' = rN(1 - N/K)$, then

$$N'' = rN'\left(1 - \frac{N}{K}\right) - N'\frac{rN}{K} = rN'\left(1 - \frac{2N}{K}\right)$$

Hence

$$\left|\frac{2N'}{N''}\right| = \left|\frac{2}{r\left(1 - \frac{2N}{K}\right)}\right|$$

The minimum value of $|2N'/N''|$ during the course of the integration depends on several factors. A graph of $|2/(1 - 2N/K)|$ as a function of N/K is shown below.

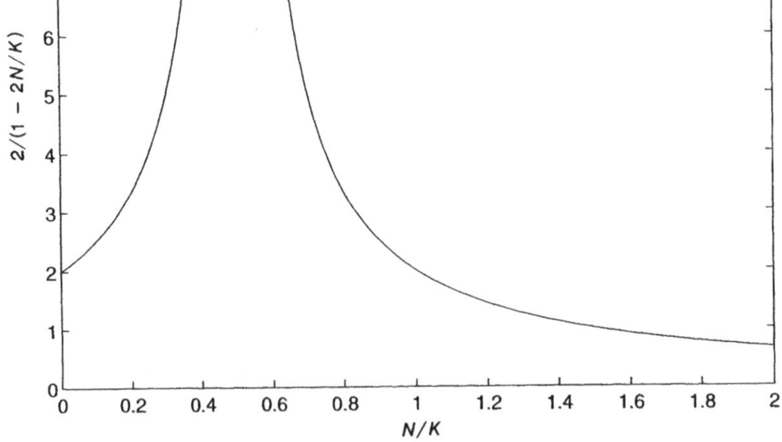

There are three possibilities if we do not assume that the integration proceeds until $N \to K$:

1. Both the initial and the final values of N are less than $0.5K$ or greater than K. In this case the minimum value of $|2N'/N''|$ occurs at $N = N_0$, that is, at the start of the integration. The minimum value is

$$\left|\frac{2}{r(2N_0/K - 1)}\right|$$

2. The initial (and hence final) value of N falls in the range $0.5K$ to K. In this case the minimum value of $|2N'/N''|$ occurs at the end of the integration. As $N \to K$, the value of $|2N'/N''| \to 2/r$.
3. The initial value of N is less than $0.5K$, and the final value of N lies in the range $0.5K$ to K. In this case the minimum value of $|2N'/N''|$ occurs either at the beginning or at the end of the integration. If N_f is the final value of N, the minimum value of $|2N'/N''|$ is the smaller of

$$\frac{2}{r\left(\dfrac{2N_f}{K} - 1\right)} \quad \text{and} \quad \frac{2}{r\left(1 - \dfrac{2N_0}{K}\right)}$$

If we assume that the integration proceeds until $N \to K$, then there are only two possibilities to be considered. If $N_0 > K$, the minimum value of $|2N'/N''|$ occurs at $N = N_0$ and equals

$$\frac{2}{r\left(\dfrac{2N_0}{K} - 1\right)}$$

If $N_0 < K$, the minimum value of $|2N'/N''|$ occurs at $N = K$ and equals $2/r$. For the nondimensionalized problem, the integration from $t = 0$ to $t = 10$ causes X to increase from 0.1 to almost 1. The minimum value of $|2X'/X''|$ is therefore 2, that is, the smaller of 2 and $2/(1 - 0.2) = 2.5$. Certainly 0.1 is much smaller than 2, but 0.5 is dubious and 1 is certainly not much smaller than 2. A graph of the results obtained by integrating the nondimensionalized finite-difference equation using the simple Euler method is shown on page 308. The results obtained with $\Delta\tau = 0.1$ (solid curve) are virtually identical with the exact solution (*). The results obtained with $\Delta\tau = 0.5$ (dashed curve) and $\Delta\tau = 1$ (dash–dot curve) deviate from the exact solution by progressively larger amounts for intermediate values of X, but all three finite-difference curves settle down to the correct equilibrium value as $\tau \to 10$. The difference between the results here and in Exercise 6.3 is the fact that in this case the

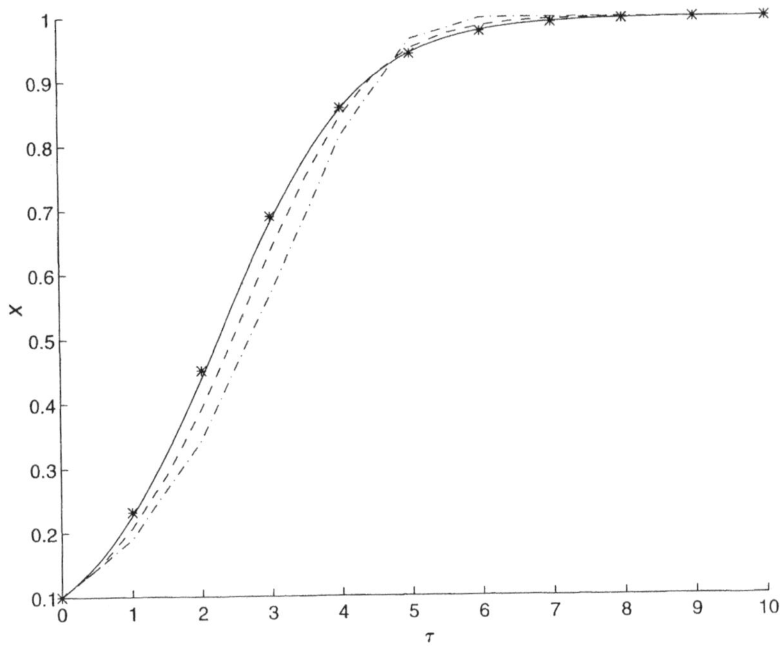

function is not constantly accelerating. The function accelerates for $X < 0.5$, during which time the Euler projections underestimate the true value of the function, as in Exercise 6.3. However, for $0.5 < X < 1$ the function is decelerating, and within this range of X the Euler projections tend to overestimate the change in the function value. This allows the Euler projections to catch up with the true function and eventually exceed it by a small amount before settling down to the equilibrium value.

6.6 The nonzero equilibrium values are $X = e/f = 0.5/0.02 = 25$ and $Y = a/c = 1/0.1 = 10$. The initial values of X' and Y' are 12 and -0.4, respectively. The expressions for X'' and Y'' are as follows:

$$X'' = (a - cY)X' - cY'X = \left[(a - cY)^2 - c(-e + fX)Y\right]X$$

$$Y'' = (-e + fX)Y' + fX'Y = \left[(-e + fX)^2 + f(a - cY)X\right]Y$$

The initial values of X'' and Y'' are 8 and 1, respectively. Hence $|2X'/X''|$ and $|2Y'/Y''|$ are 3 and -0.8, respectively. Given this information, a Δt of 0.04 is probably a reasonable time step. The results of the numerical integrations are shown in the next four figures. The first two figures show the simple Euler results, first with the prey and predator populations plotted against each other and second with both plotted against time. The plot of prey versus predator clearly shows the absence of a stable limit cycle. X and Y oscillate around the equilibrium values of 25 and 10, but the amplitude of the oscillations increases with increasing time. The second figure shows the same results with X and Y plotted against time. The third and fourth figures show the analogous results from the Runge–Kutta integration. The third figure seems to clearly show a stable limit cycle, with the prey oscillating between roughly 7 and 65 and the predator oscillating between 4 and 20. The fourth figure shows the same results with X and Y plotted against time.

The explanation for the difference in the results is that we are dealing with a function which goes through a series of maxima and minima. Whenever we approach the region of a maximum or minimum of a function F, $|2F'/F''| \to 0$, and hence it is impossible for Δt to satisfy the condition $\Delta t \ll |2F'/F''|$. The result is that the simple Euler finite-difference solution tends to overshoot (i.e., does not curve enough) each time the function passes through a maximum or minimum. The result is that a function which is supposed to be oscillating with a constant amplitude in fact oscillates with a constantly increasing amplitude. The fourth-order Runge–Kutta overcomes this problem by including additional terms in the Taylor series expansion.

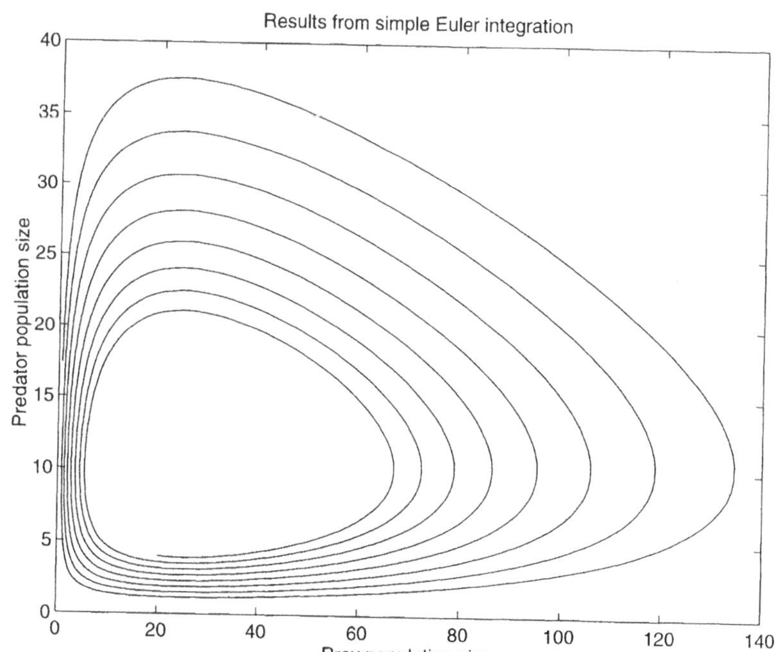

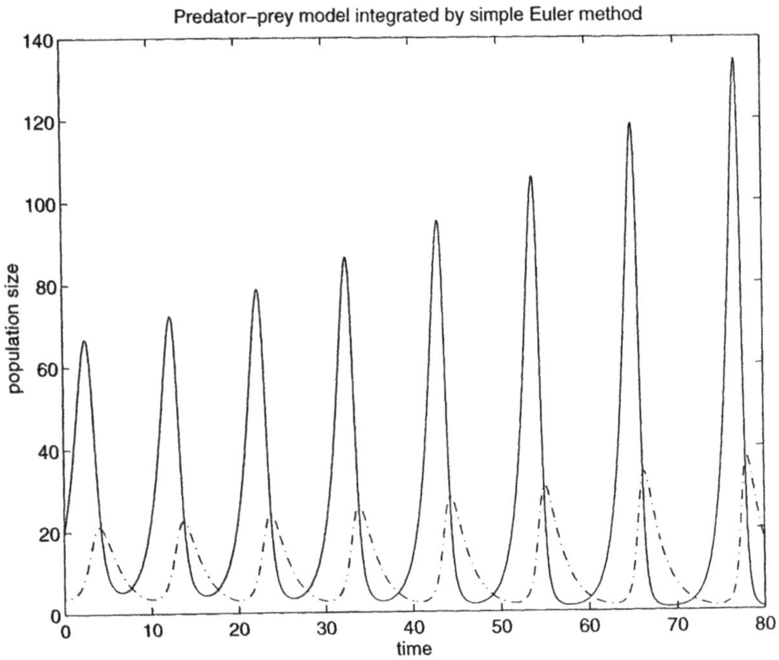
Predator–prey model integrated by simple Euler method
population size
time
0
20
40
60
80
100
120
140
0
10
20
30
40
50
60
70
80

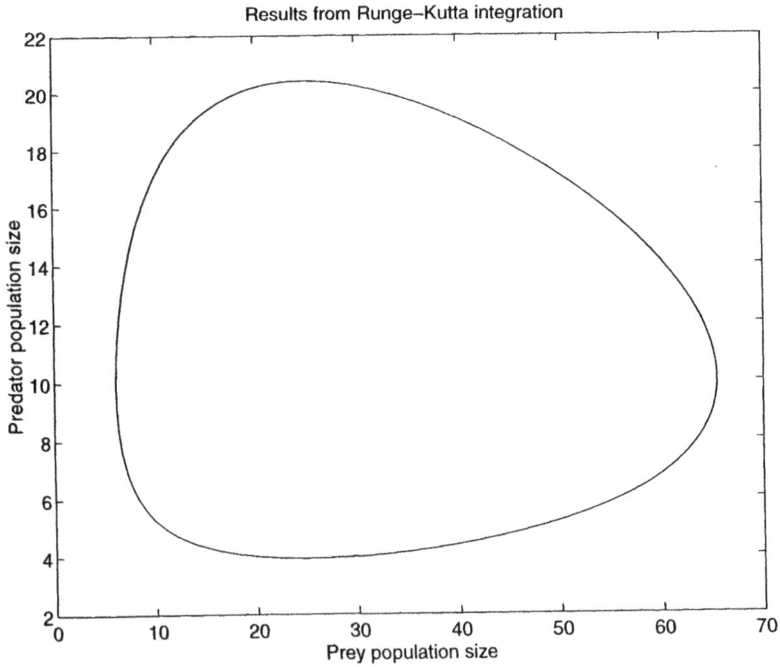
Results from Runge–Kutta integration
Predator population size
Prey population size
2
4
6
8
10
12
14
16
18
20
22
0
10
20
30
40
50
60
70

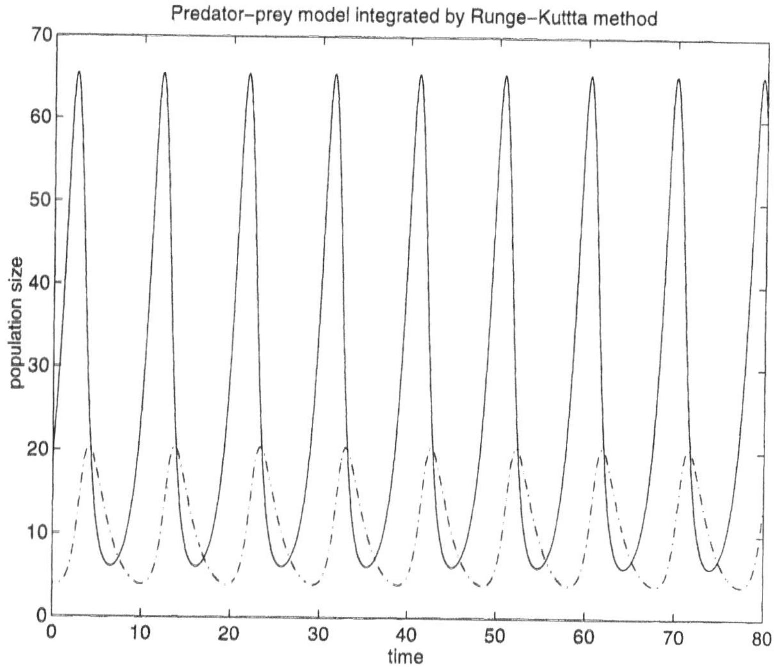

6.7 Multiplying the first equation by X_1 and subtracting it from the second equation gives

$$W_2(X_2 - X_1) = M_1 - X_1 M_0 \tag{5}$$

Multiplying the second equation by X_1 and subtracting it from the third equation gives

$$W_2 X_2(X_2 - X_1) = M_2 - X_1 M_1 \tag{6}$$

Multiplying the third equation by X_1 and subtracting it from the fourth equation gives

$$W_2 X_2^2(X_2 - X_1) = M_3 - X_1 M_2 \tag{7}$$

Dividing Equation 6 by Equation 5 gives

$$X_2 = \frac{M_1 X_1 - M_2}{M_0 X_1 - M_1} \tag{8}$$

Dividing Equation 7 by Equation 6 gives

$$X_2 = \frac{M_2 X_1 - M_3}{M_1 X_1 - M_2} \tag{9}$$

Equating the right-hand sides of Equations 8 and 9 and rearranging gives

$$(M_1 X_1 - M_2)^2 = (M_0 X_1 - M_1)(M_2 X_1 - M_3) \tag{10}$$

Rearranging this equation leads directly to

$$(M_0 M_2 - M_1^2) X_1^2 + (M_1 M_2 - M_0 M_3) X_1 + (M_1 M_3 - M_2^2) = 0 \tag{11}$$

Regardless of which root of Equation 11 we choose to assign to X_1, it follows that X_2, as defined by either Equation 8 or 9, is the other complementary root. This follows from the fact that the product of the two roots must equal $(M_1 M_3 - M_2^2)/(M_0 M_2 - M_1^2)$ and their sum must equal $(M_0 M_3 - M_1 M_2)/(M_0 M_2 - M_1^2)$.

6.8 The answers to three significant figures are as follows:

Trapezoid	505
Midpoint	492
Legendre–Gauss	497
Exact	496

CHAPTER 7

7.1 One would expect river runoff to be positively correlated with the temperature of the lower atmosphere because evaporation rates are positively correlated with temperature. In short, an increase in temperature would speed up the hydrologic cycle.

7.2 An alternative formulation that does not involve exponentiating the function $RCO_2/(1 + RCO_2)$ would be

$$0.65 + 0.35 \frac{2RCO_2}{1 + RCO_2}$$

7.3 By eliminating F_{bg} from Equations 7.4 and 7.5, it is straightforward to show that

$$F_{bc} - F_{wc} = -\frac{y_c - F_{wc}(\delta_{bc} - \delta_c)}{\alpha_c}$$

where

$$y_c = (F_{mc} + F_{wg} + F_{mg})(\delta_{bc} - \alpha_c) - \delta_c F_{mc} - \delta_g(F_{wg} + F_{mg})$$

Substituting for $F_{bc} - F_{wc}$ in Equation 7.14, we obtain

$$-\frac{y_c + F_{wc}(\delta_{bc} - \delta_c)}{\alpha_c} = \frac{f_B y_B}{\alpha_c}$$

where

$$y_B = \alpha_c f_R f_E f_D^{0.65} F_{wsi}(0)$$

Substituting the right-hand side of Equation 7.6 for F_{wc} then gives

$$y_b f_B + y_{bb} f_{BB} + y_c = 0$$

where

$$y_b = (\delta_{bc} - \delta_c) f_{LA} f_D f_E k_{wc} C$$

To derive Equation 7.43, we rewrite Equations 7.4 and 7.5 as follows:

$$F_{mg} + F_{mc} = F_{bc} + F_{bg} - F_{wc} - F_{wg}$$

$$\delta_g F_{mg} + \delta_c F_{mc} = \delta_{bc} F_{bc} + (\delta_{bc} - \alpha_c) F_{bg} - \delta_c F_{wc} - \delta_g F_{wg}$$

Multiplying the first of these equations by δ_c and subtracting from the second gives

$$F_{mg}(\delta_g - \delta_c) = F_{bc}(\delta_{bc} - \delta_c) + F_{bg}(\delta_{bc} - \alpha_c - \delta_c) - F_{wg}(\delta_g - \delta_c)$$

Dividing both sides of this equation by $(\delta_g - \delta_c)$ then leads directly to Equation 7.43.

7.4 The results are shown in the next five graphs.

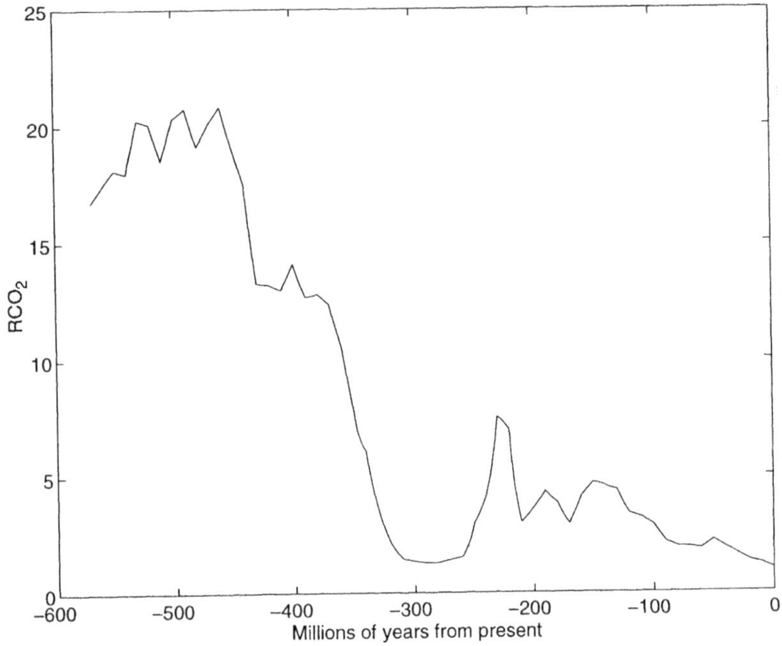

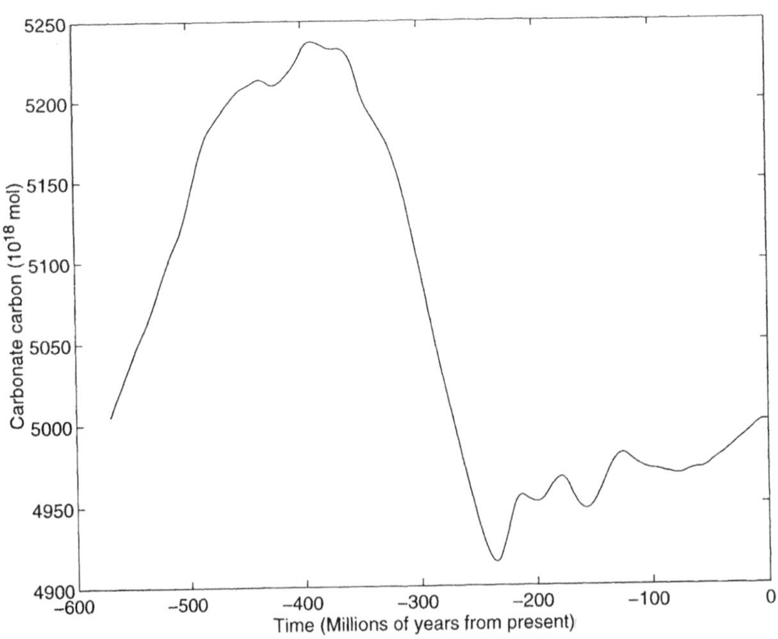

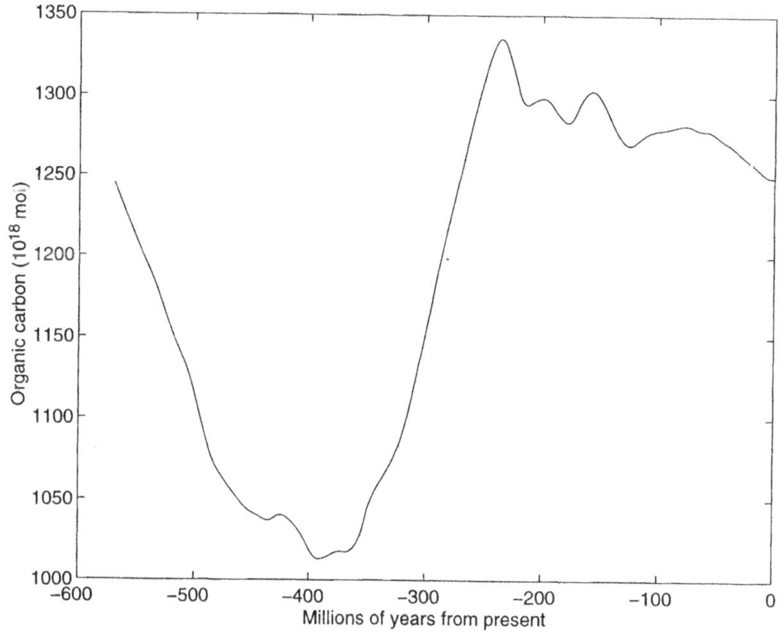
1350
1300
1250
1200
1150
1100
1050
1000
Organic carbon (10^18 mol)
-600
-500
-400
-300
-200
-100
0
Millions of years from present

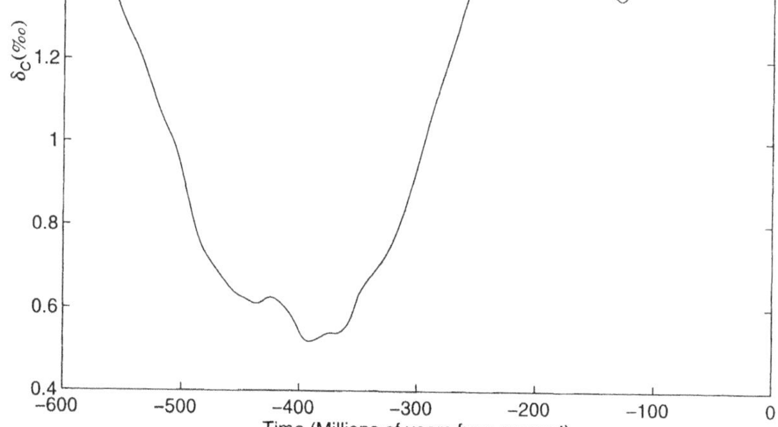
1.8
1.6
1.4
1.2
1
0.8
0.6
0.4
δC(‰)
-600
-500
-400
-300
-200
-100
0
Time (Millions of years from present)

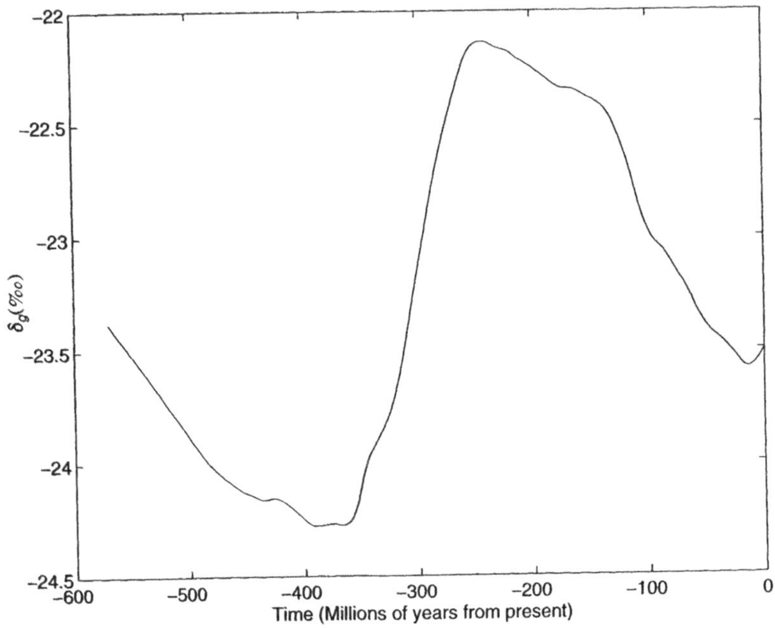

7.5 The graph is shown below. The dash–dot line is the result if R_{CO_2} is assumed to equal 1 at all times. According to the model, the Earth

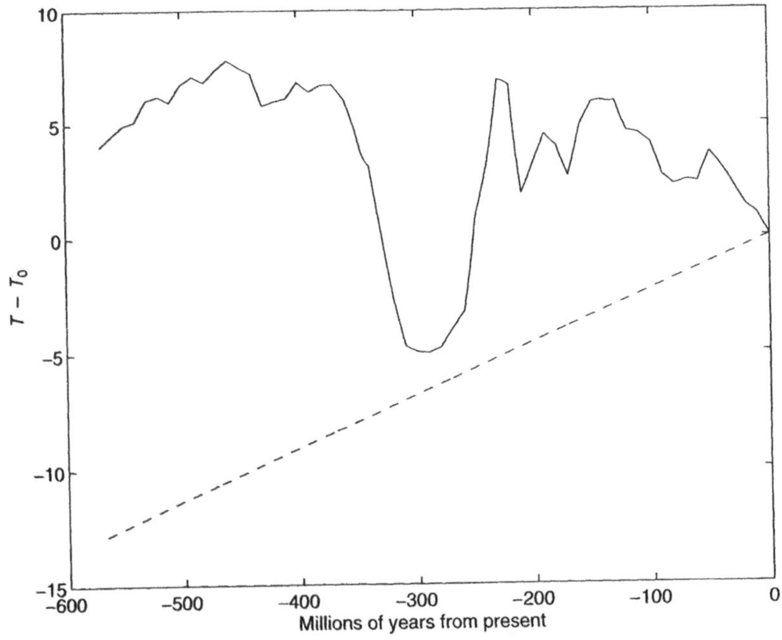

would have been 12.9° C cooler 570 million years ago if the CO_2 concentration in the atmosphere had been the same then as now. This is, of course, the paradox of the pale sun. According to the model, the earth was in fact about 5° C warmer 570 million years ago, because the CO_2 concentration in the atmosphere at that time was more then 15 times what it is now.

7.6 **(a)** The graph for the case $W_s = 0$ is shown below. The buildup of CO_2 in earlier times is less than the estimates made by the standard model. The explanation is that the rate of weathering, which consumes atmospheric CO_2, is positively correlated with temperature as discussed in the text. If the output from the sun were the same 570 million years ago as it is today, then the temperature of the earth, according to the model, would have been about 10° C warmer than it is today, and weathering processes would have consumed CO_2 at a much higher rate.

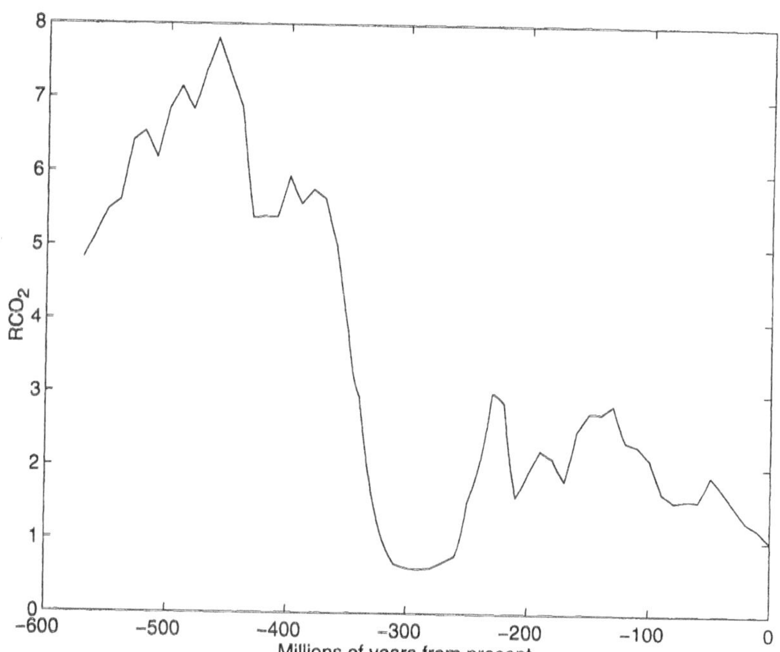

(b) The graph is shown in the upper panel on page 318. Here again, the buildup of CO_2 is much less in earlier times than predicted by the standard model. The explanation again relates to weathering, which consumes atmospheric CO_2. If land plants have not had the

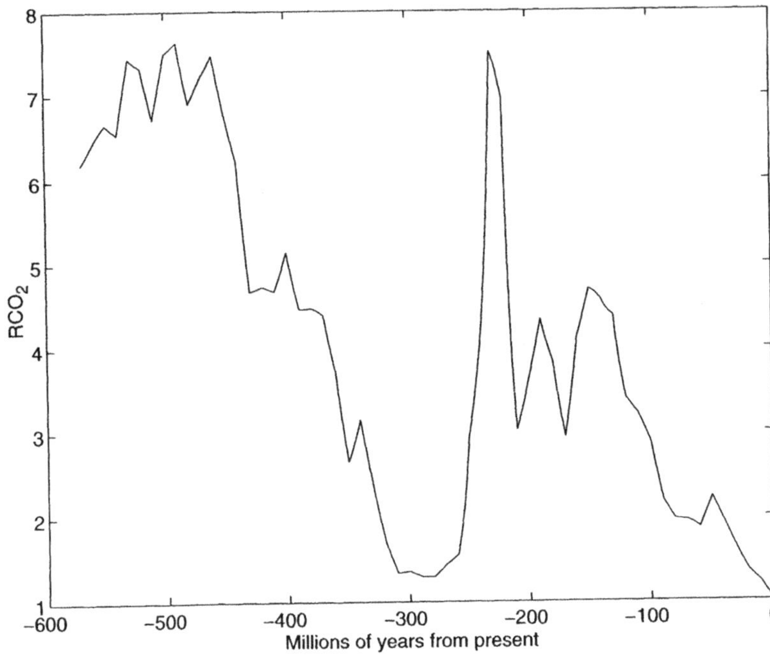
RCO2
8
7
6
5
4
3
2
1
-600
-500
-400
-300
-200
-100
0
Millions of years from present

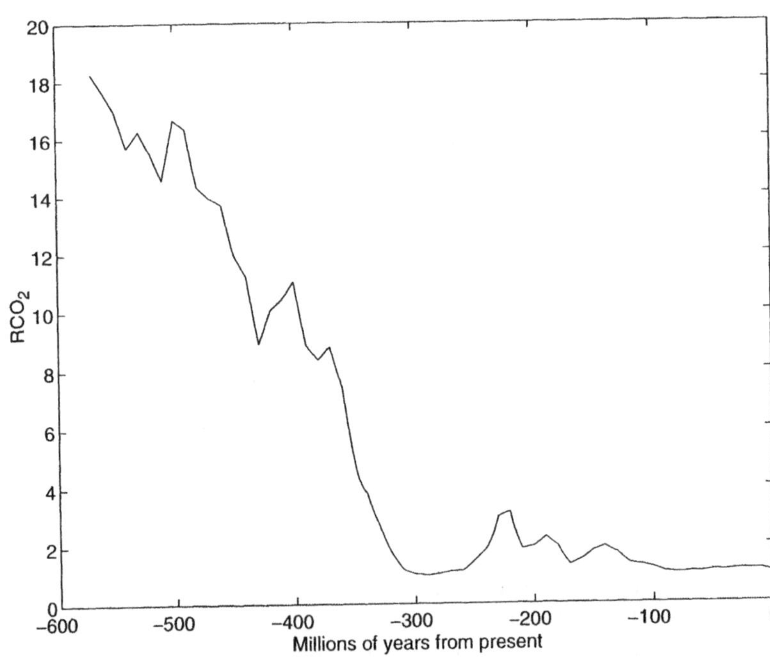
RCO2
20
18
16
14
12
10
8
6
4
2
0
-600
-500
-400
-300
-200
-100
0
Millions of years from present

postulated impact on weathering rates, the implication is that weathering rates in times past were not a great deal less than they are now; that is, they were substantially greater in times past than postulated in the standard model.

(c) The graph is shown in the lower panel on page 318. The results in this case are not dramatically different from the standard model. The reason is that in the standard model the degassing rate due to tectonics is assumed to have been greater at almost all times in the past ($f_G > 1$). Assuming $f_G = 1$ at all times therefore implies less degassing than in the standard model. On the other hand, in the standard model mean continental relief is assumed to have been less at almost all times in the past ($f_R < 1$). Assuming $f_R = 1$ at all times therefore implies a higher rate of release of CO_2 to the ocean–atmosphere–biosphere system via weathering of organic matter. The two effects therefore tend to cancel out.

7.7 The trick here is to pick off the extra terms in the expression for L that do not appear in the case $m = n - 1$. The extra terms in L are $-R_n f_n - k$(extra terms in X_{n+1}). The extra terms in X_{n+1} are

$$\frac{1}{D + k}(S_n f_n + \text{extra terms in } F)$$

The extra term in F is $f_2(Q_n - 1)\prod_{i=2}^{n-1} Q_i$. Putting all this together, we see that the extra terms in L are as follows:

$$-R_n f_n - \frac{k}{D + k} S_n f_n - \frac{k}{D + k} f_2 (Q_n - 1) \prod_{i=2}^{n-1} Q_i$$

We now note that

$$f_n = \frac{F}{Q_n} = f_2 \prod_{i=2}^{n-1} Q_i$$

and that $S_n = 1 - Q_n - R_n$. It follows that the extra term in the expression for L is

$$-f_2 \prod_{i=2}^{n-1} Q_i \left[R_n + \frac{k}{D + k}(1 - R_n - Q_n + Q_n - 1) \right]$$

$$= -R_n f_2 \frac{D}{D + k} \prod_{i=2}^{n-1} Q_i$$

The extra term in the expression for the f ratio is the extra term in L divided by f_2 and hence equals

$$-\frac{D}{D+k}R_n\prod_{i=2}^{n-1}Q_i$$

We therefore have

$$f \text{ ratio} = \left(\frac{D}{D+k}\right)\left(1-\sum_{j=3}^{n-1}R_j\prod_{k=2}^{j-1}Q_k\right) - \frac{D}{D+k}R_n\prod_{i=2}^{n-1}Q_i$$

$$= \left(\frac{D}{D+k}\right)\left(1-\sum_{j=3}^{n}R_j\prod_{k=2}^{j-1}Q_k\right)$$

7.8 If $I = I_0 e^{-kz}$, the average light intensity between the surface and depth Z is given by the equation

$$\bar{I} = \frac{1}{Z}\int_0^Z I_0 e^{-kz}\,dz = \frac{I_0}{kZ}(1-e^{-kZ})$$

The value of kz when $I = 0.001 I_0$ is $-\ln(0.001) = 6.9$. Hence

$$\bar{I} = \frac{I_0}{6.9}(1-e^{-6.9}) = 0.145 I_0$$

7.9 The equilibrium point is not globally stable if we allow negative values of X. This, of course, makes no sense biologically, but the question is certainly a legitimate one mathematically. Once X becomes negative, both terms in the logistic equation, rX and $-r(X^2/K)$, become negative, and X becomes even more negative.

7.10 The nonzero equilibrium values of X and Y are $X = 1$ and $Y = 4$. The community matrix is $\begin{pmatrix} -2 & -1 \\ 16 & -12 \end{pmatrix}$. The eigenvalues of the community matrix are -10 and -4. Eigenvectors corresponding to these eigenvalues are $\begin{pmatrix} 1 \\ 8 \end{pmatrix}$ and $\begin{pmatrix} 1 \\ 2 \end{pmatrix}$, respectively, or any scalar times these vectors.

The condition for local stability in the finite-difference situation is that both eigenvalues be negative (which they are) and that both eigenvalues satisfy the condition $\lambda\,\Delta t > -2$. Since the eigenvalues in this case are -10 and -4, this latter condition implies $\Delta t < 0.2$ and $\Delta t < 0.5$, respectively. The former condition is the stricter of the two.

Plots of X and Y versus time are shown below for $\Delta t = 0.1$, 0.15, and 0.2 with initial values of $X = 2$ and $Y = 3$. For $\Delta t = 0.1$ and 0.15, X and Y settle down to their equilibrium values of 1 and 4. This is also the case if we integrate with a Δt of 0.2 using the fourth-order

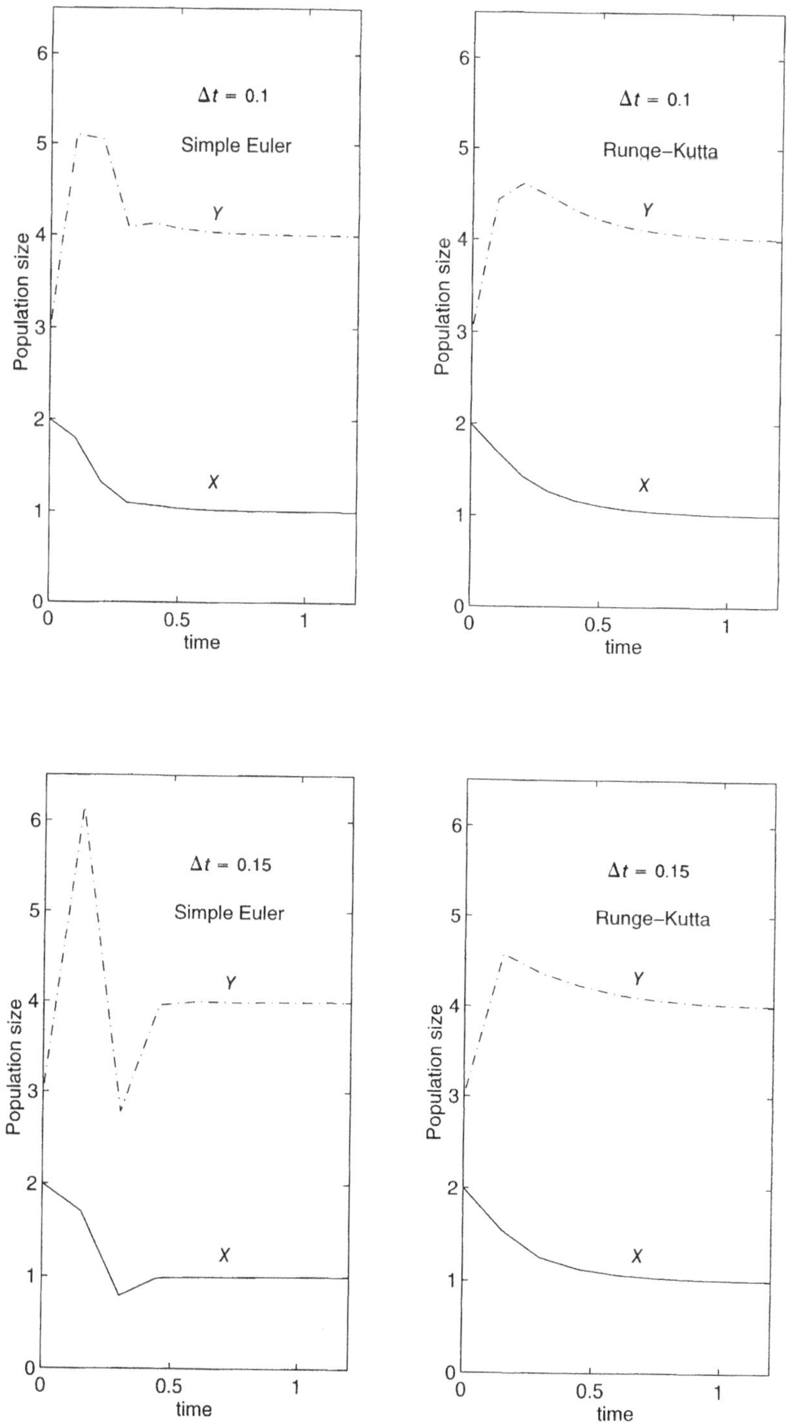

Δt = 0.1
Simple Euler
Y
X
Population size
time
0
0.5
1
2
3
4
5
6
Δt = 0.1
Runge–Kutta
Y
X
Population size
time
Δt = 0.15
Simple Euler
Y
X
Population size
time
Δt = 0.15
Runge–Kutta
Y
X
Population size
time

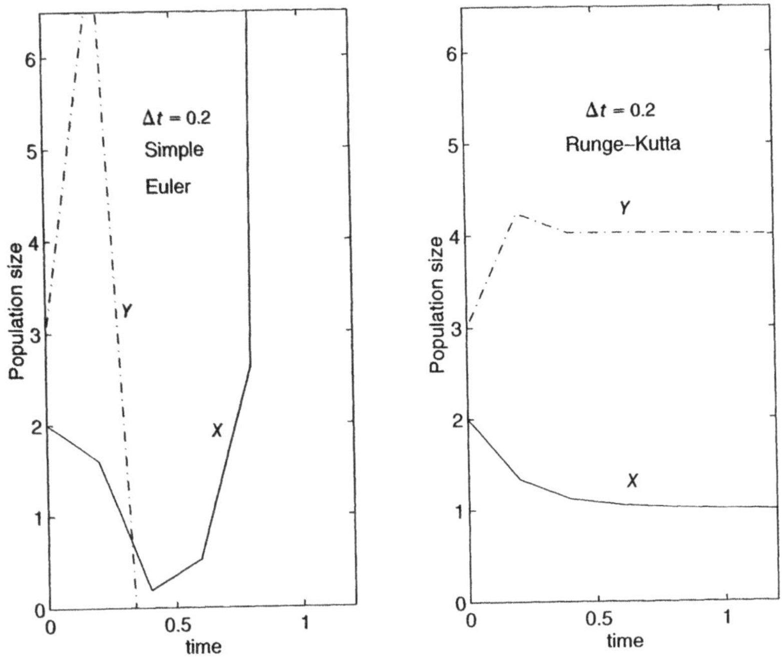

Runge–Kutta. However, X and Y blow up if we try to integrate by simple Euler using a Δt of 0.2. This is, of course, consistent with our analysis of the stability characteristics of the simple finite-difference analog of the differential equations describing this system.

7.11 The community matrix is

$$\begin{pmatrix} -1.0119 & -0.1635 & 0.1373 & 0.1380 & 0.1313 \\ 1.0119 & 0 & -0.2615 & 0 & 0 \\ 0 & 0.0681 & 0 & -0.2760 & 0 \\ 0 & 0 & 0.0229 & 0 & 0 \\ 0 & 0.0954 & 0.1013 & 0.1380 & -0.2268 \end{pmatrix}$$

The eigenvalues are as follows:

$$\begin{array}{l} -0.7565 \\ -0.3625 \\ -0.0414 + 0.0781i \\ -0.0414 - 0.0781i \\ -0.0369 \end{array}$$

7.12 We know that $\text{frac}_m = 1 - (P_{m-1}/X_{m-1})$. If X_{m-1} and frac_m are input parameters to the model, then it follows that $P_{m-1} = X_{m-1}(1 -$

frac_m). From Equation 7.66 it follows that

$$\frac{P_m}{X_m} = 1 - \text{frac}_{m+1} = 1 - \frac{L}{A_{m+1}X_{m+1}(f\ \text{ratio})}\prod_{k=1}^{m} Q_k$$

Assuming the f ratio, L, Q_k, frac_m, and X_{m+1} are input to the model, it follows that

$$A_{m+1} = \frac{L}{\text{frac}_{m+1}X_{m+1}(f\ \text{ratio})}\prod_{k=1}^{m} Q_k$$

7.13 The real parts of the eigenvalues with the least negative real parts are as follows:

Fractional Growth Rate	Real Part of Eigenvalue
0.3	−0.0541
0.4	−0.0453
0.5	−0.0369
0.6	−0.0266
0.7	−0.0219
0.8	0.0027

A graph of these is shown below. The system becomes less stable as the fraction of maximum growth rate increases, and becomes unstable when that fraction reaches approximately 0.8.

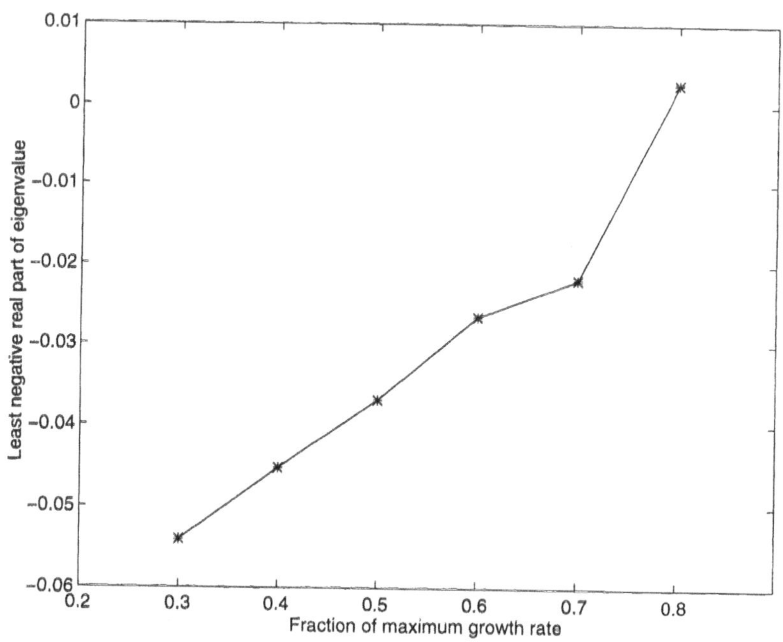

7.14 The results of the integrations for $\Delta t = 1$, 2, and 3 are shown below. In the first two cases the functions all settle down nicely to their equilib-

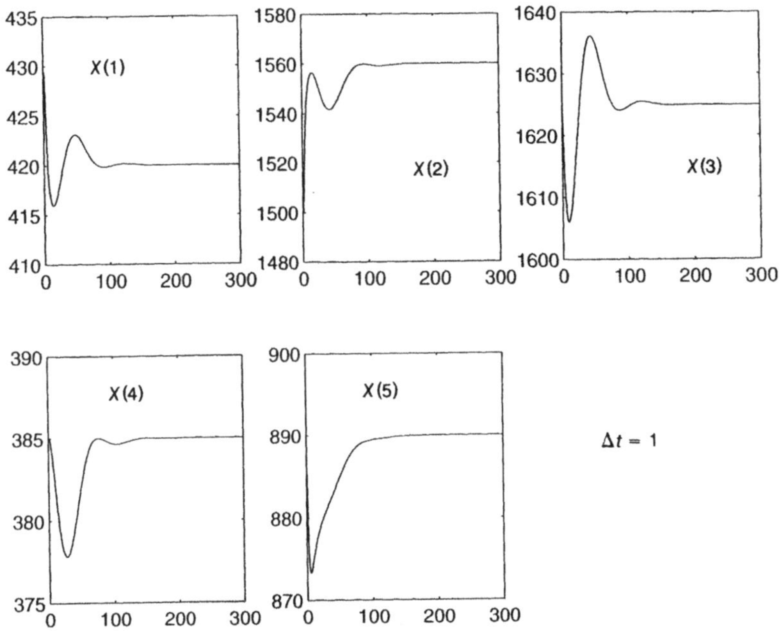

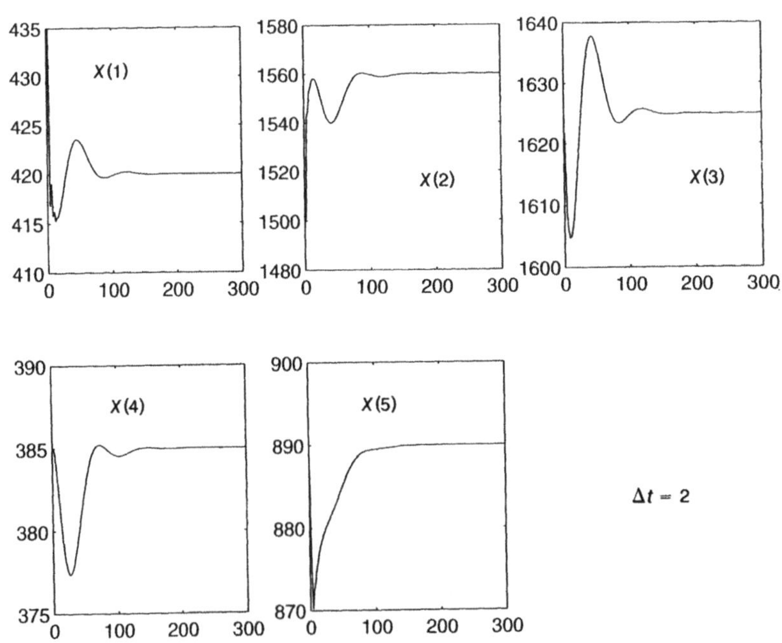

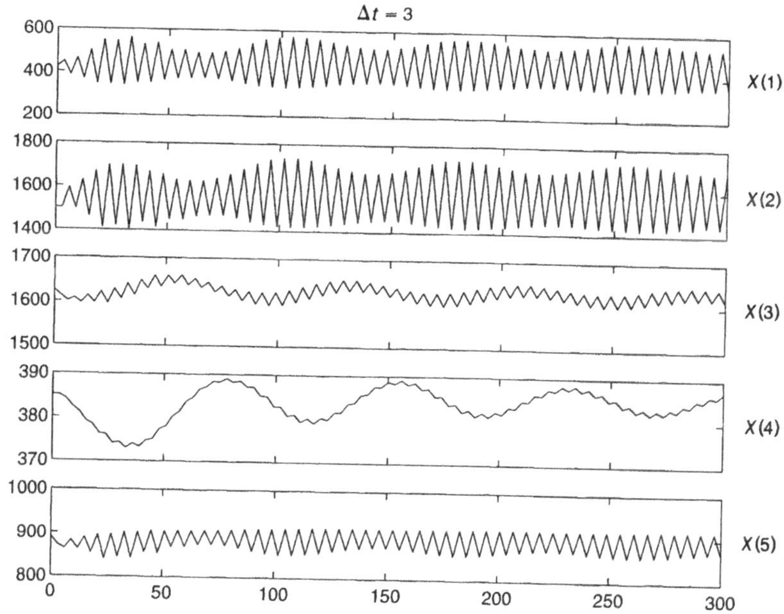

rium values, and the behavior of the functions is very similar in both cases. For $\Delta t = 3$ the functions never settle down. The reason for the behavior at $\Delta t = 3$ is, of course, due to the fact that one of the eigenvalues of the community matrix is -0.7565. The stability criterion that $\alpha\,\Delta t > -2$ is not satisfied for this eigenvalue, because $(-0.7565)(3) = -2.2695 < -2$.

7.15 In this case

$$\text{frac}_m = \left(1 - \frac{P_{m-1}}{X_{m-1}}\right)^n$$

Hence, as before, if X_{m-1} and frac_m are input parameters to the model, it follows that $P_{m-1} = X_{m-1}\sqrt[n]{1 - \text{frac}_m}$. The analog of Equation 7.66 becomes

$$\text{frac}_{m+1} = \frac{L}{A_{m+1}X_{m+1}(f\ \text{ratio})}\prod_{k=1}^{m} Q_k$$

This follows from the fact that

$$f_{m+1} = f_2\prod_{k=1}^{m} Q_k = \frac{L}{(f\ \text{ratio})}\prod_{k=1}^{m} Q_k$$

and $f_{m+1} = \text{frac}_{m+1} A_{m+1} X_{m+1}$. Assuming the f ratio, L, Q_k, frac_m, and X_{m+1} are input to the model, it follows that

$$A_{m+1} = \frac{L}{\text{frac}_{m+1} X_{m+1}(f \text{ ratio})} \prod_{k=1}^{m} Q_k$$

To derive Equation 7.102, it is only necessary to differentiate $[1 - (P_{m-1}/X_{m-1})]^n$ with respect to X_{m-1}. The result is

$$\frac{\partial}{\partial X_{m-1}} \left(1 - \frac{P_{m-1}}{X_{m-1}}\right)^n = n\left(1 - \frac{P_{m-1}}{X_{m-1}}\right)^{n-1} \frac{P_{m-1}}{X_{m-1}^2}$$

$$= \left(1 - \frac{P_{m-1}}{X_{m-1}}\right)^n \left[n \frac{P_{m-1}}{X_{m-1}^2} \left(\frac{1}{1 - \dfrac{P_{m-1}}{X_{m-1}}} \right) \right]$$

$$= \left(1 - \frac{P_{m-1}}{X_{m-1}}\right)^n \left(n \frac{P_{m-1}}{X_{m-1}(X_{m-1} - P_{m-1})} \right)$$

Shown below is the graph obtained by repeating Exercise 7.13 with $n = 0.5$, 1, and 2. With the exception of the results when the growth

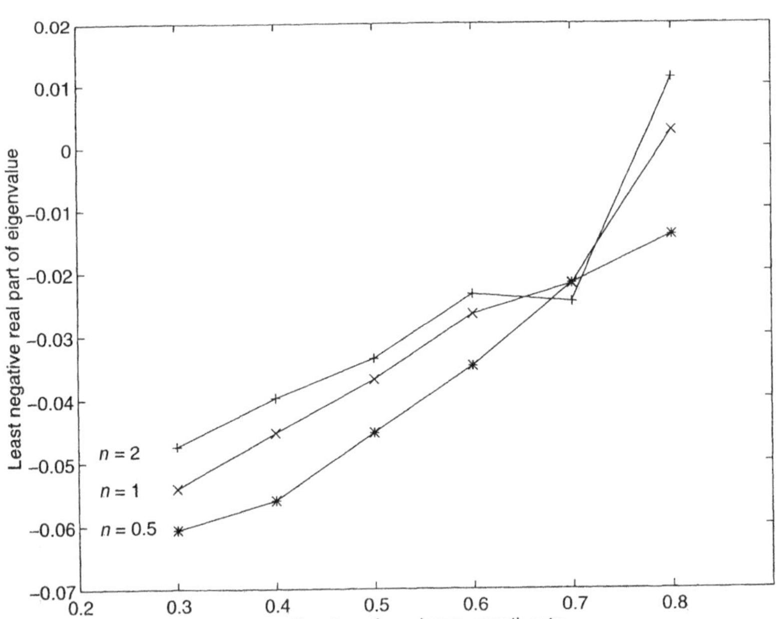

rates are 70% of their maximum values, the results all indicate that the smaller the value of n, the more stable the steady state. In other words, hyperbolas that rise rapidly normally confer more stability on the system than hyperbolas that rise gradually.

7.16 The eigenvalues are given in the question. The elements of the community matrix are as follows:

$$\begin{pmatrix} -1.0019 & -0.1635 & 0.1373 & 0.1863 & 0.0483 & 0.1104 \\ 1.0019 & 0 & -0.2615 & 0 & 0 & 0 \\ 0 & 0.0681 & 0 & -0.2760 & 0 & 0 \\ 0 & 0 & 0.0229 & 0 & -0.0966 & 0 \\ 0 & 0 & 0 & 0.0338 & 0 & 0 \\ 0 & 0.0954 & 0.1013 & 0.0559 & 0.0483 & -0.2059 \end{pmatrix}$$

An important point for the purposes of solving this problem is that the values of k and M will not be the values given in Table 7.4 but instead must be recalculated using Equations 7.67 and 7.68, respectively.

CHAPTER 8

8.1 The answers are obtained by plugging into Equations 8.10 and 8.11:

$$\mu(t) = \alpha + \beta E(t)$$

$$R(t, \Delta t) = \alpha(\alpha + \beta \Delta t) + \beta(2\alpha + \beta \Delta t) E(t) + \beta^2 E(t^2)$$

8.2 In the case of a sine wave, the process will be stationary only if the phase angle is random. Given a random phase angle, the process will then be ergodic only if the amplitude and period are constant. Clearly $\mu_x(j)$ will be 0 for all j, regardless of whether the phase angle, amplitude, and period are random or constant. However, if the amplitude and/or period are random, then $C_{xx}(\Delta t, j)$ will depend on j. The following table summarizes these conclusions.

Amplitude	Phase Angle	Period	Stationary	Ergodic
Constant	Constant	Random	No	No
Constant	Random	Constant	Yes	Yes
Random	Constant	Constant	No	No
Constant	Random	Random	Yes	No
Random	Constant	Random	No	No

8.3 Let $\Delta(f)$ be the Fourier transform of $\delta(t)$. Then, by definition,

$$\Delta(f) = \int_{-\infty}^{\infty} \delta(t) \exp(-2\pi i f t)\, dt = e^0 = 1$$

8.4

$$\delta(t) = \int_{-\infty}^{\infty} \Delta(f)\exp(2\pi ift)\,df = \int_{-\infty}^{\infty} \exp(2\pi ift)\,df$$

8.5

$$G * H(f) = \int_{-\infty}^{\infty} G(x)H(f-x)\,dx$$

$$= \int_{-\infty}^{\infty} dx \int_{-\infty}^{\infty} g(y)\exp(-2\pi ixy)\,dy \int_{-\infty}^{\infty} h(z)\exp[-2\pi i(f-x)z]\,dz$$

$$= \int_{-\infty}^{\infty} dy \int_{-\infty}^{\infty} g(y)h(z)\exp(-2\pi ifz)\,dz \int_{-\infty}^{\infty} \exp[2\pi ix(z-y)]\,dx$$

$$= \int_{-\infty}^{\infty} dy \int_{-\infty}^{\infty} g(y)h(z)\exp(-2\pi ifz)\,\delta(z-y)\,dz$$

$$= \int_{-\infty}^{\infty} g(z)h(z)\exp(-2\pi ifz)\,dz$$

8.6 Using the fact that $e^{ikx} = \cos(kx) + i\sin(kx)$ and letting $x = 2\pi n/N$, we have

$$\sum_{k=-(N-1)/2}^{(N-1)/2} G_k \exp(ikx)$$

$$= G_0 + \sum_{k=-(N-1)/2}^{-1} G_k[\cos(kx) + i\sin(kx)]$$

$$+ \sum_{k=1}^{(N-1)/2} G_k[\cos(kx) + i\sin(kx)]$$

$$= G_0 + \sum_{k=1}^{(N-1)/2} \{G_{-k}[\cos(kx) - i\sin(kx)] + G_k[\cos(kx) + i\sin(kx)]\}$$

$$= G_0 + \sum_{k=1}^{(N-1)/2} \{(G_k + G_{-k})\cos(kx) + i(G_k - G_{-k})\sin(kx)\}$$

8.7 The Fourier transform of the product of two functions g and h is given by

$$(GH)_m = \frac{1}{N}\sum_n g(n\,\Delta t)h(n\,\Delta t)\exp\left(-\frac{2\pi inm}{N}\right)$$

where the summation here and in the following equations is understood to extend from $-(N-1)/2$ to $(N-1)/2$. Using Equation 8.55, we see that this equation becomes

$$(GH)_m = \frac{1}{N}\sum_n \sum_p \sum_s G_p H_s \exp\left[\frac{2\pi i n(p+s-m)}{N}\right]$$

$$= \frac{1}{N}\sum_p \sum_s G_p H_s \sum_n \exp\left[\frac{2\pi i n(p+s-m)}{N}\right]$$

$$= \sum_p G_p H_{m-p}$$

8.8 If $|f_a' - n/\Delta t| \leq 1/(2\,\Delta t)$, where n is some integer, then $|f_a'\,\Delta t - n| \leq 1/2$. The second inequality says that $f_a'\,\Delta t$ differs from an integer n by less than or equal to 1/2. It follows that the integer n equals $f_a'\,\Delta t$ rounded to the nearest integer.

8.9 The four values of f_a' are 24/25.8, 24/12.4, 24/24.1, and 24/12.05 per day. Multiplying these frequencies by $\Delta t = 17$ days and rounding to the nearest integer gives n values of 16, 33, 17, and 34, respectively. The four aliased frequencies are therefore $|24/25.8 - 16/17|$, $|24/12.4 - 33/17|$, $|24/24.1 - 17/17|$, and $|24/12.05 - 34/17|$. These frequencies are 0.0109, 0.0057, 0.0041, and 0.0083 per day and correspond to periods of $91\frac{3}{8}$, $175\frac{2}{3}$, 241, and $120\frac{1}{2}$ days, respectively.

8.10 Since $\text{Box}(t/\tau) = 1$ for $-1/2 < t/\tau < 1/2$ and 0 elsewhere, it follows that $\text{Box}(t/\tau) = 1$ for $-\tau/2 < t < \tau/2$ and 0 elsewhere. The Fourier transform of $\text{Box}(t/\tau)$ therefore equals

$$\int_{-\tau/2}^{\tau/2} \exp(-2\pi i f t)\, dt = 2\int_0^{\tau/2} \cos(2\pi f t)\, dt$$

$$= \frac{1}{\pi f}\sin(2\pi f t)\Big|_0^{\tau/2} = \frac{1}{\pi f}\sin(\pi f \tau)$$

8.11 The Fourier transform of $\cos(2\pi f_0 t)$ equals

$$\int_{-\infty}^{\infty} \cos(2\pi f_0 t)\exp(-2\pi i f t)\, dt$$

$$= \frac{1}{2}\int_{-\infty}^{\infty} [\exp(2\pi i f_0 t) + \exp(-2\pi i f_0 t)]\exp(-2\pi i f t)\, dt$$

$$= \frac{1}{2}\int_{-\infty}^{\infty} \{\exp[2\pi i(f_0 - f)t] + \exp[-2\pi i(f_0 + f)t]\}\, dt$$

Now letting $x = -t$, we see that the Fourier transform of $\cos(2\pi f_0 t)$ equals

$$\frac{1}{2}\int_{-\infty}^{\infty}\{\exp[2\pi i(f - f_0)x] + \exp[2\pi i(f_0 + f)x]\}\, dx$$

$$= \frac{1}{2}[\delta(f - f_0) + \delta(f + f_0)]$$

8.12 The result follows from the results of Exercises 8.10 and 8.11 and the convolution theorem. First, we note that Black(t) is equal to Box(t/τ) times $0.42 + 0.5\cos(2\pi f_0 t) + 0.08\cos(4\pi f_0 t)$, where $f_0 = 1/\tau$. We have already shown in Exercise 8.10 that the Fourier transform of Box(t/τ) $= \tau\,\text{sinc}(f\tau)$. For the other two terms, we make use of the fact that the Fourier transform of the product of two functions is equal to the convolution of their Fourier transforms. Hence the Fourier transform of Box(t/τ) times $\cos(2\pi f_0 t)$ equals

$$\int_{-\infty}^{\infty}\tau\,\text{sinc}(s\tau)\frac{1}{2}[\delta(f - s + f_0) + \delta(f - s - f_0)]\, ds$$

$$= \frac{\tau}{2}\{\text{sinc}[(f + f_0)\tau] + \text{sinc}[(f - f_0)\tau]\}$$

$$= \frac{\tau}{2}\{\text{sinc}[(f\tau + 1)] + \text{sinc}[(f\tau - 1)]\}$$

Similarly, the Fourier transform of Box(t/τ) times $\cos(4\pi f_0 t)$ equals

$$\int_{-\infty}^{\infty}\tau\,\text{sinc}(s\tau)\frac{1}{2}[\delta(f - s + 2f_0) + \delta(f - s - 2f_0)]\, ds$$

$$= \frac{\tau}{2}\{\text{sinc}[(f + 2f_0)\tau] + \text{sinc}[(f - 2f_0)\tau]\}$$

$$= \frac{\tau}{2}\{\text{sinc}[(f\tau + 2)] + \text{sinc}[(f\tau - 2)]\}$$

The theorem then follows by noting that $\text{sinc}[(f\tau - 1)] = \text{sinc}[(1 - f\tau)]$ and $\text{sinc}[(f\tau - 2)] = \text{sinc}[(2 - f\tau)]$, since $\text{sinc}(x)$ is an even function of x.

8.13 The Fourier transform of $c_1g(t) + c_2h(t)$ equals

$$\int_{-\infty}^{\infty} [c_1 g(x) + c_2 h(x)] \exp(-2\pi i f x)\, dx$$

$$= c_1 \int_{-\infty}^{\infty} G(s)\, ds \int_{-\infty}^{\infty} \exp[2\pi i x(s-f)]\, dx$$

$$+ c_2 \int_{-\infty}^{\infty} H(s)\, ds \int_{-\infty}^{\infty} \exp[2\pi i x(s-f)]\, dx$$

$$= c_1 \int_{-\infty}^{\infty} G(s)\, \delta(s-f)\, ds + c_2 \int_{-\infty}^{\infty} H(s)\, \delta(s-f)\, ds = c_1 G(f) + c_2 H(f)$$

8.14 Using Equation 8.40, we see that the Fourier transform of $g(t+T)$ equals $\int_{-\infty}^{\infty} g(x+T)\exp(-2\pi i f x)\, dx$. Now let $y = x + T$. Then $x = y - T$ and $dx = dy$. Hence

$$\int_{-\infty}^{\infty} g(x+T)\exp(-2\pi i f x)\, dx$$

$$= \int_{-\infty}^{\infty} g(y)\exp[-2\pi i f(y-T)]\, dy$$

$$= \exp(2\pi i f T) \int_{-\infty}^{\infty} g(y)\exp(-2\pi i f y)\, dy = G(f)\exp(2\pi i f T)$$

8.15 We are given $R_{gg}(x) = \int_{-\infty}^{\infty} g^*(t) g(t+x)\, dt$. Now if $g(t) = \int_{-\infty}^{\infty} G(s)\exp(2\pi i s t)\, ds$, then $g^*(t) = \int_{-\infty}^{\infty} G^*(f)\exp(-2\pi i f t)\, df$. Hence

$$R_{gg}(x) = \int_{-\infty}^{\infty} G^*(f)\, df \int_{-\infty}^{\infty} G(s)\exp(2\pi i x s)\, ds \int_{-\infty}^{\infty} \exp[2\pi i t(s-f)]\, dt$$

$$= \int_{-\infty}^{\infty} G^*(f)\, df \int_{-\infty}^{\infty} G(s)\exp(2\pi i x s)\, \delta(s-f)\, ds$$

$$= \int_{-\infty}^{\infty} |G(f)|^2 \exp(2\pi i x f)\, df$$

8.16 If $G(f) = \int_{-\infty}^{\infty} g(t)\exp(-2\pi i f t)\, dt$, then

$$G^*(f) = \int_{-\infty}^{\infty} g(t)\exp(2\pi i f t)\, dt$$

$$= \int_{-\infty}^{\infty} g(t)\exp[-2\pi i(-f)t]\, dt = G(-f)$$

8.17 The idea here is to show that interchanging β_0 and β_1 in an MA(1) model does not change the autocorrelation coefficient function. For the model $X_t = \mu + \beta_1 Z_t + \beta_0 Z_{t-1}$, we have

$$\begin{aligned} R_{xx}(k) &= E(X_t X_{t+k}) \\ &= E[(\mu + \beta_1 Z_t + \beta_0 Z_{t-1})(\mu + \beta_1 Z_{t+k} + \beta_0 Z_{t+k-1})] \\ &= \mu^2 + \beta_1^2 E(Z_t Z_{t+k}) + \beta_0^2 E(Z_{t-1} Z_{t+k-1}) \\ &\quad + \beta_0 \beta_1 E(Z_t Z_{t+k-1} + Z_{t-1} Z_{t+k}) \end{aligned}$$

The remaining expectation values are clearly 0, since the Z's are random numbers with expectation values of 0. The expectation values of the product of two Z's will be nonzero only if the subscripts on the Z's are the same, since the Z's are independent random numbers. Given this fact, we can easily see that there are only three values of k that will lead to a nonzero contribution from the product Z's:

$$k = 0 \qquad R_{xx}(0) = \mu^2 + \beta_1^2 E(Z_t Z_t) + \beta_0^2 E(Z_{t-1} Z_{t-1}) = \mu^2 + (\beta_0^2 + \beta_1^2)\sigma^2$$

$$k = 1 \qquad R_{xx}(1) = \mu^2 + \beta_0 \beta_1 E(Z_t Z_t) = \mu^2 + \beta_0 \beta_1 \sigma^2$$

$$k = -1 \quad R_{xx}(-1) = \mu^2 + \beta_0 \beta_1 E(Z_{t-1} Z_{t-1}) = \mu^2 + \beta_0 \beta_1 \sigma^2$$

Since the expectation value of the process is independent of time, it follows that

$$\begin{aligned} C_{xx}(0) &= R_{xx}(0) - \mu^2 = (\beta_0^2 + \beta_1^2)\sigma^2 \\ C_{xx}(1) &= R_{xx}(1) - \mu^2 = \beta_0 \beta_1 \sigma^2 \\ C_{xx}(-1) &= R_{xx}(-1) - \mu^2 = \beta_0 \beta_1 \sigma^2 \end{aligned}$$

It follows immediately that

$$\rho_{xx}(0) = 1$$

$$\rho_{xx}(1) = \rho_{xx}(-1) = \frac{\beta_0 \beta_1}{\beta_0^2 + \beta_1^2}$$

8.18 We can use essentially the same approach here as we used in solving Exercise 8.17:

$$\begin{aligned} R_{xx}(k) &= E(X_t X_{t+k}) \\ &= E[(\mu + \beta_0 Z_t + \beta_1 Z_{t-1} + \beta_2 Z_{t-2}) \\ &\qquad \times (\mu + \beta_0 Z_{t+k} + \beta_1 Z_{t+k-1} + \beta_2 Z_{t-2})] \\ &= \mu^2 + \beta_0^2 E(Z_t Z_{t+k}) + \beta_1^2 E(Z_{t-1} Z_{t+k-1}) + \beta_2^2 E(Z_{t-2} Z_{t+k-2}) \\ &\quad + \beta_0 \beta_1 E(Z_t Z_{t+k-1} + Z_{t-1} Z_{t+k}) + \beta_0 \beta_2 E(Z_t Z_{t+k-2} + Z_{t-2} Z_{t+k} \\ &\quad + \beta_1 \beta_2 E(Z_{t-1} Z_{t+k-2} + Z_{t-2} Z_{t+k-1}) \end{aligned}$$

In this case we can see that there are five values of k that will lead to a nonzero contribution from the product Z's:

$$k = 0 \qquad R_{xx}(0) = \mu^2 + \beta_0^2 E(Z_t Z_t) + \beta_1^2 E(Z_{t-1} Z_{t-1}) + \beta_2^2 E(Z_{t-2} Z_{t-2})$$

$$= \mu^2 + (\beta_0^2 + \beta_1^2 + \beta_2^2)\sigma^2$$

$$k = 1 \qquad R_{xx}(1) = \mu^2 + \beta_0 \beta_1 E(Z_t Z_t) + \beta_1 \beta_2 E(Z_{t-1} Z_{t-1})$$

$$= \mu^2 + \beta_1(\beta_0 + \beta_2)\sigma^2$$

$$k = 2 \qquad R_{xx}(2) = \mu^2 + \beta_0 \beta_2 E(Z_t Z_t) = \mu^2 + \beta_0 \beta_2 \sigma^2$$

$$k = -1 \qquad R_{xx}(-1) = \mu^2 + \beta_0 \beta_1 E(Z_{t-1} Z_{t-1}) + \beta_1 \beta_2 E(Z_{t-2} Z_{t-2})$$

$$= \mu^2 + \beta_1(\beta_0 + \beta_2)\sigma^2$$

$$k = -2 \qquad R_{xx}(-2) = \mu^2 + \beta_0 \beta_2 E(Z_{t-2} Z_{t-2}) = \mu^2 + \beta_0 \beta_2 \sigma^2$$

It is clear from this analysis that interchanging β_0 and β_2 has no effect on the autocorrelation function and hence on the autocorrelation coefficient function.

8.19 In this case we have $X_t = \sum_{m=0}^{\infty} \alpha_1^m Z_{t-m}$. Hence (skipping a few steps)

$$R_{xx}(k) = \mu^2 + E\left(\sum_{m=0}^{\infty} \alpha_1^m Z_{t-m} \sum_{n=0}^{\infty} \alpha_1^n Z_{t+k-n} \right)$$

The expectation values will be 0 unless $m = n - k$ or, equivalently, $n = m + k$. Hence

$$R_{xx}(k) = \mu^2 + \sum_{m=0}^{\infty} \alpha_1^m \alpha_1^{m+k} E(Z_{t-m}^2) = \mu^2 + \sigma^2 \sum_{m=0}^{\infty} \alpha_1^{2m+k}$$

For $R_{xx}(-k)$ we have

$$R_{xx}(-k) = \mu^2 + E\left(\sum_{m=0}^{\infty} \alpha_1^m Z_{t-m} \sum_{n=0}^{\infty} \alpha_1^n Z_{t-k-n} \right)$$

The expectation values will be 0 unless $m = k + n$. Hence

$$R_{xx}(-k) = \mu^2 + \sum_{n=0}^{\infty} \alpha_1^n \alpha_1^{n+k} E(Z_{t-k-n}^2) = \mu^2 + \sigma^2 \sum_{n=0}^{\infty} \alpha_1^{2n+k}$$

8.20 We begin with $\sum_{k=1}^{(N-1)/2} \alpha_1^k [\exp(-2\pi ifk/N) + \exp(2\pi ifk/N)]$. Letting $z = \alpha_1 \exp(-2\pi if/N)$, we can see that this summation can be written as $\sum_{k=1}^{(N-1)/2}(z^k + z^{*k})$, where z^* is the complex conjugate of z. Using the standard recipe for the sum of a geometric series, this summation becomes

$$\frac{z}{1-z}(1 - z^{(N-1)/2}) + \frac{z^*}{1-z^*}(1 - z^{*(N-1)/2})$$

We can now write

$$1 + \sum_{k=1}^{(N-1)/2} \alpha_1^k \left[\exp\left(-\frac{2\pi ifk}{N}\right) + \exp\left(\frac{2\pi ifk}{N}\right)\right]$$

$$= 1 + \frac{z}{1-z}(1 - z^{(N-1)/2}) + \frac{z^*}{1-z^*}(1 - z^{*(N-1)/2})$$

$$= \frac{1}{(1-z)(1-z^*)}\{(1-z)(1-z^*) + z(1-z^*)[1 - z^{(N-1)/2}]$$

$$+z^*(1-z)[1 - z^{*(N-1)/2}]\}$$

$$= \frac{(1-z)(1-z^*) + z(1-z^*) + z^*(1-z)}{(1-z)(1-z^*)}$$

$$- \frac{1}{(1-z)(1-z^*)}\mathrm{Re}\left[z(1-z^*)z^{(N-1)/2}\right]$$

The second term in this expression clearly goes to 0 as $N \to \infty$, since it is proportional to α_1 raised to the $(N+1)/2$ power, and we are assuming $|\alpha_1| < 1$. The first term equals

$$\frac{1 - zz^*}{|1 - z|^2} = \frac{1 - \alpha_1^2}{|1 - \alpha_1 \exp(-2\pi fi/N)|^2}$$

It then follows immediately that

$$\lim_{N\to\infty} \frac{1}{N}\frac{\sigma^2}{1-\alpha_1^2}\left[1 + \sum_{k=1}^{(N-1)/2} \alpha_1^k \left[\exp\left(-\frac{2\pi ifk}{N}\right) + \exp\left(\frac{2\pi ifk}{N}\right)\right]\right]$$

$$= \frac{\sigma^2}{N|1 - \alpha_1 \exp(-2\pi if/N)|^2}$$

8.21 The solution is

$$R_{xx}(0) = \mu^2 + \frac{\sigma^2(1-\alpha_2)}{(1+\alpha_2)\left[(1-\alpha_2)^2 - \alpha_1^2\right]}$$

8.22 Our first job is to show that for any MA(q) process, $R_{xx}(k) = R_{xx}(-k)$. We have already shown this to be the case for autoregressive processes in Exercise 8.19. For an MA(q) process we have

$$R_{xx}(k) = E(X_t X_{t+k})$$
$$= E\left[\left(\mu + \sum_{n=0}^{q} \beta_n Z_{t-n}\right)\left(\mu + \sum_{m=0}^{q} \beta_m Z_{t+k-m}\right)\right]$$

With the exception of μ^2, the only nonzero expectation values will involve products of Z's that have the same subscript. This means $n = m - k$ or $m = n + k$. Hence

$$R_{xx}(k) = \mu^2 + E\left(\sum_{n=0}^{q-k} \beta_n \beta_{n+k} Z_{t-n} Z_{t-n}\right) = \mu^2 + \sigma^2 \sum_{n=0}^{q-k} \beta_n \beta_{n+k}$$

Similarly,

$$R_{xx}(-k) = E(X_t X_{t-k})$$
$$= E\left[\left(\mu + \sum_{n=0}^{q} \beta_n Z_{t-n}\right)\left(\mu + \sum_{m=0}^{q} \beta_m Z_{t-k-m}\right)\right]$$

In this case the nonzero expectation values occur when $n = k + m$. Hence

$$R_{xx}(-k) = \mu^2 + E\left(\sum_{m=0}^{q-k} \beta_m \beta_{m+k} Z_{t-k-m} Z_{t-k-m}\right)$$
$$= \mu^2 + \sigma^2 \sum_{m=0}^{q-k} \beta_m \beta_{m+k} = R_{xx}(k)$$

Having shown that $R_{xx}(-k) = R_{xx}(k)$ for an MA(q) process, the idea now is to show that the expectation value of $|X_f|^2$ equals the Fourier transform of the autocorrelation function. It is convenient first to expand $|\sum_{k=0}^{q} \beta_k \exp(-2\pi ikf/N)|^2$. The expansion gives

$$\left|\sum_{k=0}^{q} \beta_k \exp\left(-\frac{2\pi ikf}{N}\right)\right|^2$$

$$= \left[\sum_{k=0}^{q} \beta_k \exp\left(-\frac{2\pi ikf}{N}\right)\right]\left[\sum_{m=0}^{q} \beta_m \exp\left(\frac{2\pi imf}{N}\right)\right]$$

$$= \left[\beta_0 + \beta_1 \exp\left(-\frac{2\pi if}{N}\right) + \cdots + \beta_q \exp\left(-\frac{2\pi iqf}{N}\right)\right]$$

$$\times\left[\beta_0 + \beta_1 \exp\left(\frac{2\pi if}{N}\right) + \cdots + \beta_q \exp\left(\frac{2\pi iqf}{N}\right)\right]$$

$$= \sum_{k=0}^{q} \beta_k^2 + \sum_{n=0}^{q-1} \sum_{p=n+1}^{q} \beta_n \beta_p \left\{\exp\left[\frac{2\pi i(n-p)f}{N}\right] + \exp\left[-\frac{2\pi i(n-p)f}{N}\right]\right.$$

$$= \sum_{k=0}^{q} \beta_k^2 + 2\sum_{n=0}^{q-1} \sum_{p=n+1}^{q} \beta_n \beta_p \cos\left[\frac{2\pi(n-p)f}{N}\right]$$

Now looking at Exercise 8.22 and Equation 8.93, we note that all $R_{xx}(k)$ include a term μ^2. Noting that f is some integer between 0 and $(N-1)/2$, it follows that

$$\sum_{k=-(N-1)/2}^{(N-1)/2} \mu^2 \exp\left(-\frac{2\pi ikf}{N}\right) = \mu^2 \sum_{k=-(N-1)/2}^{(N-1)/2} \exp\left(-\frac{2\pi ikf}{N}\right) = 0$$

Hence we need only concern ourselves with the contribution to $R_{xx}(k)$ from $\sigma^2\sum_{m=0}^{q-k} \beta_m \beta_{m+k}$, and this contribution occurs only for $|k| \le q$. For $k = 0$ the contribution is $\sigma^2\sum_{m=0}^{q} \beta_m^2$. The nonzero contribution for $k > 0$ to $(2/N)\sum_{k=1}^{(N-1)/2} R_{xx}(k)\cos(2\pi kf/N)$ will be

$$\frac{2}{N}\sum_{k=1}^{q} R_{xx}(k)\cos\left(\frac{2\pi kf}{N}\right)$$

$$= \frac{2}{N}\sigma^2 \sum_{k=1}^{q} \sum_{m=0}^{q-k} \beta_m \beta_{m+k} \cos\left(\frac{2\pi kf}{N}\right)$$

Now let $p = m + k$, in which case $k = p - m$. With this substitution,

$$\frac{2\sigma^2}{N}\sum_{k=1}^{q}\sum_{m=0}^{q-k}\beta_m\beta_{m+k}\cos\left(\frac{2\pi kf}{N}\right)$$
$$=\frac{2\sigma^2}{N}\sum_{m=0}^{q-1}\sum_{p=m+1}^{q}\beta_m\beta_p\cos\left[\frac{2\pi(p-m)f}{N}\right]$$

We conclude that

$$\frac{1}{N}\sum_{k=-(N-1)/2}^{(N-1)/2}R_{xx}(k)\exp\left(-\frac{2\pi ikf}{N}\right)$$
$$=\frac{\sigma^2}{N}\left\{\sum_{m=0}^{q}\beta_m^2+2\sum_{m=0}^{q-1}\sum_{p=m+1}^{q}\beta_m\beta_p\cos\left[\frac{2\pi(p-m)f}{N}\right]\right\}$$
$$=\frac{\sigma^2}{N}\left|\sum_{m=0}^{q}\beta_m\exp\left(-\frac{2\pi imf}{N}\right)\right|^2$$

8.23 Let $\{X_F\}$ and $\{X_L\}$ be the first and last $N-1$ observations, respectively. Then

$$\hat{\mu}=\frac{\langle X_F\rangle\langle X_F X_L\rangle-\langle X_L\rangle\langle X_F^2\rangle}{\langle(X_F-\langle X_F\rangle)(X_L-X_F)\rangle}$$

where $\langle X\rangle$ indicates the average of $N-1$ values of X.

8.24 If $\sigma^2 = 1$, then $V(X_t)=\sum_{m=0}^{q}\beta_m^2$. Hence

$$V(X_t)\left(1+2\sum_{k=1}^{q}\rho_{xx}(k)\right)=\sum_{m=0}^{q}\beta_m^2\left(1+2\sum_{k=1}^{q}\rho_{xx}(k)\right)$$
$$=\sum_{m=0}^{q}\beta_m^2+2\sum_{k=1}^{q}\sum_{n=0}^{q-k}\beta_n\beta_{n+k}$$

Now let $p = n + k$. With this substitution we see that

$$V(X_t)\left(1+2\sum_{k=1}^{q}\rho_{xx}(k)\right)=\sum_{m=0}^{q}\beta_m^2+2\sum_{n=0}^{q-1}\sum_{p=n+1}^{q}\beta_n\beta_p$$
$$=\left(\sum_{k=0}^{q}\beta_k\right)^2$$

INDEX

Printed and bound by CPI Group (UK) Ltd, Croydon, CR0 4YY
16/09/2026
14963456-0005